The Practical Approach in Chemistry Series

SERIES EDITORS

L. M. Harwood
Department of Chemistry
University of Reading

C. J. Moody
Department of Chemistry
University of Exeter

The Practical Approach in Chemistry Series

Organocopper reagents
Edited by Richard J. K. Taylor

Macrocycle synthesis
Edited by David Parker

High-pressure techniques in chemistry and physics
Edited by Wilfried B. Holzapfel and Neil S. Isaacs

Preparation of alkenes
Edited by Jonathan M. J. Williams

Transition metals in organic synthesis
Edited by Susan E. Gibson (née Thomas)

Matrix-isolation techniques
Ian R. Dunkin

Lewis acid reagents
Edited by Hisashi Yamamoto

Organozinc reagents
Edited by Paul Knochel and Philip Jones

Lewis Acid Reagents

A Practical Approach

Edited by

HISASHI YAMAMOTO

Graduate School of Engineering
Nagoya University

OXFORD
UNIVERSITY PRESS

1999

OXFORD
UNIVERSITY PRESS

Great Clarendon Street, Oxford OX2 6DP

Oxford University Press is a department of the University of Oxford
and furthers the University's aim of excellence in research, scholarship,
and education by publishing worldwide in

Oxford New York

Athens Auckland Bangkok Bogotá Buenos Aires Calcutta
Cape Town Chennai Dar es Salaam Delhi Florence Hong Kong Istanbul
Karachi Kuala Lumpur Madrid Melbourne Mexico City Mumbai
Nairobi Paris São Paulo Singapore Taipei Tokyo Toronto Warsaw
and associated companies in Berlin Ibadan

Oxford is a registered trade mark of Oxford University Press

Published in the United States
by Oxford University Press Inc., New York

A catalogue record for this book is available from the British Library

Library of Congress Cataloging-in-Publication Data
(Data applied for)

ISBN 0 19 850099 8 (Hbk)

Typeset by Footnote Graphics, Warminster, Wilts
Printed in Great Britain by Bookcraft Ltd, Midsomer Norton, Avon

Preface

Lewis acids are becoming a powerful tool in many different modern reactions, such as the Diels-Alder reactions, Ene reactions, Sakurai reactions, and Aldol synthesis. In fact, the importance and practicality of Lewis acid reagents as valuable means of obtaining a variety of organic molecules is now fully acknowledged by chemists in the synthetic organic society. This prominence is due to the explosive development of newer and even more efficient methods during the last decade, and the numbered publications on these reagents is actually increasing exponentially each year. Research on asymmetric synthesis has become more important and popular in the total synthesis of natural products, pharmaceuticals, and agricultural agents, and Lewis acid chemistry plays a major role in this arena.

Comprehensive coverage of the literature on each area of Lewis acid is not necessarily provided here. Rather, the aim of the book is to furnish a detailed and accessible laboratory guide useful for researchers who are not familiar with the benefits of Lewis acids. It includes information on reagent purification, reaction equipment and conditions, work-up procedures, and other expert advice. The primary goal is thus to dispel the mystery surrounding Lewis acid reagents and to encourage more scientists to use these powerful synthetic tools to maximum effect. The book contains 14 independently referenced chapters describing a variety of Lewis acids using different metals. Each metal has different characteristic features of reagent preparation and practicality which clearly described in that chapter.

I would like to thank Professor K. Ishihara for helping to check parts of the manuscripts and for useful suggestions. I would also like to express my personal gratitude to all of the invited contributors who carefully honored the deadlines and thus made the editorial job much easier.

It is my strong hope that this book will be found an invaluable reference for graduate students as well as chemists at all levels in both academic and industrial laboratories.

Nagoya   H. Y.
October 1998

Contents

Contents

4. Magnesium(II) and other alkali and alkaline earth metals 65

Akira Yanagisawa

5. Zinc(II) reagents 77

Nobuki Oguni

6. Chiral titanium complexes for enantioselective catalysis 93

Koichi Mikami and Masahiro Terada

Contents

Contributors

TAKUZO AIDA

Department of Chemistry and Biotechnology, Graduate School of Engineering, The University of Tokyo, Hongo, Bunkyo, Tokyo 113-8656, Japan

SHIN-ICHI FUKUZAWA

Department of Applied Chemistry, Chuo University, 1113127 Kasuga, Bunkyo-ku, Tokyo 152, Japan

AKIRA HOSOMI

Department of Chemistry, University of Tsukuba, Tsukuba, Ibaraki 305-8571, Japan

KAZUAKI ISHIHARA

Research Center for Advanced Waste and Emission Management, Nagoya University Furo-cho, Chikusa, Nagoya 464-8603, Japan

SHŪ KOBAYASHI

Graduate School of Pharmaceutical Sciences, The University of Tokyo, Hongo, Bunkyo-ku, Tokyo 113-0033, Japan

KEIJI MARUOKA

Graduate School of Science, Hokkaido University, Supporo 060-0810, Japan

KOICHI MIKAMI

Department of Chemical Technology, Tokyo Institute of Technology, Meguro-ku, Tokyo 152, Japan

KATSUKIYO MIURA

Department of Chemistry, University of Tsukuba, Tsukuba, Ibaraki 305-8571, Japan

YUKIHIRO MOTOYAMA

School of Materials Science, Toyohashi University of Technology, Tempaku-cho, Toyahashi, 441 Japan

TAKESHI NAKAI

Department of Chemical Technology, Tokyo Institute of Technology, Meguro-ku, Tokyo 152-8552, Japan

Contributors

HISAO NISHIYAMA

School of Materials Science, Toyohashi University of Technology, Tempaku-cho, Toyohashi 441, Japan

NOBUKI OGUNI

Department of Chemistry, Faculty of Science, Yamaguchi University, 1677-1 Yoshida, Yamaguchi City 753, Japan

MASAYA SAWAMURA

Department of Chemistry, Graduate School of Science, The University of Tokyo, Hongo, Bunkyo, Tokyo 113, Japan

KEISUKE SUZUKI

Department of Chemistry, Tokyo Institute of Technology, Meguro-ku, Tokyo 152-8551, Japan

DAISUKE TAKEUCHI

Department of Chemistry and Biotechnology, Graduate School of Engineering, The University of Tokyo, Hongo, Bunkyo, Tokyo 113-8656, Japan

MASAHIRO TERADA

Department of Chemical Technology, Tokyo Institute of Technology, Meguro-ku, Tokyo 152, Japan

KATSUHIKO TOMOOKA

Department of Chemical Technology, Tokyo Institute of Technology, Meguro-ku, Tokyo 152-8552, Japan

HISASHI YAMAMOTO

Graduate School of Engineering, Nagoya University Furo-cho, Chikusa, Nagoya 464-01, Japan

AKIRA YANAGISAWA

Graduate School of Engineering, Nagoya University, Furo-cho, Chikusa, Nagoya 464-01, Japan

1

Introduction

HISASHI YAMAMOTO

Are enzymatic reactions really good models for laboratory chemical reactions? An enzyme is a giant molecule, large enough to support a substrate, whereas chemical reagents are composed of much smaller molecules. Still, the much smaller molecular apparatus of human-made reagents is expected to induce reactions with selectivities comparable to those of a large enzyme. Clearly, the design of new reagents requires careful abstraction and simplification of the true mechanism of an enzyme, much like the design of an aircraft might be based on the aerodynamics of a bird.

A case in point is the important role of hydrogen bonding during enzymatic reactions. In the course of such processes, the giant template of the enzyme will specify quite accurately the position and direction of a proton for hydrogen bonding, before and after the reaction. However, a proton by itself cannot behave in this fashion. A perfect sphere, it has no directional selectivity for hydrogen bonding outside the domain of the enzyme, thus it is unable to act as a 'delicate finger' in an ordinary organic reaction as it does in the enzymatic transformation. It is natural to wonder whether an appropriate substitute for the proton might induce human-made reactions capable of selectivities comparable to those afforded by enzymes.

An excellent candidate as a proton substitute is a Lewis acid. The observation that organoaluminium, organolithium, organoboron and many other organometallic compounds immediately ignite when exposed to air, reflecting the high affinity of these metals for oxygen, inspired us to devise a new series of reagents based on those metals: true 'Designer Lewis Acids' for organic synthesis. For example, since an organoaluminium compound would have three ligands around the metal, the structural design of such a catalyst could be quite flexible. The goal, then, was to engineer an artificial proton of a special shape, which could be utilized as an effective tool for chemical reactions, by harnessing the high reactivity of the metal atom towards oxygen. Such a concept was initially researched by examining the influence of specially designed Lewis acid compounds.

In the *Encyclopedia of reagents for organic synthesis* edited by Paquette,

Hisashi Yamamoto

the reagent function index listed the following metals as being used as Lewis acid reagents:[1]

Aluminum, Antimony, Boron, Cadmium, Cerium, Cobalt, Copper, Europium, Germanium, Hafnium, Iron, Lanthanum, Lithium, Magnesium, Molybdenum, Nickel, Palladium, Phosphorus, Silicon, Silver, Sulfur, Thallium, Tin, Titanium, Vanadium, Ytterbium, Zinc, Zirconium

A truly varied group of elements are used as the Lewis acid reagent and each metal has its own characteristic features. We therefore decided in this book to classify these reagents according to their metals.

It need not be pointed out that Lewis acid-promoted carbon–carbon bond formation reaction is one of the most important processes in organic synthesis. Classically, Friedel–Crafts reaction, ene reaction, Diels–Alder reaction, and Mukaiyama aldol synthesis are catalysed with ordinary Lewis acids such as $AlCl_3$, $TiCl_4$, $BF_3 \cdot OEt_2$, or $SnCl_4$ (Fig. 1.1). These classical Lewis acids activate the functional groups of substrates, and the reactions proceed in relatively low stereo-, regio-, or chemoselectivities. On co-ordination with well-designed ligand(s), a Lewis acid exhibits substantially new reactivity. Furthermore, a designer Lewis acid leads to an isolation of monomeric Lewis acid species whose structural features can be easily understood and easily extended to designer chiral catalysts for asymmetric syntheses. Thus, metal ligand tunings are the most essential component in the design of Lewis acid reagents.[2–4]

Lewis acid-mediated reactions can be classified as follows. The complex between substrate and Lewis acid rearranges to produced the product (type 1).

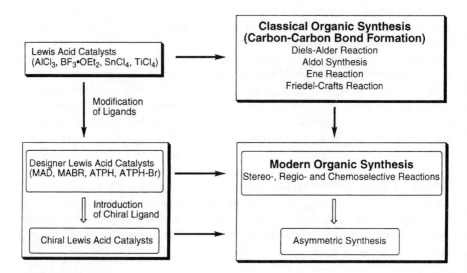

Fig. 1.1 Role of Lewis acid in organic synthesis.

2

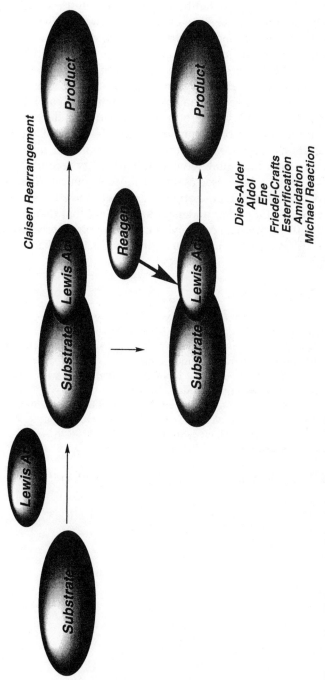

Fig. 1.2 Classification of Lewis acid promoted reactions.

3

Claisen rearrangement promoted by Lewis acid catalyst is a typical example of type 1. On the other hand, some complexes between Lewis acids and substrates are stable enough and the formed complexes react with a variety of reagents from outside the system to generate the product (type 2). The reaction between the Lewis acid activated unsaturated carbonyl compounds with dienes, Diels–Alder reaction, is an example of type 2 (see Fig. 1.2).

A Lewis basic carbonyl group can be activated through co-ordination with a metal-centred Lewis acid, with profound reactivity and stereochemical consequences. In the context of asymmetric synthesis, many of the Lewis acid-mediated reactions are known to proceed with improved stereoselectivities as compared to their non-catalysed counterparts; very recently, a number of chiral Lewis acids have been used as remarkably efficient catalysts for carbonyl addition processes. Although the origins of many of the effects brought about by Lewis acids are still poorly understood, it is clear that the conformational preferences of the Lewis acid carbonyl complex are ultimately responsible for determining the stereochemical course of Lewis acid-mediated reactions.[5,6]

References

1. Paquette, L. A. (ed.) *Encyclopedia of reagents for organic synthesis*, John Wiley & Sons, Chichester, New York, Brisbane, Toronto, Singapore, **1995**.
2. Schinzer, D. (ed.) *Selectivities in Lewis acid promoted reaction*, Kluwer Academic Publishers, Dordrecht, Boston, London, **1988**.
3. Sasntelli, M.; Pons, J.-M., *Lewis acids and selectivity in organic synthesis*, CRC Press, Boca Raton, New York, London, Tokyo, **1996**.
4. *Methods of Organic Chemistry* (Houben-Weyl), Additional and Suppl. Vol. of the 4th Edn, Vol. E 21b; *Stereoselective Synthesis* (ed. G. Helmchen, R. W. Hoffmann, J. Mulzer, E. Schaumann), Thieme, Stuttgart, **1995**.
5. Shambayati, S.; Crowe, W. E.; Schreiber, S. L. *Angew. Chem., Int. Ed. Engl.* **1990**, *29*, 256.
6. Denmark, S. E.; Almstead, N. G. *J. Am. Chem. Soc.*, **1993**, **115**, 3133.

2

Synthetic utility of bulky aluminium reagents as Lewis acid receptors

KEIJI MARUOKA

1. Introduction

Organoaluminium compounds, little known until the 1950s, have been widely accepted and increasingly important in the field of industry and in the laboratory,[1-13] particularly after K. Ziegler and colleagues discovered the direct synthesis of trialkylaluminiums and their brilliant application to the polymerization of olefins.[14,15] The chemistry of organoaluminium compounds has been understood in terms of the Lewis acidity of their monomeric species, which is directly related to the tendency of the aluminium atom to complete electron octets. Organoaluminium compounds possess a strong affinity for various heteroatoms in organic molecules, particularly oxygen. In fact, bond strength of aluminium and electronegative atoms such as oxygen is extremely strong; the bond energy of the Al–O bond is estimated to be 138 kcal mol^{-1}. In view of this high bond strength, most organoaluminium compounds are particularly reactive with oxygen and often ignite spontaneously in air. Accordingly, they easily generate 1:1 co-ordination complexes even with neutral bases such as ethers, which is in marked contrast with lithium and magnesium derivatives. Utilization of this property, commonly identified with 'oxygenophilicity', in organic synthesis allows facile reactions with hetero atoms particularly oxygen- and carbonyl containing compounds.

$$Me_3Al + Et_2O \rightarrow Me_3Al \cdots OEt_2 \qquad \Delta Hf = -20.21 \text{ kcal mol}^{-1}$$

The major difference between organoaluminium compounds and more common Lewis acids such as aluminium chloride and bromide is attributable to the structural flexibility of organoaluminium reagents. Thus, the structure of an aluminium reagent is easily modified by changing one or two of its ligands. Described below are our recent practical strategies to selective organic synthesis with modified organoaluminium reagents.

2. Amphiphilic alkylations

2.1 Amphiphilic carbonyl alkylations

Organoaluminium compounds are endowed with high oxygenophilic character, and hence are capable of forming long-lived monomeric 1:1 complexes with carbonyl substrates. For example, the reaction of benzophenone with Me_3Al in a 1:1 molar ratio gives a yellow, long-lived monomeric 1:1 species

$$Ph_2C=O \xrightarrow[25\,°C]{Me_3Al} [Ph_2C=O \cdots AlMe_3] \xrightarrow{80\,°C} Ph_2\overset{\overset{\displaystyle Me}{|}}{C}-OAlMe_2$$
$$\text{yellow complex}$$

which decomposed unimolecularly to dimethylaluminum 1,1-diphenylethoxide during some minutes at 80 °C or many hours at 25 °C.[16] This unique property may be utilized for stereoselective activation of the carbonyl group. Among various organoaluminium derivatives examined, exceptionally bulky, oxygenophilic organoaluminium reagents such as methylaluminium bis(2,6-di-*tert*-butyl-4-alkylphenoxide) (MAD and MAT), have shown excellent diastereofacial selectivity in carbonyl alkylation.[17,18] Thus, treatment of 4-*tert*-butyl cyclohexanone with MAD or MAT in toluene produced a 1:1 co-ordination complex which on subsequent treatment with methyl-lithium or Grignard reagents in ether at −78 °C afforded the equatorial alcohol almost exclusively (Scheme 2.1). Methyl-lithium or Grignard reagents solely undergo preferential equatorial attack yielding axial alcohols as the major product. MAD and MAT have played a crucial role in the stereoselective synthesis of hitherto inaccessible equatorial alcohols from cyclohexanones as shown in Table 2.1.

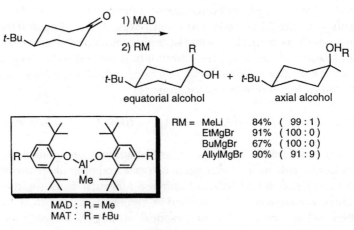

RM = MeLi   84% (99 : 1)
      EtMgBr   91% (100 : 0)
      BuMgBr   67% (100 : 0)
      AllylMgBr 90% (91 : 9)

MAD : R = Me
MAT : R = *t*-Bu

Scheme 2.1

Table 2.1 Stereoselective alkylation of cyclic ketones

Entry	Alkylation agent	Chemical yield (ax/eq ratio)

1	MeLi	75% (79 : 21)
2	MAD/MeLi	84% (1 : 99)
3	MAT/MeLi	92% (0.5 : 99.5)
4	EtMgBr	95% (48 : 52)
5	MAD/EtMgBr	91% (0 : 100)
6	BuMgBr	58% (56 : 44)
7	MAD/BuMgBr	67% (9 : 100)

8	MeLi	73% (92 : 8)
9	MAD/MeLi	84% (14 : 86)
10	MAT/MeLi	80% (10 : 90)

11	MeLi	80% (83 : 17)
12	MAD/MeLi	69% (9 : 91)
13	MAT/MeLi	95% (3 : 97)
14	BuMgBr	86% (79 : 21)
15	MAD/BuMgBr	75% (1 : 99)

16	MeLi	77% (75 : 25)
17	MAD/MeLi	82% (1 : 99)

This approach has been quite useful in the stereoselective alkylation of steroidal ketones. Reaction of 3-cholestanone with MeLi gave predominantly 3β-methylcholestan-3α-ol (axial alcohol), whereas amphiphilic alkylation of the ketone with MAD/MeLi or MAT/MeLi afforded 3α-methylcholestan-3β-ol (equatorial alcohol) exclusively (Scheme 2.2). In addition, unprecedented *anti*-Cram selectivity was achievable in the MAD- or MAT-mediated alkylation of α-chiral aldehydes possessing no ability to be chelated.

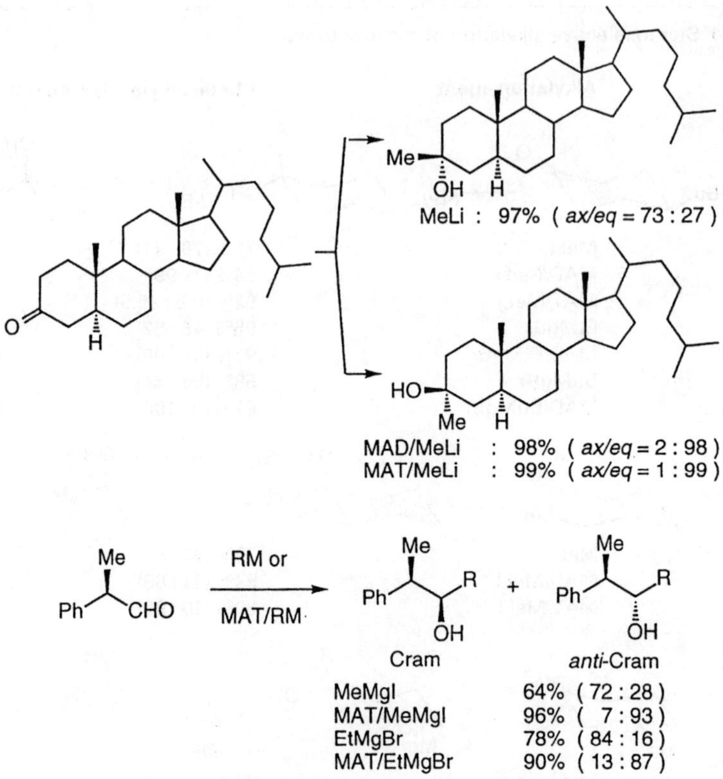

Me̅-
OH   H
MeLi : 97% (*ax/eq* = 73 : 27)

HO̅-
Me   H
MAD/MeLi : 98% (*ax/eq* = 2 : 98)
MAT/MeLi : 99% (*ax/eq* = 1 : 99)

Me
Ph   CHO
RM or
MAT/RM

Me
Ph   R
OH
Cram

Me
Ph   R
OH
anti-Cram

MeMgI	64%	(72 : 28)
MAT/MeMgI	96%	(7 : 93)
EtMgBr	78%	(84 : 16)
MAT/EtMgBr	90%	(13 : 87)

Scheme 2.2

In contrast to the facile MAD- or MAT-mediated alkylation of cyclic ketones with primary organolithium or Grignard reagents, reduction takes precedence over alkylation with hindered alkylation agents such as *t*-butylmagnesium chloride in the presence of MAD[18] (Scheme 2.3). This amphiphilic reduction system appears to be complementary to the existing methodologies using L-Selectride for obtaining axial selectivity.

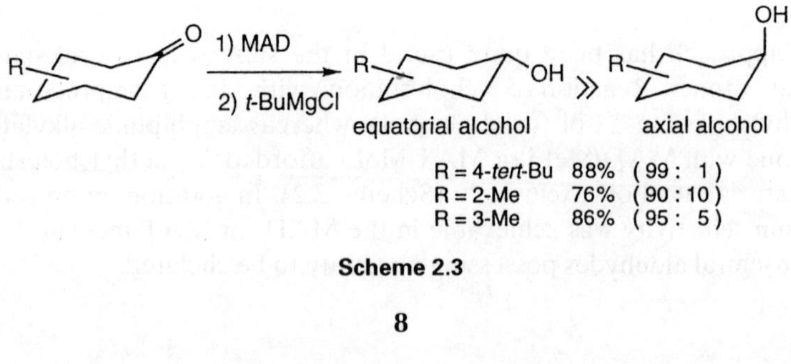

R   1) MAD   R   OH » R   OH
         2) *t*-BuMgCl
                        equatorial alcohol   axial alcohol

R = 4-*tert*-Bu	88%	(99 : 1)
R = 2-Me	76%	(90 : 10)
R = 3-Me	86%	(95 : 5)

Scheme 2.3

Protocol 1.

Synthesis of equatorial 4-*tert*-butyl-1-methylcyclohexanol. Amphiphilic alkylation of 4-*tert*-butylcyclohexanone with MAD/MeLi system

Caution! Carry out all procedures in a well-ventilated hood, and wear disposable vinyl or latex goves and chemical-resistant safety goggles.

Equipment

- Magnetic stirrer
- Three-necked, round-bottomed flask (300 mL) A three-way stopcock is fitted to the top of the flask and connected to a vacuum/argon source
- Teflon-coated magnetic stirring bar
- Medium-gauge needle

- All-glass syringe with a needle-lock Luer (volume appropriate for quantity of solution to be transferred)
- Vacuum/inert gas source (argon source may be an argon balloon)

Materials

- 2,6-Di-*tert*-butyl-4-methylphenol (FW 220.4), 6.61 g, 30 mmol   irritant
- Trimethylaluminum in hexane (FW 72.1), 2 M solution in hexane, 7.5 mL, 15 mmol   **Pyrophoric, moisture sensitive**
- Dry toluene, 110 mL   **flammable, toxic**
- MeLi in ether (FW 22.0), 1.5 M solution in ether, 10 mL, 15 mmol   **flammable, moisture sensitive**
- 4-*tert*-butylcyclohexanone (FW 154.3), 1.54 g, 10 mmol
- 1M HCl, 50 mL   **toxic**
- Technical ether for extraction, 100 mL   **flammable, toxic**
- Silica gel for flash chromatography 300 g, Merck Kieselgel 60 (Art. 9385)   **irritant dust**
- Ether for flash chromatography   **flammable, toxic**
- *n*-Hexane for flash chromatography   **flammable, irritant**

1. Clean all glass wares, syringes, needles, and stirring bar and dry for at least 2 h in a 100°C electric oven before use.
2. Assemble the flask, stirring bar, and stop cocks under argon while the apparatus is still hot.
3. Support the assembled flask using a clamp and a stand with a heavy base.
4. Dry the apparatus with an electric heat gun under vacuum (1–2 mm Hg) for 5 min, then back-full the flask with argon. Repeat to a total of three times.
5. Place 2,6-di-*tert*-butyl-4-methylphenol and flush with argon.
6. Charge dry toluene (100 mL).
7. Stir the mixture, degassed under vacuum, and replaced by argon.
8. Support the bottle containing trimethylaluminium in hexane using a clamp and a stand with a heavy base.
9. Fill a syringe with trimethylaluminium in hexane from the bottle containing trimethylaluminium using argon pressure. Apply the argon pressure to fill the syringe slowly with the required volume. Transfer the reagent in the syringe to the reaction flask at room temperature.
10. Stir the resulting solution at this temperture for 1 h to give methyl-

9

Protocol 1. *Continued*

aluminium bis(2,6-di-*tert*-butyl-4-methylphenoxide) (MAD) almost quantitatively. During this operation, nearly 2 equivalents of methane gas are
evolved per 1 equiv of trimethylaluminium.

11. Cool the reaction vessel to −78°C in a dry ice–methanol bath.

12. Transfer 4-*tert*-butylcyclohexanone, which is dissolved in 10 mL of dry
    toluene, to the syringe, and add to the reaction flask over 5 min at −78°C.
    Then add MeLi in ether over 10–15 min at −78°C. Stir the whole mixture at
    −78°C for 1 h in order to complete the alkylation.

13. Place 1M HCl solution and the stirring bar in the Erlenmeyer flask. While
    1M HCl solution is stirred vigorously, add the reaction mixture slowly at
    0°C to avoid excessive foaming on hydrolysis.

14. Remove the ice bath and stir the entire mixture vigorously at 25°C for 20 min.

15. Transfer the mixture to a separating funnel and separate the two layers.
    Extract the water layer with ether twice (2 × 50 mL).

16. Combine the ethereal extracts in a 500 mL flask. Dry over anhydrous mag-
    nesium sulfate, and filter through filter paper. Concentrate the filtrate under
    reduced pressure by means of a rotary evaporator.

17. Purify the oily residue by column chromatography on silica gel (ether/
    hexane as eluants) to give 625 mg (84%) of a mixture of axial and equatorial
    4-*tert*-butylcyclohexanols (ratio = 1:99) as a colourless oil, the ratio of
    which is determined by capillary GLC analysis. Each isomer is character-
    ized by ^{1}H NMR analysis.

2.2 Amphiphilic conjugate alkylations

Conjugate addition to α,β-unsaturated carbonyl compounds is generally
effected by soft organometallics as often seen in organocopper chemistry. In
contrast, the use of organolithiums alone has never been developed to a useful level because of their hard nucleophilic character. This difficulty has been
successfully overcome by using aluminium tris(2,6-diphenylphenoxide)
(ATPH) (Scheme 2.4) as a carbonyl stabilizer.[19] This type of conjugate alkyla-

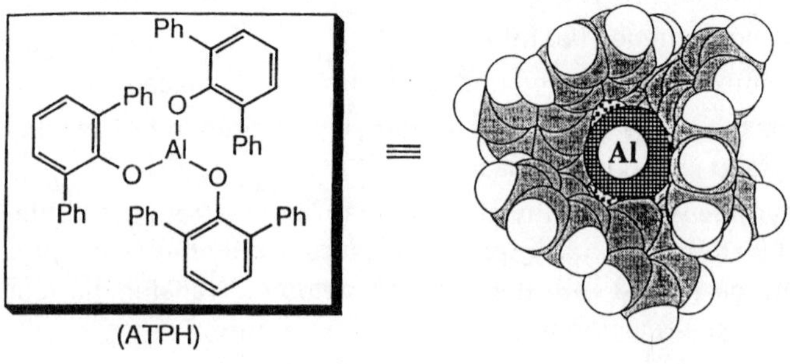

(ATPH)

Scheme 2.4

10

tion can be classified as an amphiphilic conjugate alkylation that is markedly different from previously known nucleophilic and/or electrophilic conjugate alkylations (Scheme 2.5).

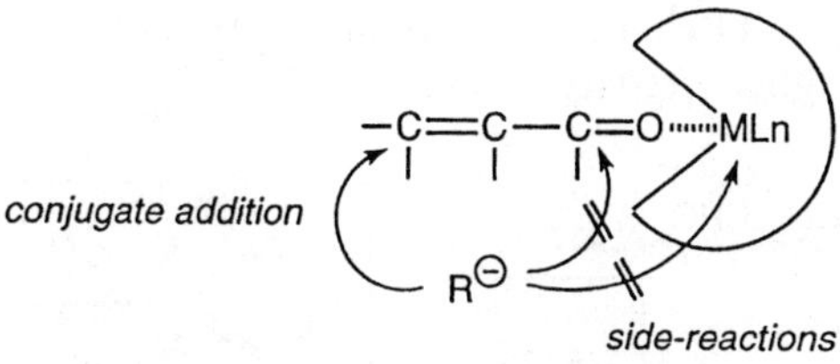

Scheme 2.5. Amphiphilic alkylation system.

This methodology is particularly effective for the conjugate alkylation to α,β-unsaturated aldehydes, which, among various conjugate acceptors, are prone to be more susceptible to 1,2 addition with a number of nucleophiles than α,β-unsaturated ketones, esters, and amides (Scheme 2.6). In addition, conjugate addition of lithium alkynides and thermally unstable lithium carbenoids, which are very difficult to achieve in organocopper chemistry, are realized with this amphiphilic conjugate alkylation system (Scheme 2.7).

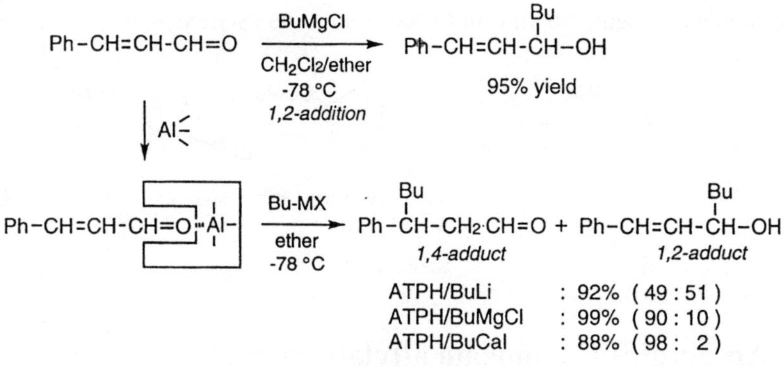

Scheme 2.6

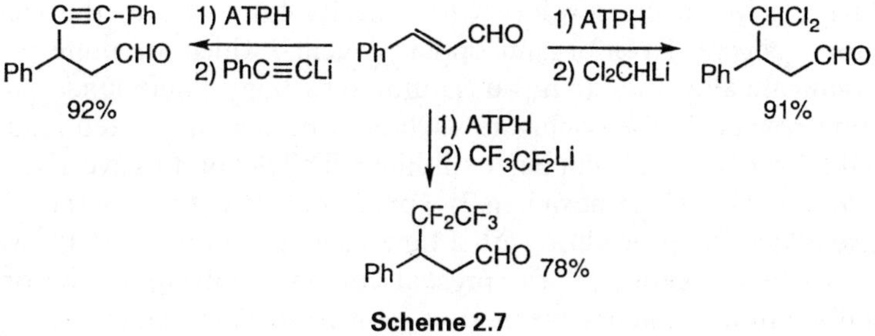

Scheme 2.7

This amphiphilic conjugate alkylation has been successfully applied for the nucleophilic alkylation to electron-deficient arenes based on the unprecedented conjugate addition of organolithiums to aromatic aldehydes and ketones by complexation with ATPH.[20] Thus, initial complexation of benzaldehyde or acetophenone with ATPH and subsequent addition of organolithiums affords 1,6-adducts with high selectivity (Scheme 2.8).

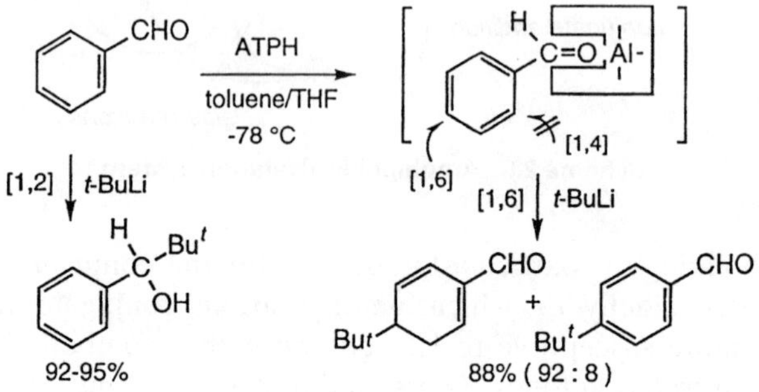

CHO
ATPH
toluene/THF
-78 °C
[1,2]
t-BuLi
[1,6]
[1,4]
[1,6]
t-BuLi
92-95%
88% (92 : 8)

Scheme 2.8

The ratio of dearomatization to aromatization products is highly dependent on the choice of solvents and quenching methods as exemplified by the amphiphilic conjugate alkylation to acetophenone (Scheme 2.9).

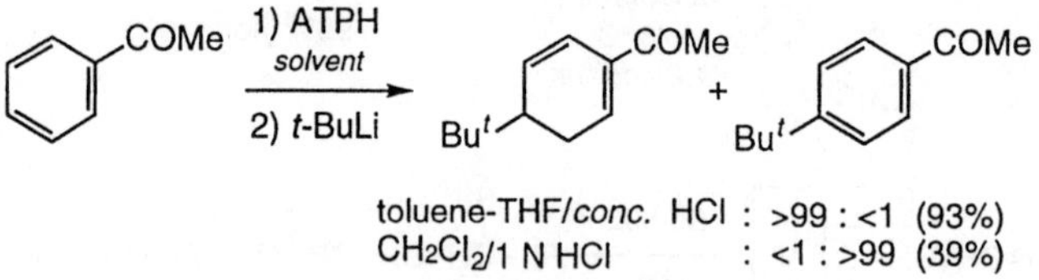

COMe
1) ATPH
solvent
2) t-BuLi
toluene-THF/conc. HCl : >99 : <1 (93%)
CH2Cl2/1 N HCl : <1 : >99 (39%)

Scheme 2.9

2.3 Amphiphilic conjugate allylations

Conjugate allylation to α,β-unsaturated aldehydes is an extremely difficult, hitherto unattainable transformation in organic synthesis, and no effective procedure has yet been developed to a useful level due to the lack of a satisfactory reagent. Even organocopper reagents, which are quite powerful in the conjugate alkylation to α,β-unsaturated carbonyl compounds, gave disappointing results for the conjugate allylation. In fact, attempted reaction of cinnamaldehyde with allylcopper or lithium diallylcuprate gave rise to 1,2-adduct, *trans*-1-phenyl-1,5-hexadien-3-ol predominantly. The new, amphiphilic conjugate alkylation procedure with a Lewis acid receptor, ATPH,[19] was also found to be less effective for the present conjugate allylation, and only the ATPH/allyl-lithium system gave modest 1,4-selectivity (Scheme 2.10).

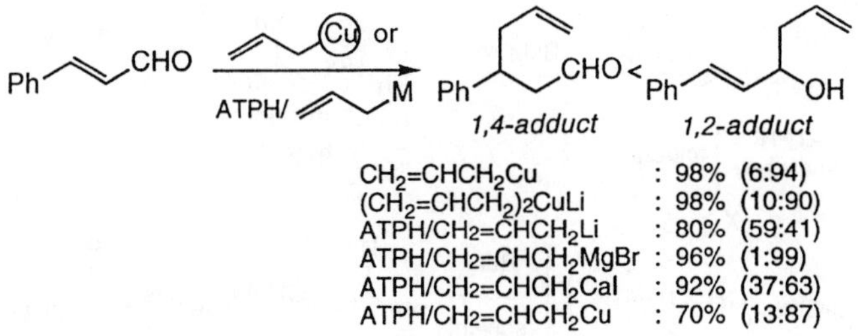

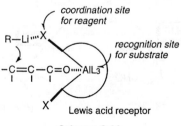

Ph
CHO
Cu
or
ATPH/
M
Ph
CHO <
Ph
OH
1,4-adduct
1,2-adduct
CH2=CHCH2Cu : 98% (6:94)
(CH2=CHCH2)2CuLi : 98% (10:90)
ATPH/CH2=CHCH2Li : 80% (59:41)
ATPH/CH2=CHCH2MgBr : 96% (1:99)
ATPH/CH2=CHCH2CaI : 92% (37:63)
ATPH/CH2=CHCH2Cu : 70% (13:87)

Scheme 2.10

This tendency is contradictory, for example, to our previous observation of the ATPH/Bu-M system for the conjugate alkylation to cinnamaldehyde, where the 1,4-selectivity is enhanced by changing nucleophiles (Bu–M) from BuLi (1.4-/1.2-ratio = 50:50) to BuMgCl (90:10) and BuCaI (98:2).[19] Considering the wide availability and versatility of organolithium reagents,[21–24] a new Lewis acid receptor possessing appropriate co-ordination sites for alkyllithium nucleophiles has been devised (Scheme 2.11). Among various functionalized ATPH derivatives as a Lewis acid receptor, *p*-F-ATPH (Scheme 2.12) was found to be highly effective for this transformation, which clearly demonstrated the synthetic utility of the strong lithium/fluorine participation in selective organic synthesis.[25] First, the 1,4-selectivity for the conjugate alkylation to cinnamaldehyde was examined with the modified ATPH/BuLi system in model experiments. Selected results are shown in Table 2.2. *p*-(MeO)-ATPH and *p*-(MeS)-ATPH showed slightly better selectivity than ATPH (Table 2.2, entries 2 and 3). The 1,4-selectivity was further enhanced by designing *p*-Cl-ATPH and *p*-F-ATPH (entries 4 and 5). Significant solvent and temperature effects on the 1,4-selectivity were also observed (entries 6–9), and eventually the optimum reaction condition was achieved using 1,2-dimethoxyethane (DME) solvent for BuLi at lower temperature under the influence of *p*-F-ATPH in toluene, giving the 1,4-adduct with 95% selectivity (entry 9). Here, the chelation of BuLi with DME is quite appropriate to increase the steric size of the nucleophile (BuLi), while still maintaining the co-ordination ability of Li^+ to fluorine atoms of *p*-F-ATPH. This molecular recognition system is highlighted by the first successful conjugate addition of

coordination site
for reagent
R—Li
X
recognition site
for substrate
AlL3
X
Lewis acid receptor

Scheme 2.11

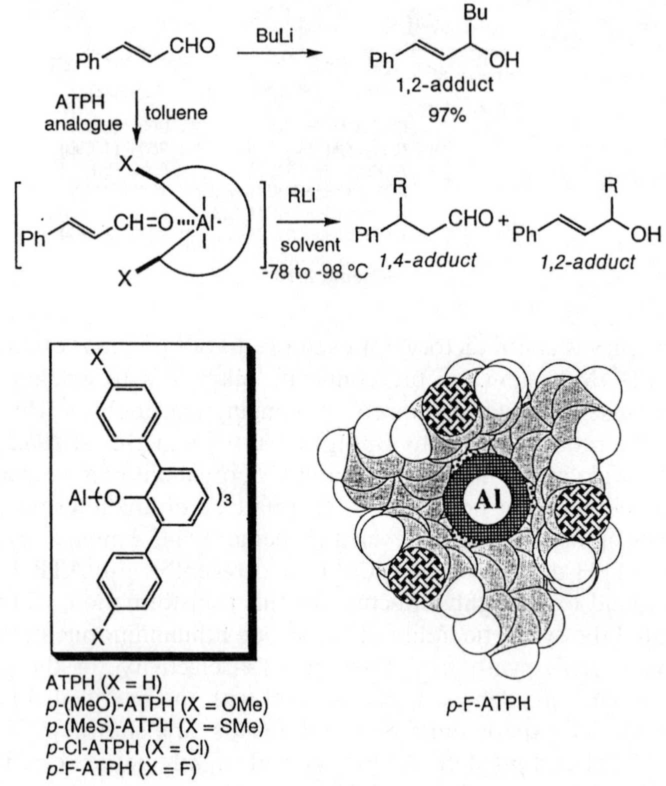

Scheme 2.12

allyl-lithium reagents to α,β-unsaturated aldehydes by complexation with the modified Lewis acid receptor, *p*-F-ATPH (entry 13).

Furthermore, the conjugate addition of prenyl-lithium to cinnamaldehyde proceeded equally well with excellent selectivity under optimized reaction conditions, where the α/γ ratio of the conjugate adducts was profoundly influenced by the solvent effect (Scheme 2.13).

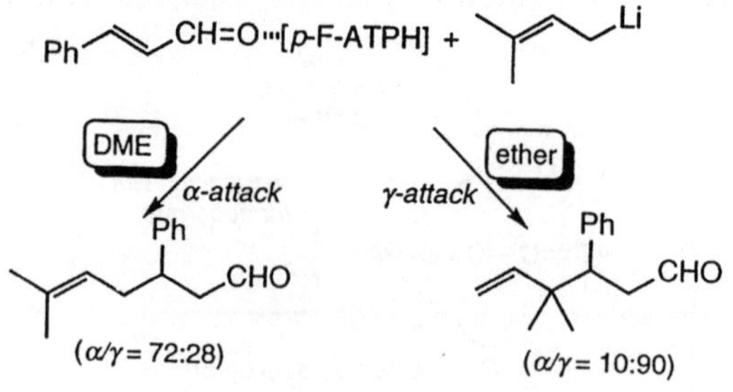

Scheme 2.13

14

Table 2.2 Conjugate addition of RLi to cinnamaldehyde with modified ATPH

Entry	ATPH analogue	RLi/solvent	Temp (°C)	Yield (1,4/1,2 ratio)
1	ATPH	BuLi/hexane	−78	92% (50 : 50)
2	p-(MeO)-ATPH	BuLi/hexane	−78	80% (55 : 45)
3	p-(MeS)-ATPH	BuLi/hexane	−78	91% (57 : 43)
4	p-Cl-ATPH	BuLi/hexane	−78	92% (63 : 37)
5	p-F-ATPH	BuLi/hexane	−78	87% (76 : 24)
6		BuLi/ether	−78	90% (79 : 21)
7		BuLi/THF	−78	82% (86 : 14)
8		BuLi/DME	−78	75% (90 : 10)
9		BuLi/DME	−98	83% (95 : 5)
10	p-F-ATPH	Allyl-Li-ether	−78	94% (77 : 23)
11		Allyl-Li/THF	−78	89% (50 : 50)
12		Allyl-Li/DME	−78	75% (90 : 10)
13		Allyl-Li-DME	−98	83% (95 : 5)

3. Regio- and stereocontrolled Diels–Alder reaction

The Diels–Alder reaction is undoubtedly the best known and most thoroughly investigated of all cycloaddition reactions because of sustained interest in its mechanism and its exceptionally broad application to regio- and stereo-defined synthesis. The rate of this reaction is usually accelerated by the presence of certain Lewis acids. As revealed by the space-filling model, the exceptionally bulky aluminium reagent, MAD (Scheme 2.14), in addition to its Lewis acidity, provides an exceptionally bulky molecular cleft, which may feature a complementary size, shape, and co-ordination capacity for structurally similar ester substrates. Indeed, MAD allows the discrimination of two structurally different ester carbonyls of unsymmetrical fumarates such as *t*-butyl methyl fumarate, where the sterically less hindered methoxycarbonyl moiety binds selectively to the bulky molecular cleft of MAD as revealed by low-temperature ^{13}C NMR spectroscopy. This finding has been successfully applied to the regio- and stereocontrolled Diels–Alder reaction of unsymmetrical fumarates.[26] Thus, the Diels–Alder reaction of the *t*-butyl methyl

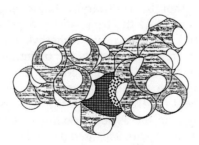

Scheme 2.14. MAD.

15

fumarate/MAD complex with cyclopentadiene at −78°C resulted in stereo-selective formation of the cycloadduct **1** (Scheme 2.15) almost exclusively. In addition, treatment of the 1:1 co-ordination complex with 2-substituted 1,3-butadiene (X = Me, $OSiMe_3$) gave the cycloadduct **2** (X = Me, $OSiMe_3$) with high regioselectivity. In marked contrast, the cycloadditions with Et_2AlCl as an ordinary Lewis acid were found to have a total lack of selectivity.

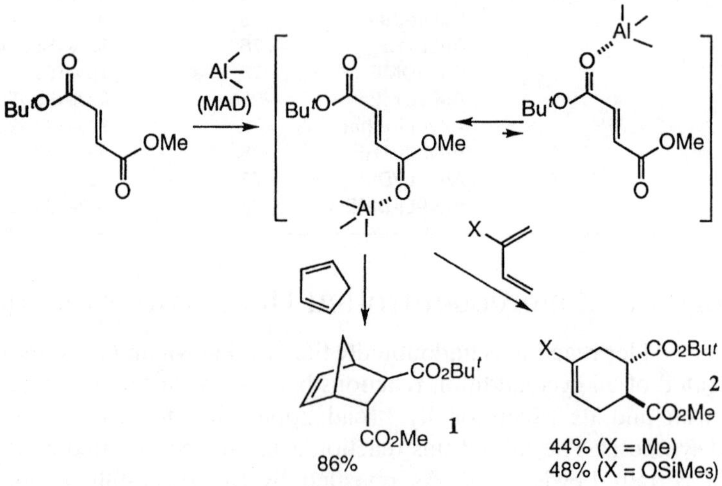

Scheme 2.15

4. Asymmetric hetero-Diels–Alder reaction

Based on the concept of the diastereoselective activation of carbonyl groups with MAD or MAT as described in the section 2.1, the first reliable, bulky chiral organoaluminium reagent, (*R*)-BINAL or (*S*)-BINAL has been devised for enantioselective activation of carbonyl groups. The sterically hindered, enantiomerically pure (*R*)-(+)-3,3'-bis(triarylsilyl)binaphthol((*R*)-**3**) required for the preparation of (*R*)-BINAL can be synthesized in two steps from (*R*)-(+)-3,3'-dibromobinaphthol by bis-triarylsilylation and subsequent intra-molecular 1,3-rearrangement of the triarylsilyl groups as shown in Scheme 2.16.[27] Reaction of (*R*)-**3** in toluene with trimethylaluminium produced the chiral organoaluminium reagent (*R*)-BINAL quantitatively. Its molecular weight, found cryoscopically in benzene, corresponds closely to the value calculated for the monomeric species. The modified chiral organoaluminium reagents, (*R*)-BINAL and (*S*)-BINAL were shown to be highly effective as chiral Lewis acid catalysts in the asymmetric hetero Diels–Alder reaction.[28] Reaction of various aldehydes with activated diene **4** (Scheme 2.17) under the influence of a catalytic amount of BINAL (5–10 mol%) at −20°C, after ex-

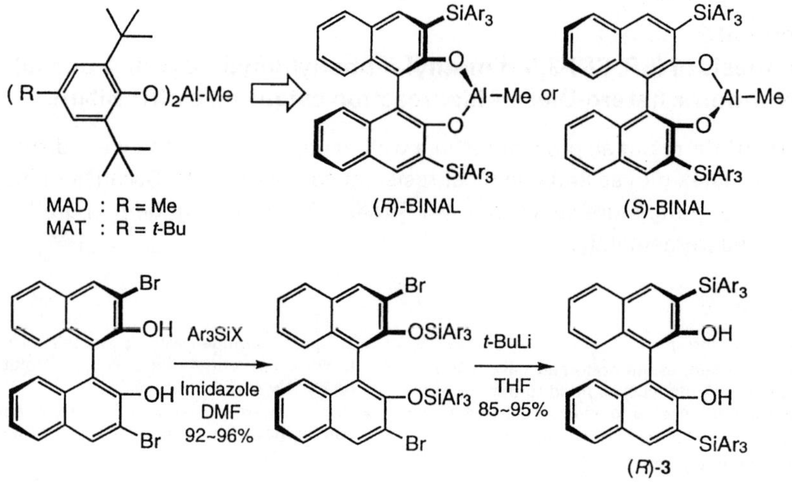

MAD : R = Me
MAT : R = t-Bu

(R)-BINAL

(S)-BINAL

Scheme 2.16

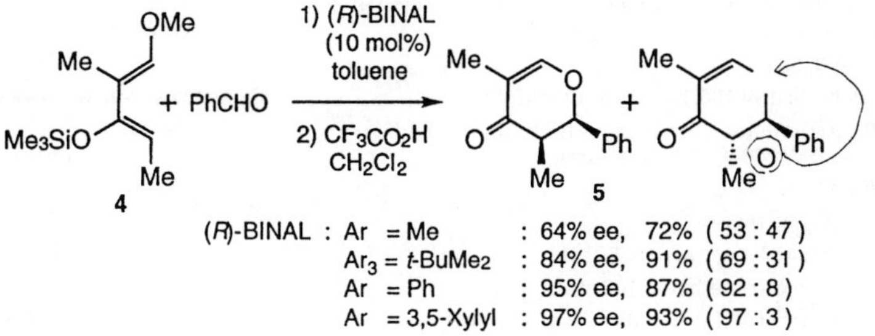

(R)-BINAL : Ar = Me : 64% ee, 72% (53 : 47)
Ar₃ = t-BuMe₂ : 84% ee, 91% (69 : 31)
Ar = Ph : 95% ee, 87% (92 : 8)
Ar = 3,5-Xylyl : 97% ee, 93% (97 : 3)

Scheme 2.17

posure of the resulting hetero Diels–Alder adducts to trifluoroacetic acid, gave predominantly *cis*-dihydropyrone **5** in high yield with excellent enantio-selectivity. The enantioface differentiation of prochiral aldehydes is control-lable by judicious choice of the size of trialkylsilyl moiety in BINAL, thereby allowing the rational design of the catalyst for asymmetric induction. In fact, switching the triarylsilyl substituent (Ar = Ph or 3,5-xylyl) to the *tert*-butyldimethylsilyl or trimethylsilyl group led to a substantial loss of enantio as well as cis selectivity in the hetero Diels–Alder reaction of benzaldehyde and activated diene **4**. In marked contrast, the chiral organoaluminium reagent derived from trimethylaluminium and (R)-(+)-3,3′-dialkylbinaphthol (alkyl = H, Me, or Ph) could be utilized, but only as a stoichiometric reagent and results were disappointing both in terms of reactivity and enantioselectivity for this hetero Diels–Alder reaction.

17

Protocol 2.
Synthesis of (5*R*, 6*R*)-3,5-dimethyl-6-phenyldihydropyran-4-one (5).
Asymmetric hetero-Diels–Alder reaction catalysed by (*R*)-BINAL

Caution! Carry out all procedures in a well-ventilated hood, and wear disposable vinyl or latex gloves and chemical-resistant safety goggles. Because of its high oxygen and moisture sensitivity, (*R*)-BINAL reagent should be prepared *in situ* and used immediately.

Equipment

- Magnetic stirrer
- Three-necked, round-bottomed flask (200 mL). A three-way stopcock is fitted to the top of the flask and connected to a vacuum/argon source
- Teflon-coated magnetic stirring bar
- Medium-gauge needle
- All-glass syringe with a needle-lock Luer (volume appropriate for quantity of solution to be transferred)
- Vacuum/inert gas source (argon or nitrogen source may be an argon or a nitrogen balloon)

Materials

- (*R*)-(+)-3,3′-Bis(triphenylsilyl)binaphthol (FW 800), 0.88 g, 1.1 mmol — **irritant**
- Trimethylaluminum in hexane (FW 72.1), 1 M solution in hexane, 1 mL, 1 mmol — **pyrophoric, moisture sensitive**
- Dry toluene, 50 mL — **flammable, toxic**
- Benzaldehyde (FW 106.1), 1.06 g, 10 mmol — **highly toxic, cancer suspecting agent**
- (1*E*,3*Z*)-2,4-Dimethyl-1-methoxy-3-(trimethylsiloxy)-1,3-butadiene (FW 200.4), 2.20 g, 11 mmol — **moisture sensitive**
- 10% HCl, 50 mL — **toxic**
- 1 M NaHCO₃, 50 mL
- Technical ether for extraction, 50 mL — **flammable, toxic**
- Technical dichloromethane, 100 mL — **toxic, irritant**
- Trifluoroacetic acid (FW 114.0), 1.37 g, 12 mmol — **corrosive, toxic**
- Silica gel for flash chromatography, 200 g, Merck Kieselgel 60 (Art. 9385) — **irritant dust**
- Ether for flash chromatography — **flammable, toxic**
- *n*-Hexane for flash chromatography — **flammable, irritant**

1. Clean all glassware, syringes, needles, and stirring bar and dry at least 2 h in an electric oven at 100°C before use.
2. Assemble the flask, stirring bar, and stop cocks under argon while the apparatus is still hot.
3. Support the assembled flask using a clamp and a stand with a heavy base.
4. Dry the apparatus with an electric heat gun under vacuum (1–2 mm Hg) for 5 min, then back-fill the flask with argon. Repeat to a total of three times.
5. Place (*R*)-(+)-3,3′-bis(triphenylsilyl)binaphthol in the flask and flush with argon.
6. Add dry toluene (50 mL).
7. Stir the mixture, degassed under vacuum, and replaced by argon.

8. Support the bottle containing trimethylaluminium in hexane using a clamp and a stand with a heavy base.

9. Fill a syringe with trimethylaluminium in hexane from the bottle containing trimethylaluminium using argon pressure. Apply the argon pressure to fill the syringe slowly with the required volume. Transfer the reagent in the syringe to the reaction flask at room temperature.

10. Stir the resulting solution at room temperature for 1 h to give (*R*)-BINAL (Ar = Ph) almost quantitatively. During this operation, nearly 2 equivalents of methane gas are evolved per 1 equiv of trimethylaluminium.

11. Cool the reaction vessel to a temperature of −20°C in a dry ice/*o*-xylene bath. *o*-Xylene is recommended as refrigerant in place of carbon tetrachloride (toxic and cancer suspecting agent).

12. Add benzaldehyde and (1*E*,3*Z*)-2,4-dimethyl-1-methoxy-3-(trimethylsiloxy)-1,3-butadiene sequentially at −20°C. Stir the mixture at −20°C for 2 h in order to complete the cycloaddition.

13. Place 10% HCl solution and the stirring bar in the Erlenmeyer flask. While stirring the diluted HCl solution vigorously, add the reaction mixture slowly at 0°C to avoid excessive foaming on hydrolysis.

14. Remove the ice bath and stir the entire mixture vigorously at 25°C for 20 min.

15. Transfer the mixture to a separating funnel and separate the two layers. Extract the water layer with ether twice (2 × 25 mL).

16. Combine the ethereal extracts to a 300 mL flask. Dry the layer over anhydrous magnesium sulfate, and filter through a filter paper. Concentrate the filtrate under reduced pressure by means of a rotary evaporator.

17. Dilute the oily residue with dichloromethane.

18. Add trifluoracetic acid at 0°C, and stir the mixture at 0°C for 1 h.

19. Transfer the mixture to a separating funnel containing diluted Nat/CO₃ solution, and separate the two layers. Extract the water layer with dichloromethane twice (2 × 25 mL).

20. Combine the extracts in a flask. Dry the layer over anhydrous magnesium sulfate, and filter through a filter paper. Concentrate the filtrate under reduced pressure by means of a rotary evaporator.

21. Purify the oily residue by column chromatography on silica gel (ether/hexane = 1:3 as eluents) to give 1.56 g (77%, 95% ee) of (5*R*,6*R*)-3,5-dimethyl-6-phenyldihydropyran-4-one which is characterized by ¹H NMR analysis: [α]ᴅ +7.1° (*c* 1.0, chloroform).

   An interesting method for the preparation of chiral aluminium reagents has appeared recently. The chiral organoaluminium reagent, (*R*)-BINAL or (*S*)-BINAL can be generated *in situ* from the corresponding racemate (±)-BINAL by diastereoselective complexation with certain chiral ketones

(Scheme 2.18).[29] Among several terpene-derived chiral ketones, 3-bromo-camphor was found to be the most satisfactory. The hetero Diels–Alder reaction of benzaldehyde and 2,4-dimethyl-1-methoxy-3-trimethylsiloxy-1,3-butadiene (**4**) with 0.1 equiv. of (±)-BINAL (Ar = Ph) and *d*-bromocamphor at −78 °C gave rise to *cis*-adduct **5** as the major product with 82% ee. Although the level of asymmetric induction attained does not yet match that acquired with the enantiomerically pure BINAL (Ar = Ph, 95% ee), one recrystallization of the *cis*-adduct **5** of 82% ee from hexane gave essentially enantiomerically pure **5**, thereby enhancing the practicality of this method. This study demonstrates the potential for broad application of the *in situ* generated chiral catalyst via diastereoselective complexation in asymmetric synthesis.

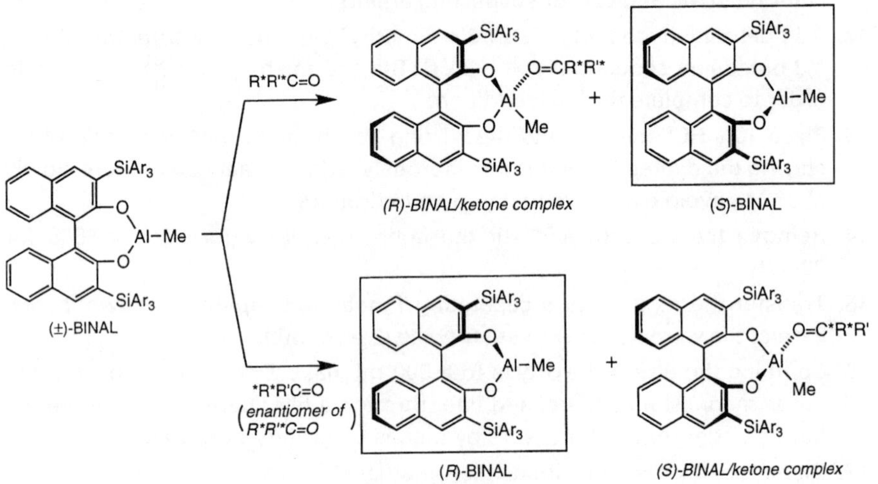

Scheme 2.18

Since the enantioselective activation of carbonyl with the chiral aluminium, (*R*)-BINAL or (*S*)-BINAL, had been demonstrated, the asymmetric ene reaction of electron-deficient aldehydes with various alkenes, by using the latter reagent, could also be considered a feasible transformation.[30] Indeed in the presence of powdered 4Å molecular sieves, the chiral aluminium reagent, (*R*)-BINAL or (*S*)-BINAL can be used as a catalyst without any loss of enantioselectivity (Scheme 2.19).

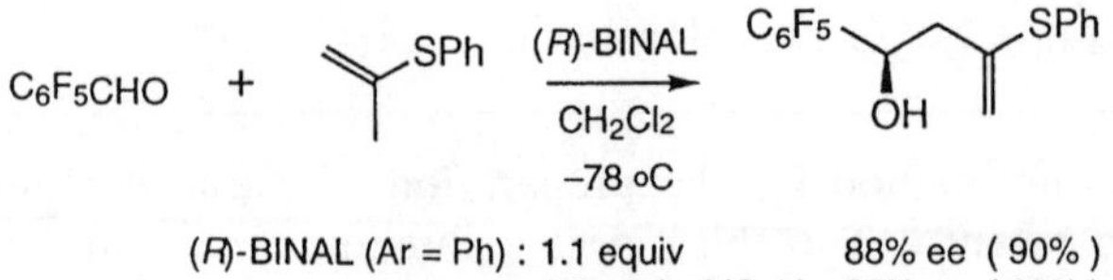

Scheme 2.19

5. Claisen rearrangement

Aliphatic Claisen rearrangements normally require high temperatures. However, in the presence of Lewis acidic organoaluminium reagents, the rearrangements have been accomplished under very mild conditions. Treatment of simple ally vinyl ether substrates with trialkylaluminiums resulted in the [3,3] sigmatropic rearrangement and subsequent alkylation on the aldehyde carbonyl group[31] (Scheme 2.20). Notably, ordinary strong Lewis acids such as $TiCl_4$, $SnCl_4$, and $BF_3 \bullet OEt_2$ could not be used for this rearrangement. The rearrangement–reduction product was obtained exclusively with *i*-Bu_3Al or DIBAH. The aluminium thiolate, Et_2AlSPh or a combination of Et_2AlCl and PPh_3 was effective for the rearrangement providing the normal Claisen products, γ,δ-unsaturated aldehydes, however, without any stereoselectivity. Accordingly, a new molecular recognition approach for the stereocontrolled Claisen rearrangement of allyl vinyl ethers has been developed based on the stereoselective activation of ether moiety using aluminium-type Lewis acid receptors. Thus, treatment of 1-butyl-2-propenyl vinyl ether with ATPH in CH_2Cl_2 afforded *E*-Claisen products predominantly (*E/Z* ratio = 94:6) (Scheme 2.21). Use of sterically more hindered aluminium tris(2-α-naphthyl-6-phenylphenoxide) (ATNP) exhibited better selectivity (*E/Z* ratio = 98:2).[32]

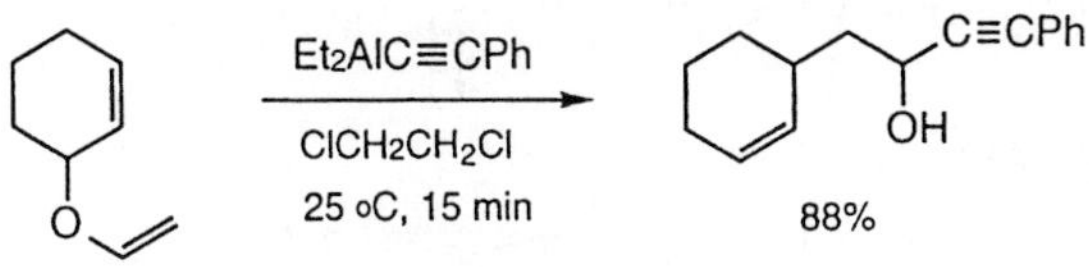

Scheme 2.20

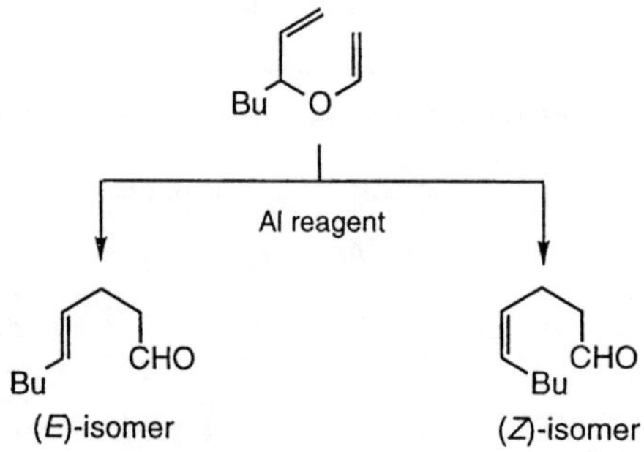

MABR	41%	(*E/Z* = 91 : 9)
Et₂AlSPh	84%	(*E/Z* = 39 : 61)
Et₂AlCl + PPh₃	81%	(*E/Z* = 43 : 57)
ATPH	87%	(*E/Z* = 94 : 6)
ATNP	90%	(*E/Z* = 98 : 2)

Scheme 2.21

21

In marked contrast, use of exceptionally bulky, Lewis acidic receptor, methyl-aluminium bis(2,6-di-*tert*-butyl-4-bromophenoxide) (MABR) resulted in pre-dominant formation of *Z*-Claisen products (*E/Z* ratio = 9:91), which was very difficult to attain by the conventional methodologies including thermal Claisen rearrangement and its variants (Carroll, the orth ester, Eschenmoser, and Ireland rearrangements).[33]

In a similar manner, the concept of the enantioselective activation of car-bonyl groups with the bulky, chiral aluminiums, (*R*)-BINAL or (*S*)-BINAL, has also been extended to the enantioselective activation of an ether oxygen, which gave rise to the first successful example of the asymmetric Claisen re-arrangement of allylic vinyl ethers **6** catalyzed by (*R*)-BINAL or (*S*)-BINAL[34,35] (Scheme 2.22). This method provides an easy asymmetric synthesis of various acylsilanes **7** or **8** (X = SiR$_3$) and acylgermane **7** (X = GeMe$_3$) with high enantiomeric purity (Table 2.3). Among the various trialkylsilyl substituents

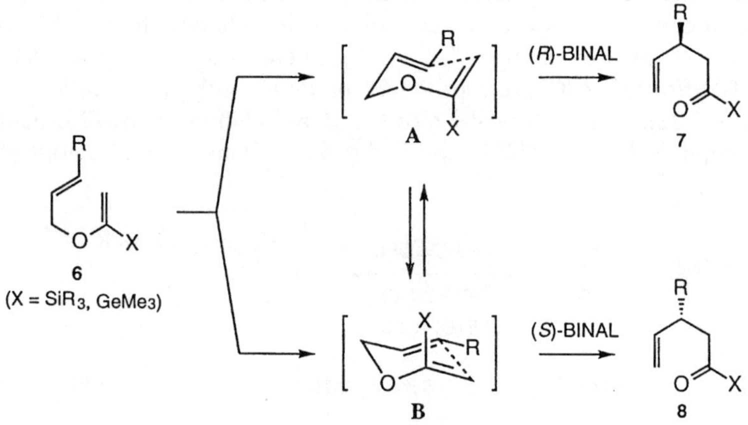

Scheme 2.22

Table 2.3 Asymmetric Claisen rearrangement of allylic vinyl ether **6** (Scheme 2.22)

Allyl vinyl ether	(R)-BINAL	Product	Yield	Optical yield
6 (R=Ph, X=SiMe$_3$)	Ar$_3$=Bu^{t}Me$_2$	**7**	22%	14% ee
6 (R=Ph, X=SiMe$_3$)	Ar=Ph	**7**	86%	80% ee
6 (R=Ph, X=SiMe$_3$)	Ar$_3$=Bu^{t}Ph$_2$	**7**	99%	88% ee
6 (R=Ph, X=SiMe$_3$)	Ar=Ph[a]	**8**	85%	80% ee
6 (R=Ph, X=SiMe$_2$Ph)	Ar=Ph	**7**	65%	85% ee
6 (R=Ph, X=SiMe$_2$Ph)	Ar$_3$=Bu^{t}Ph$_2$	**7**	76%	90% ee
6 (R=cyclohexyl X=SiMe$_3$)	Ar=Ph	**7**	79%	61% ee
6 (R=cyclohexyl X=SiMe$_3$)	Ar$_3$=Bu^{t}Ph$_2$	**7**	84%	71% ee
6 (R=Ph, X=GeMe$_3$)	Ar=Ph	**7**	73%	91% ee
6 (R=Ph, X=GeMe$_3$)	Ar$_3$=Bu^{t}Ph$_2$	**7**	68%	93% ee

[a]Use of (*S*)-BINAL as catalyst.

of BINAL, use of a bulkier *t*-butyldiphenylsilyl group gives rise to the highest enantioselectivity. Conformational analysis of two possible chairlike transition-state structures of an allyl vinyl ether substrate **6** reveals that the chiral organoaluminium reagent BINAL can discriminate between these two conformers **A** amd **B**, which differ from each other only in the orientation of α-methylene groups of ethers.

Notably, the asymmetric Claisen rearrangement of *cis*-allylic α-(trimethyl-silyl)vinyl ethers with (R)-BINAL produced optically active acylsilanes with the same absolute configuration as those produced from *trans*-allylic α-(trimethylsilyl)vinyl ethers[36] (Scheme 2.23).

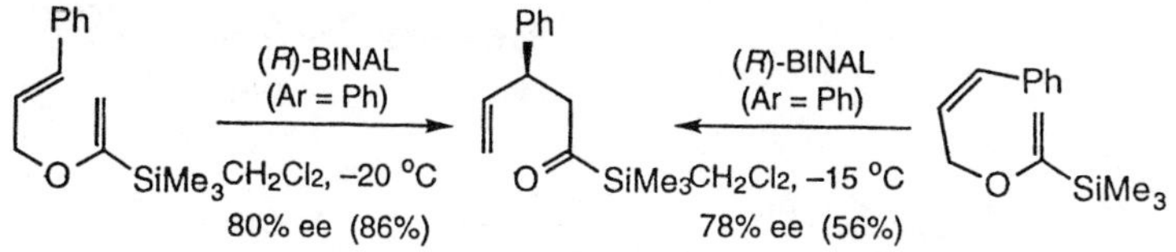

Scheme 2.23

The bulky, chiral organoaluminium reagents of type (R)-BINAL or (S)-BINAL is only applicable for allyl vinyl ether substrates possessing bulky α-silyl and α-germyl substituents. Later, the asymmetric Claisen rearrangement of simple allyl vinyl ether substrates has been developed with the designing of chiral ATPH analogues, aluminium tris((R)-1-α-naphthyl-3-phenylnaphth-oxide) ((R)-ATBN) or aluminium tris((R)-3-(p-fluorophenyl)-1-α-naphthyl-naphthoxide) ((R)-ATBN-F) with high enantioselectivity[32] (Scheme 2.24).

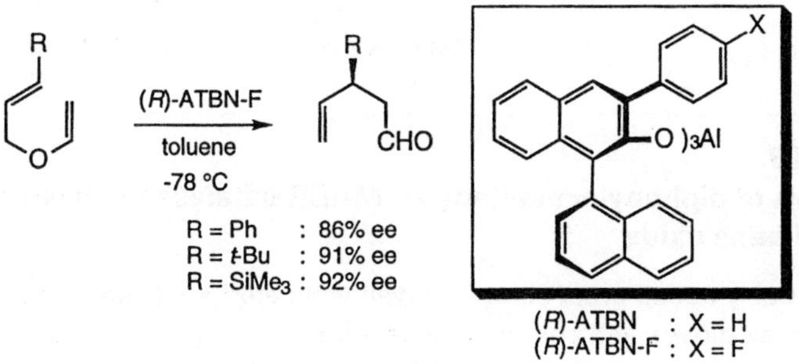

Scheme 2.24

6. Epoxide rearrangement

The exceptional bulkiness of the modified organoaluminium reagent, MABR, can also be utilized for the rearrangement of epoxy substrates under very mild conditions with high efficiency and selectivity. So far, $BF_3 \bullet OEt_2$ catalyst is regarded as a reliable Lewis acid catalyst for the epoxide rearrangement.

However, attempted rearrangement of *tert*-butyldimethylsilyl ether of epoxy citronellol with $BF_3 \cdot OEt_2$ resulted in the formation of a number of products. In contrast, treatment of this substrate with MABR under mild conditions (-78 to $-20\,°C$) gave the desired aldehyde almost quantitatively (Scheme 2.25). In addition, certain epoxy substrates can be rearranged with catalytic use of MABR.[37]

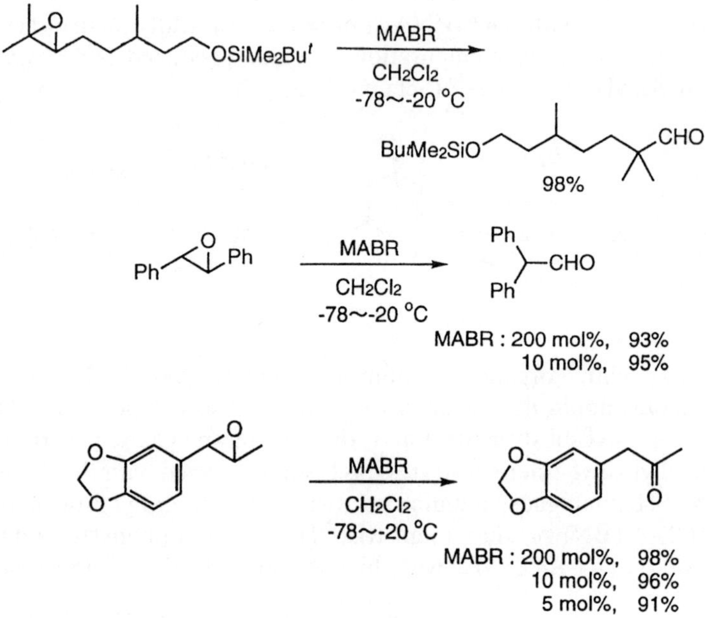

Scheme 2.25

Protocol 3.
Synthesis of diphenylacetaldehyde. MABR-catalysed rearrangement of *trans*-stilbene oxide

Caution! Carry out all procedures in a well-ventilated hood, and wear disposable vinyl or latex gloves and chemical-resistant safety goggles. Because of its high oxygen and moisture sensitivity, MABR reagent should be prepared *in situ* and used immediately.

Equipment

- Magnetic stirrer
- All-glass syringe with a needle-lock Luer (volume appropriate for quantity of solution to be transferred)
- Teflon-coated magnetic stirrer bar
- Vacuum/inert gas source (argon source may be an argon balloon)

- Medium-gauge needle
- Three-necked, round-bottomed flask (1 L) fitted with a condenser and a pressure-equalizing dropping funnel. A three-way stopcock is fitted to the top of the condenser and connected to a vacuum/argon source

Materials

- 4-Bromo-2,6-di-*tert*-butylphenol (FW 285.2), 3.42 g, 12 mmol   **irritant**
- Trimethylaluminium in hexane (FW 72.1), 2 M solution in
  hexane, 3 mL, 6 mmol   **pyrophoric, moisture sensitive**
- Dry dichloromethane, 300 mL   **toxic, irritant**
- *trans*-Stilbene oxide (FW), 11.78 g, 60 mmol
- Sodium fluoride (FW 42.0), 1.01 g, 24 mmol   **moisture sensitive, toxic**
- Water (FW 18), 324 μL, 18 mmol
- Technical dichloromethane   **toxic, irritant**
- Silica gel for flash chromatography, 500 g, Merck Kieselgel
  60 (Art. 9385)   **irritant dust**
- Ether for flash chromatography   **flammable, toxic**
- Dichloromethane for flash chromatography   **toxic, irritant**
- *n*-Hexane for flash chromatography   **flammable, irritant**

1. Clean all glassware, syringes, needles, and stirring bar and dry for at least 2 h in an electric oven at 100 °C before use.

2. Assemble the flask, stirring bar, and stop cocks under argon while the apparatus is still hot.

3. Support the assembled flask using a clamp and a stand with a heavy base.

4. Dry the apparatus with an electric heat gun under vacuum (1–2 mm Hg) for 5 min, then back-fill the flask with argon. Repeat to a total of three times.

5. Place 4-bromo-2,6-di-*tert*-butylphenol and flush with argon.

6. Add freshly distilled dichloromethane (300 mL).

7. Stir the mixture, degas under vacuum, and replace by argon.

8. Support the bottle containing trimethylaluminium in hexane using a clamp and a stand with a heavy base.

9. Fill a syringe with trimethylaluminium in hexane from the bottle containing trimethylaluminium using argon pressure. Apply the argon pressure to fill the syringe slowly with the required volume. Transfer the reagent in the syringe to the reaction flask at room temperature.

10. Stir the resulting solution at this temperature for 1 h to give methylaluminium bis(4-bromo-2,6-di-*tert*-butylphenoxide) (MABR) almost quantitatively. During this operation, nearly 2 equivalents of methane gas are evolved per 1 equiv of trimethylaluminium.

11. Cool the reaction vessel to −20 °C in a dry ice/*o*-xylene bath. *o*-Xylene is recommended as refrigerant in place of carbon tetrachloride (toxic and cancer suspecting agent).

12. Transfer *trans*-stilbene oxide, which is dissolved in 25 mL of dry dichloromethane, to the dropping funnel, and added to the reaction flask over 15–20 min at −20 °C. Stir the mixture at −20 °C for 20 min in order to complete the rearrangement.

13. Add sodium fluoride, and injected water dropwise at −20 °C. To avoid excessive foaming on hydrolysis water should be added carefully by syringe.

Protocol 3. *Continued*

14. Stir the entire mixture vigorously at −20 °C for 5 min at 0 °C for 30 min.

15. Filter the contents of the flask with the aid of three 50 mL portions of dichloromethane. The sodium fluoride–water workup offers an excellent method for large-scale preparations, and is generally applicable for product isolation in the reaction of organoaluminium compounds.

16. Evaporate the combined filtrates under reduced pressure with a rotary evaporator. Purify the oily residue by column chromatography (column diameter: 9.5 cm) on silica gel (ether/dichloromethane/hexane = 1:2:20 to 1:1:10 as eluants) to give 11.02–11.17 g (94–95%) of diphenylacetaldehyde as a colourless oil, which is characterized by ^{1}H NMR analysis.

Although the acid-catalysed rearrangement of epoxides to carbonyl compounds is a well-known transformation and a number of reagents have been elaborated for this purpose, only a few reagents have been used successfully for the rearrangement of functionalized epoxides with respect to the efficiency and selectivity of the reaction. With the stoichiometric use of MABR, however, a new, stereocontrolled rearrangement of epoxy silyl ethers leading to β-siloxy aldehydes has been developed under mild conditions. Interestingly, used in combination with the Sharpless asymmetric epoxidation of allylic alcohols, this rearrangement represents a new approach to the synthesis of various optically active β-hydroxy aldehydes,[38,39] (Scheme 2.26) which are

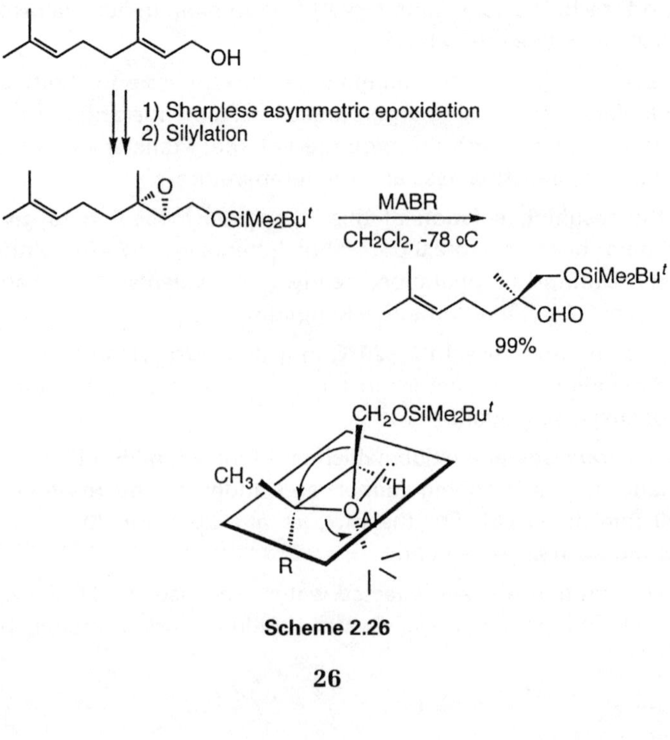

Scheme 2.26

quite useful intermediates in natural product synthesis. Based on the optical rotation sign and value of the β-siloxy aldehydes, this organoaluminium-promoted rearrangement proceeds with rigorous transfer of the epoxide chirality, and the observed stereoselectivity can be interpreted to arise from the *anti* migration of the siloxymethyl group to the epoxide moiety. Here

Table 2.4 MABR-catalysed rearrangement of optically active epoxy silyl ethers

Entry	Mol % of MABR	Conditions (°C, h)	Chemical yield

1	200	−78, 0.5	95%
2	20	−20, 0.3	82%
3	10	−20, 0.3	74%

4	200	−78, 1	99%
5	20	−78, 0.2; 0, 0.5	82%
6	10	−78, 0.2; 0. 1	74%

7	200	−78, 1; −40, 0.5	98%
8	20	−78, 0.2; 0, 1	79%
9	10	−78, 0.2; 0, 3	68%

10	200	−78, 0.5	87%
11	20	−40, 0.5	75%
12	10	−78, 0.2; −20, 0.5, 0, 0.5	71%

| 13 | 200 | −78, 0.5; −20, 2 | 88% |
| 14 | 20 | −78, 0.2; 0, 5 | 77% |

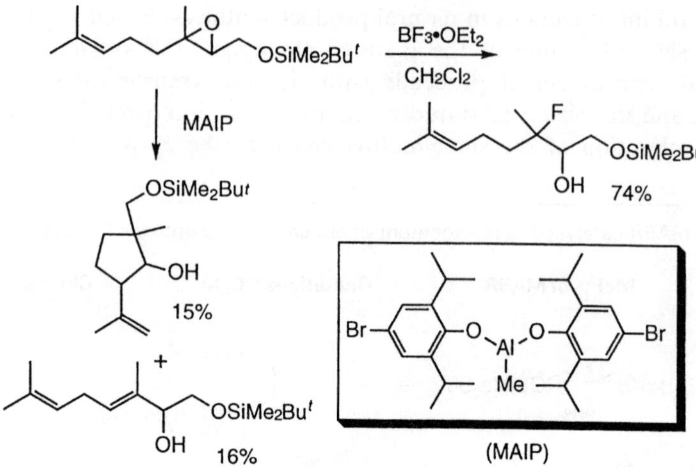

Scheme 2.26

again, the exceptional bulkiness of 2,6-di-*tert*-butyl-4-bromophenoxy ligands in MABR is essential for the smooth rearrangement of epoxy silyl ethers, and the less bulky methylaluminium bis(4-bromo-2,6-di-isopropylphenoxide) (MAIP) was found to be totally ineffective for the rearrangement of *tert*-butyldimethylsilyl ether of epoxygeraniol (Table 2.4). Again, $BF_3 \cdot OEt_2$ as an ordinary Lewis acid gave fluorohydrines as sole isolable products (Scheme 2.27).

References

1. Mole, T.; Jeffery, E. A. *Organoaluminum compounds.* Elsevier, Amsterdam **1972**.
2. Reinheckel, H.; Haage, K.; Jahnke, D. *Organometal. Chem. Rev. A* **1969**, *4*, 47.
3. Lehmkuhl, H.; Ziegler, K.; Gellert, H.; Houben-Weyl, G. *Methoden der Organischen Chemie* (4th edn) Vol. XIII, Part 4. Thieme, Stuttgart, **1970**.
4. Bruno, G. *The use of aliminum alkalys in organic synthesis.* Ethyl Corporation, Baton Rouge, FL, USA, **1970, 1973, 1980**.
5. Negishi, E. *J. Organometal. Chem. Libr.* **1982**, *1*, 93.
6. Yamamoto, H.; Nozaki, H. *Angew. Chem. Int. Ed. Engl.* **1978**, *17*, 169.
7. Negishi, E. *Organometallics in organic synthesis.* Vol. 1, Wiley, New York, **1980**, p. 286.
8. Eisch, J. J. In: *Comprehensive organometallic chemistry* (ed. G. Wilkinson, F. G. A. Stone, E. W. Abel). Pergamon Press, Oxford, **1982**, Vol. 1, p 555.
9. Zietz, Jr, J. R.; Robinson, G. C.; Leindsay, K. L. In: *Comprehensive organometallic chemistry* (ed. G. Wilkinson, F. G. A. Stone, E. W. Abel). Pergamon Press, Oxford, **1982**, Vol. 7, p 365.
10. Zweifel, G.; Miller, J. A. *Org. React.* **1984**, *32*, 375.
11. Maruoka, K.; Yamamoto, H. *Angew. Chem. Int. Ed. Engl.* **1985**, *24*, 668.

12. Maruoka, K.; Yamamoto, H. *Tetrahedron* **1988**, *44*, 5001.
13. Maruoka, K. In: *Synthesis of Organometallic Compounds* (ed. S. Komiya). Wiley, Chichester, **1997**.
14. Ziegler, K. In: *Organometallic Chemistry* (ed. H. Zeiss). Reinhold, New York, **1960**, p 194.
15. Wilke, G. In: *Coordination Polymerization* (ed. J. C. W. Chien). Academic Press, New York, **1975**.
16. Strarowieyski, K. B.; Pasynkiewics, S.; Skowronska-Ptasinska, M. *J. Organometal. Chem.* **1975**, *90*, C43.
17. Maruoka, K.; Itoh, T.; Yamamoto, H. *J. Am. Chem. Soc.* **1985**, *107*, 4573.
18. Maruoka, K.; Sakurai, M.; Yamamoto, H. *Tetrahedron Lett.* **1985**, *26*, 3853.
19. Maruoka, K.; Imoto, H.; Saito, S.; Yamamoto, H. *J. Am. Chem. Soc.* **1994**, *116*, 4131.
20. Maruoka, K.; Ito, M.; Yamamoto, H. *J. Am. Chem. Soc.*, **1995**, *117*, 9091.
21. Wakefield, B. J. *The chemistry of organolithium compounds*. Pergamon, Oxford, **1974**.
22. Beswick, M. A.; Wright, D. S. In: *Comprehensive organometallic chemistry II* (ed. E. W. Abel, F. G. A. Stone, G. Wilkinson). Pergamon, Oxford, **1995**, Vol. 1, 1.
23. Gray, M.; Tinkl, M.; Snieckus, V. In: *Comprehensive organometallic chemistry II* (ed. E. W. Abel, F. G. A. Stone, G. Wilkinson). Pergamon, Oxford, **1995**, Vol. 11, 1.
24. Wakefield, B. J. *Organolithium Method*. Academic Press, London, **1988**.
25. Ooi, T.; Kondo, Y.; Maruoka, K. *Angew. Chem. Int. Ed. Engl.*, **1997**, *36*, 1183.
26. Maruoka, K.; Saito, S.; Yamamoto, H. *J. Am. Chem. Soc.* **1992**, *114*, 1089.
27. Maruoka, K.; Ito, T.; Araki, Y.; Shirasaka, T.; Yamamoto, H. *Bull. Chem. Soc. Jpn* **1988**, *61*, 2975.
28. Maruoka, K.; Ito, T.; Shirasaka, T.; Yamamoto, H. *J. Am. Chem. Soc.* **1988**, *110*, 310.
29. Maruoka, K.; Yamamoto, H. *J. Am. Chem. Soc.* **1989**, *111*, 789.
30. Maruoka, K.; Hoshino, Y.; Shirasaka, T.; Yamamoto, H. *Tetrahedron Lett.*, **1988**, *29*, 3967.
31. Takai, K.; Mori, I.; Oshima, K.; Nozaki, H. *Bull. Chem. Soc., Jpn* **1986**, *57*, 446.
32. Maruoka, K.; Saito, S.; Yamamoto, H. *J. Am. Chem. Soc.* **1995**, *117*, 6153.
33. Nonoshita, K.; Banno, H.; Maruoka, K.; Yamamoto, H. *J. Am. Chem. Soc.* **1990**, *112*, 316.
34. Maruoka, K.; Banno, H.; Yamamoto, H. *J. Am. Chem. Soc.* **1990**, *112*, 7791.
35. Maruoka, K.; Banno, H.; Yamamoto, H. *Tetrahedron-Asymmetry* **1991**, *2*, 647.
36. Maruoka, K.; Yamamoto, H. *Synlett* **1991**, 793.
37. Maruoka, K.; Nagahara, S.; Ooi, T.; Yamamoto, H. *Tetrahedron Lett.* **1989**, *30*, 5607.
38. Maruoka, K.; Ooi, T.; Yamamoto, H. *J. Am. Chem. Soc.* **1989**, *111*, 6431.
39. Maruoka, K.; Ooi, T.; Nagahara, S.; Yamamoto, H. *Tetrahedron* **1991**, *47*, 6983.

3

Boron reagents

KAZUAKI ISHIHARA

1. Introduction

Arylboron reagents with electron-withdrawing aromatic groups and chiral boron reagents as Lewis acid catalysts are described in this chapter.

The classical boron Lewis acids, BX_3, RBX_2 and R_2BX (X = F, Cl, Br, OTf) have become popular tools in organic synthesis. In general, these are used stoichiometrically in organic transformations under anhydrous conditions, since the presence of even a small amount of water causes rapid decomposition or deactivation of the promoters. To obviate some of these inherent problems, we have demonstrated the potential of tris(pentafluorophenyl)boron (**1**), bis(pentafluorophenyl)borinic acid (**2**), bis(3,5-bis(trifluoromethyl)phenyl) borinic acid (**3**), 3,4,5-trifluorophenylboronic acid (**4**), and 3,5-bis(trifluoromethyl)phenylboronic acid (**5**) as a new class of boron catalysts.

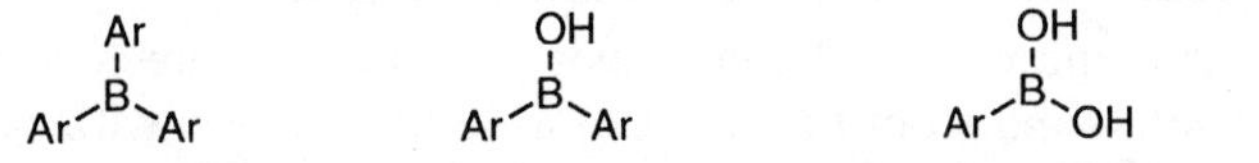

Ar   OH   OH
|
Ar–B–Ar   Ar–B–Ar   Ar–B–OH

1 (Ar=C_6F_5)   **2** (Ar=C_6F_5)   **4** (Ar=3,4,5-$F_3C_6H_2$)
                       3 (Ar=3,5-$(CF_3)_2C_6H_3$)   **5** (Ar=3,5-$(CF_3)_2C_6H_3$)

Chiral boron catalysts have been widely used as Lewis acids in the asymmetric Diels–Alder, Mukaiyama aldol, Sakurai–Hosomi allylation, and aldol-type reactions of imines. As one successful example, we have achieved highly enantioselective carbo-Diels–Alder, hetero-Diels–Alder, aldol, and allylation reactions using a common chiral (acyloxy)borane (CAB) catalyst as depicted in Scheme 3.1. These carbon–carbon bond formations are very important and useful in asymmetric synthesis.

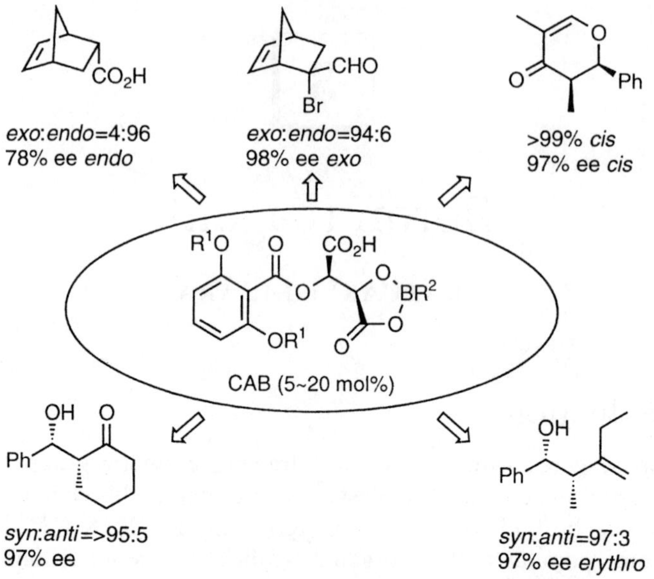
CO2H
exo:endo=4:96
78% ee endo
CHO
Br
exo:endo=94:6
98% ee exo
O
O
Ph
>99% cis
97% ee cis
R1O
O
CO2H
O
O
BR2
O
OR1
O
CAB (5~20 mol%)
OH
O
Ph
syn:anti=>95:5
97% ee
OH
Ph
syn:anti=97:3
97% ee erythro

Scheme 3.1

2. Arylboron reagents with electron-withdrawing aromatic group

2.1 Tris(pentafluorophenyl)boron as an efficient, air stable, and water-tolerant Lewis acid catalyst

1 is an air stable and water-tolerant Lewis acid catalyst, which is readily prepared as a white solid from boron trichloride by reaction with pentafluorophenyl-lithium.[1,2] This reagent does not react with pure oxygen.[1,2] It is very thermally stable, even at 270 °C, and soluble in many organic solvents.[1,2] Although **1** is available when exposed to air (not anhydrous grade), it acts better as a catalyst under anhydrous conditions.

2.1.1 Mukaiyama aldol reactions of silyl enol ethers with aldehydes or other electrophiles[3,5]

The aldol-type reactions of various silyl enol ethers with aldehydes or other electrophiles proceed smoothly in the presence of 2–10 mol% of **1**. Although the reaction is remarkably promoted by using an anhydrous solution of **1**, the reaction of silyl enol ethers with a commercial aqueous solution of formaldehyde takes place without incident. Silyl enol ethers react with chloromethyl methyl ether or trimethylorthoformate; hydroxymethyl, methoxymethyl, or dimethoxymethyl C1 groups can be introduced at the α-position of the carbonyl group (Scheme 3.2).

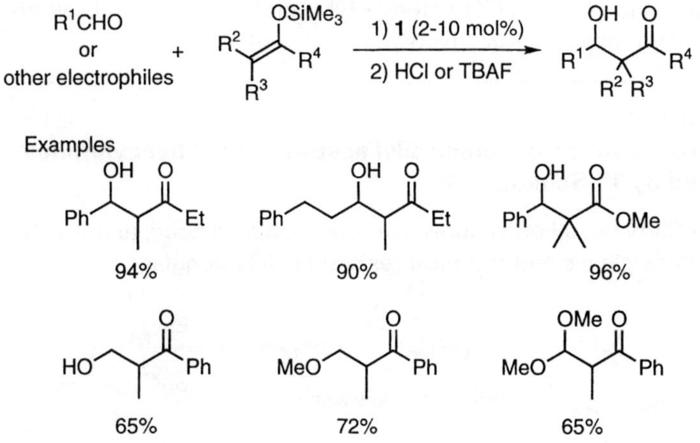

Scheme 3.2

2.1.2 Aldol-type reaction of ketene silyl acetals with *N*-benzylimines[4,5]

1 (anhydrous grade) is a highly active catalyst for the aldol-type reaction between ketene silyl acetals and *N*-benzylimines because of its stability and comparatively low value of bond energy and affinity toward nitrogen-containing compounds (Scheme 3.3). *N*-Benzylimines are useful substrates because β-benzylamino acid esters produced are readily debenzylated by hydrogenolysis over palladium on carbon. Catalysis is carried out using 0.2–10 mol% catalyst loading in toluene. The condensation proceeds smoothly even with aliphatic enolizable imines derived from primary and secondary aliphatic

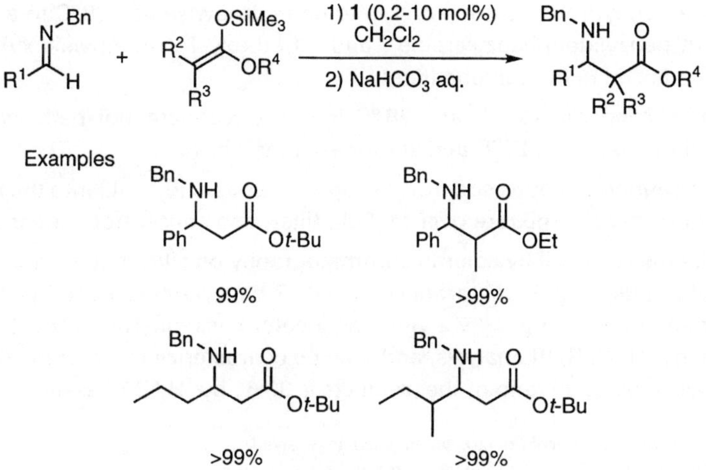

Scheme 3.3

Kazuaki Ishihara

aldehydes, and (E)- and (Z)-ketene silyl acetals give *anti* and *syn* products as major diastereomers, respectively.

Protocol 1.
Aldol-type reaction of ketene silyl acetals with *N*-benzylimines catalysed by 1[5] (Scheme 3.4)

Caution! Carry out all procedures in a well-ventilated hood, and wear disposable vinyl or latex gloves and chemical-resistant safety goggles.

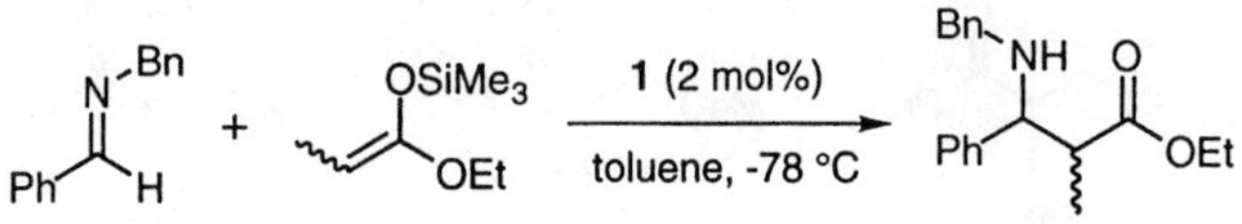

Scheme 3.4

Equipment

- Two-necked, round-bottomed flask (25 mL)
- Separating funnel (50 mL)
- Magnetic stirrer bar
- Sintered glass filter funnel
- Magnetic stirrer
- Erlenmeyer flasks (50 mL)
- Three-way stopcock
- Column for flash chromatography
- Vacuum/argon source

Materials

• Anhydrous 1 (0.247 M) in toluene,[a] 24 μL, 0.006 mmol	flammable toxic
• *N*-Benzylidenebenzylamine (FW 195.3),[5] 116.8 mg, 0.6 mmol	moisture sensitive
• 1-Ethoxy-1-(trimethylsiloxy)propene[5] (FW 174.3), 208.9 mg, 1.2 mmol	flammable
• Dry toluene[b]	flammable toxic
• Silica gel for flash chromatography, Merck 9385	irritant dust

1. Add an anhydrous solution of **1** in toluene dropwise at −78°C to a solution of *N*-benzylidenebenzylamine and 1-ethoxy-1-(trimethylsiloxy)propene (*E/Z* = 85:15) in dry toluene (6 mL).

2. After stirring for 13.5 h at −78°C in a dry ice/methanol bath, warm the reaction mixture to 25°C and stir for another 2 h.

3. After pouring aqueous sodium hydrogencarbonate (0.1 mL) into the resultant solution, dry the mixture over MgSO₄, filter, and concentrate under vacuum.

4. Purify the crude oil by column chromatography on silica gel (eluant: hexane–ethylacetate, 15:1) to produce ethyl 3-benzylamino-2-methyl-3-phenyl-propanoate (178 mg, >99% yield) as a colourless oil. The product is >98% pure by ^{1}H NMR, IR analysis, and may be characterized further by elemental analysis. *Syn/anti* ratio of the products is 36:64 by ^{1}H NMR assay.

---

[a] From Toso-Akuzo Chemical Co. Ltd, Japan; used as received.
[b] Distil toluene from calcium hydride under argon.

2.1.3 Stereoselective rearrangement of epoxides[6]

The protic or Lewis acid-promoted rearrangement of epoxides to carbonyl compounds is a well-known synthetic transformation. $BF_3 \bullet OEt_2$ appears to be the most widely used Lewis acid for rearrangement. It is often consumed or altered in the course of these reactions, and is thus a reagent rather than a catalyst, although less than an equivalent is effective in some instances. We have found that **1** is a highly efficient catalyst in the rearrangement of epoxides. The rearrangement of trisubstituted epoxides proceeds successfully in the presence of catalytic amounts of **1** (anhydrous grade) *via* a highly selective alkyl shift to give the corresponding aldehydes (Scheme 3.5). The exceptional bulkiness of **1** may be effective in the selective rearrangement of epoxides via an alkyl shift.

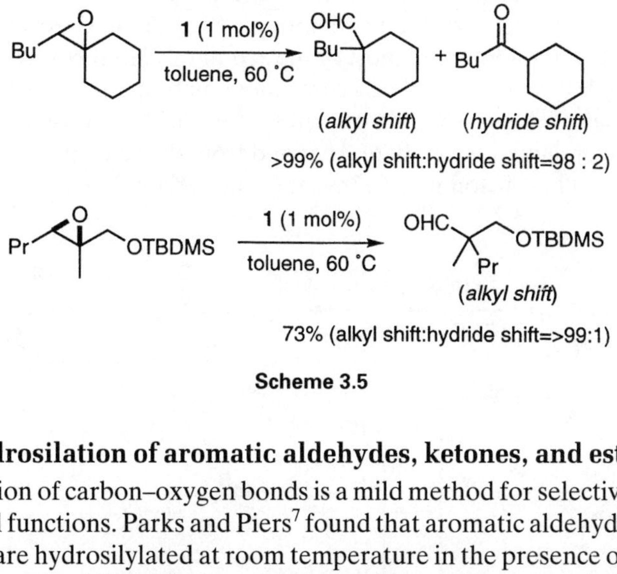

>99% (alkyl shift:hydride shift=98 : 2)

73% (alkyl shift:hydride shift=>99:1)

Scheme 3.5

2.1.4 Hydrosilation of aromatic aldehydes, ketones, and esters[7]

Hydrosilation of carbon–oxygen bonds is a mild method for selective reduction of carbonyl functions. Parks and Piers[7] found that aromatic aldehydes, ketones, and esters are hydrosilylated at room temperature in the presence of 1–4 mol% of **1** and 1 equiv of Ph_3SiH. The reduction takes place by an unusual nucleophilic/electrophilic mechanism: the substrate itself serves to nucleophilically activate the Si–H bond, while hydride transfer is facilitated by **1** (Scheme 3.6).

2.2 Diarylborinic acids as efficient catalysts for selective dehydration of aldols[8]

Diarylborinic acids with electron-withdrawing aromatic groups are effective for Mukaiyama aldol condensation.[8] The catalytic activities of diarylborinic acids **2** and **3** are much higher than those of the corresponding arylboronic acids. Recently, we developed the selective dehydration of β-hydroxy carbonyl compounds catalysed by diarylborinic acids.[8] The dehydration is strongly promoted in tetrahydrofuran (THF). In most cases, the reaction proceeds smoothly, and

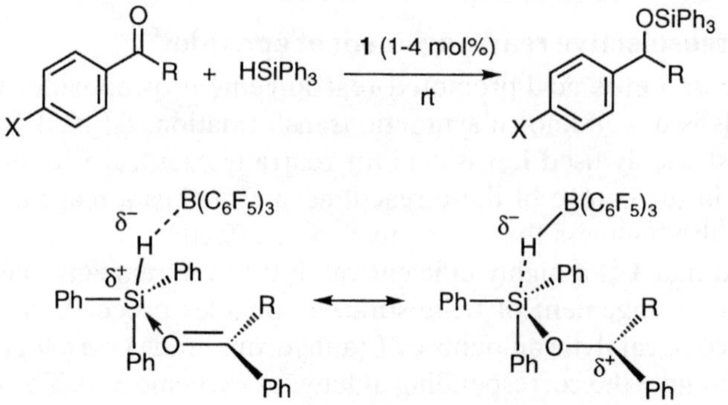

Scheme 3.6

α,β-enones are obtained as *E* isomers in high yields. In the reaction of α-substituted-β-hydroxy carbonyl compounds, α,β-enones are preferentially obtained from *anti*-aldols, and most of the *syn*-aldols are recovered. Thus, the present dehydration is a useful and convenient method for isolating pure *syn*-aldol from *syn*- and *anti*-isomeric mixtures. The transformation to α,β-enones occurs *via* an enolate intermediate **7** derived from the selective deprotonation of a pseudo-axial α-proton perpendicular to the carbonyl face of a cyclic intermediate **6** (Scheme 3.7).

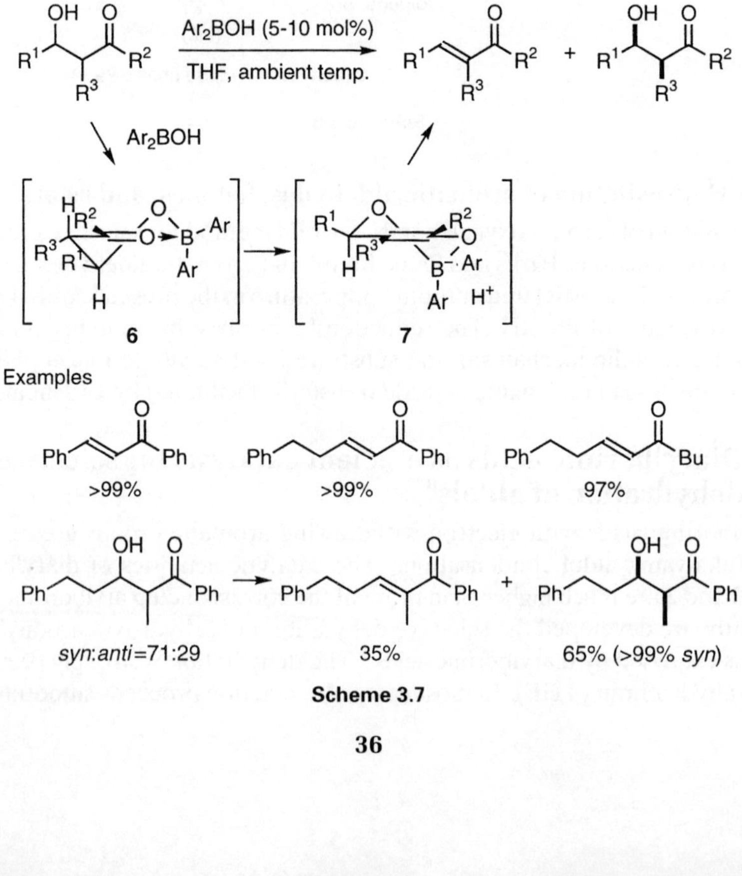

Scheme 3.7

36

Protocol 2.
Dehydration of aldols catalysed by 2[8] (Scheme 3.8)

Caution! Carry out all procedures in a well-ventilated hood, and wear disposable vinyl or latex gloves and chemical-resistant safety goggles.

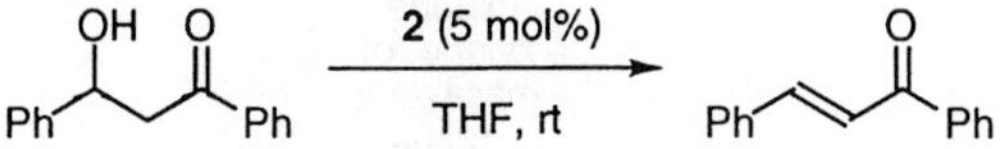

Scheme 3.8

Equipment

- 3 Round-bottomed flask (5 mL)
- Separating funnel (50 mL)
- Magnetic stirrer bar
- Sintered glass filter funnel
- Magnetic stirrer

- Erlenmeyer flasks (50 mL)
- Pressure-equalized addition funnel (10 mL)
- Column for flash chromatography
- Three-way stopcock
- Vacuum/argon source

Materials

• 2[a] (FW 361.9), 3.6 mg, 0.01 mmol	**Toxic, white solid**
• 1,3-Diphenyl-3-hydroxy-1-propanone (FW 226.3), 45 mg, 0.2 mmol	**irritant**
• Dry THF[b]	**flammable irritant**
• Silica gel for flash chromatography, Merck 9385	**irritant dust**

1. Add **2** to a solution of 1,3-diophenyl-3-hydroxy-1-propanone in THF (1 mL) and stir the reaction mixture at room temperature for 2 h.

2. Pour the reaction mixture into 1M NaOH aq., and extract with CH_2Cl_2 three times. Dry the combined organic extracts over $MgSO_4$ filter, and concentrate in vacuum.

3. Purify the crude product by column chromatography on silica gel using a mixture of hexane and ethyl acetate (5/1) as eluent to give chalcone (40 mg, 0.19 mmol, 96%). The product is >98% pure by ^{1}H NMR, IR analysis, and may be characterized further by elemental analysis.

[a] **2** can be prepared by hydrolysis of the known bis(pentafluorophenyl)boron chloride. Chambers, R. D.; Chivers, T. *J. Chem. Soc.* **1965**, 3933.
[b] Distil THF from sodium and benzophenone under argon.

2.3 3,4,5-Trifluorophenylboronic acid as an amidation catalyst[9]

There are several different routes to carboxamides. In most cases, a carboxylic acid is converted into a more reactive intermediate, e.g. acid chloride, which is then allowed to react with an amine. For practical reasons, it is preferable to form the reactive intermediate *in situ*. We have found that arylboronic acids with electron-withdrawing groups **4** and **5** act as highly efficient catalysts in the amidation between carboxylic acids and amines (Scheme 3.9). The catalytic amidation of optically active aliphatic α-hydroxycarboxylic acids with benzylamine proceeds with no measurable loss of enantiomeric purity.

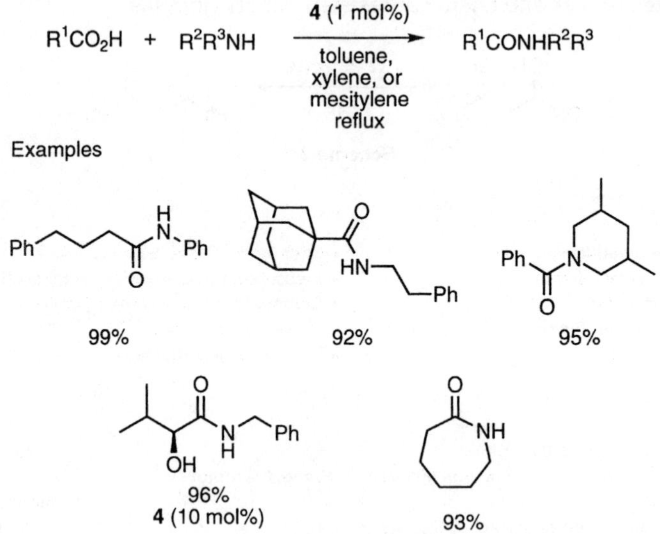

$$R^1CO_2H \; + \; R^2R^3NH \xrightarrow[\substack{\text{toluene,}\\\text{xylene, or}\\\text{mesitylene}\\\text{reflux}}]{\textbf{4} \text{ (1 mol\%)}} R^1CONHR^2R^3$$

Examples

99%   92%   95%

96%
4 (10 mol%)   93%

Scheme 3.9

Protocol 3.
Amidation reaction of carboxylic acids and amines catalysed by 4[9] (Scheme 3.10)

Caution! Carry out all procedures in a well-ventilated hood, and wear disposable vinyl or latex gloves and chemical-resistant safety goggles.

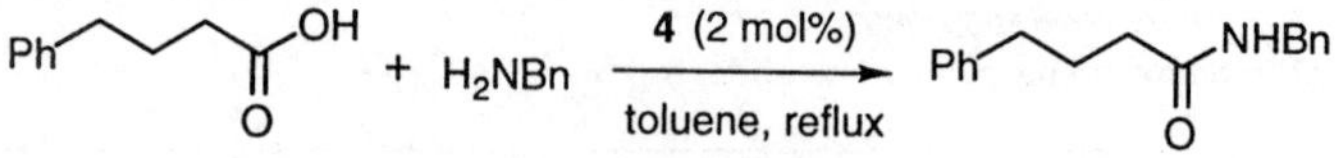

Scheme 3.10

Equipment

- Round-bottomed flask (50 mL)
- Separating funnel (50 mL)
- Magnetic stirrer bar
- Sintered glass filter funnel
- Magnetic stirrer
- Erlenmeyer flasks (50 mL)

- Pressure-equalized addition funnel (10 mL)
- Column for flash chromatography
- Three-way stopcock
- Vacuum/argon source
- Reflux condenser
- Oil bath

Materials

- **4**[a] (FW 175.9), 8.8 mg, 0.05 mmol   **toxic, white solid**
- 4-Phenylbutyric acid (FW 164.2), 821 mg, 5 mmol   **irritant**
- Benzylamine (FW 107.16), 546 μL, 5 mmol   **flammable**
- Dry toluene[a]   **flammable toxic**
- 4-Å molecular sieves ca. 4 g   **pellets**

1. Add 4-phenylbutyric acid, benzylamine, and **4** in 25 mL of toluene to a dry, round-bottomed flask fitted with a stirrer bar and a pressure-equalized addition funnel (containing a cotton plug and 4-Å molecular sieves and functioning as a Soxhlet extractor) surmounted by a reflux condenser.
2. Bring the mixture to reflux by the removal of water.
3. After 18 h, cool the resulting mixture to ambient temperature, wash with aqueous ammonium chloride and aqueous sodium hydrogencarbonate successively, and extract the product with ethyl acetate. Dry the combined organic layers over magnesium sulfate. Evaporate the solvent, and purify the residue by column chromatography on silica gel (eluant: hexane–ethyl acetate = 3:1) to give N-benzyl-4-phenylbutyramide (1.221 g, 96% yield). The product is >98% pure by ^{1}H NMR, IR analysis, and may be characterized further by elemental analysis.

[a] Distil toluene from calcium hydride under argon.

---

3. Chiral boron reagents as Lewis acids

3.1 Catalytic enantioselective carbo-Diels–Alder reactions

The asymmetric Diels–Alder reaction is now of great interest because of its potential to introduce some asymmetric centres simultaneously during carbon–carbon bond formation.

Kaufmann and Boese[10] developed asymmetric catalyst **8** derived from $H_2BBr\text{-}SMe_2$ and 1,1-binaphthol. The diborate structure with a propeller-like shape has been established by X-ray analysis. The reaction between methacrolein and cyclopentadiene catalysed by **8** gives the cycloadducts with 97% exo selectivity and 90% ee.

8

Using another promising approach[11–16] it was found that an (acyloxy)-borane $RCO_2BR'_2$ behaves as a Lewis acid, and the chiral (acyloxy)borane

Kazuaki Ishihara

(CAB) **9** is an excellent asymmetric catalyst for the Diels–Alder reaction between cyclopentadiene and acrylic acid[11] or methacrolein[12,13] (Scheme 3.11). The α-substituent on the dienophile increases the enantioselectivity. When there is a β-substitution on the dienophile, the cycloadduct is almost racemic, but for a substrate having substituents at both α- and β-positions, high ee's have been observed. According to NOE studies, the effective shielding of the si-face of the co-ordinated α,β-enal arises from π-stacking of the 2,6-di-iso-propoxybenzene ring and the co-ordinated aldehyde.

CAB **9** (R¹=Me, R²=H)
CAB **10** (R¹=i-Pr, R²=H
or o-PhOC₆H₄)

Examples

9 (10 mol%)
exo/endo: 4/96
endo: 78% ee

9 (10 mol%)
exo/endo: 89/11
exo: 96% ee

9 (10 mol%)
exo/endo: 4/96
exo: 92% ee

Scheme 3.11

Protocol 4.
Enantioselective Diels–Alder reaction catalysed by CAB 9[13] (Scheme 3.12)

Caution! Carry out all procedures in a well-ventilated hood, and wear disposable vinyl or latex gloves and chemical-resistant safety goggles.

Scheme 3.12

Equipment

- Three-necked round-bottomed flask (100 mL)
- Separating funnel (500 mL)
- Magnetic stirrer bar
- Sintered glass filter funnel
- Magnetic stirrer
- Erlenmeyer flasks (500 mL)

- Pressure-equalized addition funnel (10 mL)
- Column for flash chromatography
- Three-way stopcock
- Vacuum/argon source
- Rubber septum

40

Materials

- Mono (2,6-dimethoxybenzoyl)tartaric acid[13] (FW 314.2),
  1.57 g, 5 mmol — **irritant**
- Borane–THF (1.40 M)[a], 3.57 mL, 5 mmol — **flammable liquid, moisture-sensitive**
- Methacrolein (FW 70.09)[b], 4.14 mL, 50 mmol — **flammable liquid, corrosive**
- 2,3-Dimethyl-1,3-butadiene (FW 82.15)[c], 8.49 mL, 75 mmol — **flammable liquid**
- Dry dichloromethane[d] — **flammable toxic**
- 4-Å molecular sieves ca. 4 g — **pellets**

1. Equip a three-necked round-bottomed flask containing a magnetic stirrer bar with a rubber septum and three-way stopcock with an argon inlet. Repeat flushing with dry argon to displace the air.

2. Charge the flask with mono(2,6-dimethoxybenzoyl)tartaric acid and 50 mL of dry dichloromethane, and cool in an ice bath.

3. Through the septum, with a syringe, add dropwise a borane–THF solution at 0 °C over a period of 30 min.

4. Stir the reaction mixture for 15 min at 0 °C and then cool to −78 °C in a dry ice/methanol bath.

5. Add freshly distilled methacrolein to this solution via a syringe dropwise.

6. After the addition is complete, introduce 2,3-dimethyl-1,3-butadiene to the solution at the same temperature and stir the mixture for 12 h.

7. Pour the cold reaction mixture into 150 mL of ice-cold saturated sodium bicarbonate and extract the product with three 200-mL portions of hexane. Wash the combined organic phase with brine (2 × 200 mL), dry over sodium sulfate, filter, and concentrate at atmospheric pressure. Distil the residue at reduced pressure to afford (1R)-1,3,4-trimethyl-3-cyclohexene-1-carboxaldehyde (6.53 g, 86%) as a colourless liquid, b.p. 92–93 °C (23 mm). The product is >98% pure by ^{1}H NMR, IR analysis, and may be characterized further by elemental analysis.[e]

---

[a] Titrate borane–THF complex which can be obtained from Toso-Akuzo Chemical Company, Ltd. in Japan before use. Vigorous evolution of hydrogen is observed during addition of borane–THF solution to the reaction mixture.

[b] Dry methacrolein from Tokyo Kasei Kogyo Company, Ltd with calcium sulfate and distil through a 20-cm Vigreux column under argon prior to use.

[c] Distil 2,3-dimethyl-1,3-butadiene, from Tokyo Kasei Kogyo Company Ltd, before use.

[d] Distil dichloromethane from calcium hydride under argon.

[e] The optical purity of this adduct is 95% as determined by 200 MHz ^{1}H NMR spectroscopy and GC analysis (capillary column PEG, 0.25 mm × 25 m, purchased from Gaskuro Kogyo Company, Ltd in Japan) after conversion into the corresponding chiral acetal as follows. Stir a solution of the adduct, (2R,4R)-2,4-pentanediol (1.2 equiv, obtained from Wako Pure Chemical Industries), triethyl ortho-formate (1.2 equiv.), and p-toluenesulfonic acid monohydrate (as a 5 mM solution) in dry benzene at ambient temperature for 3 h. Pour the mixture into saturated sodium bicarbonate and extract the product with ether. Dry the combined organic phases over sodium sulfate and concentrate on a rotary evaporator. Purify the residue by flash column chromatography on silica gel using hexane–ethyl acetate (25:1) as eluant to give the acetal quantitatively.

Kazuaki Ishihara

In 1991, Helmchen *et al.*[17,19] and our group[18] found that *N*-sulfonyl derivatives of α-amino acid react with diborane, giving complexes formulated as CAB (Scheme 3.13). These CAB complexes catalyse asymmetric cycloadditions between α,β-enals and dienes.

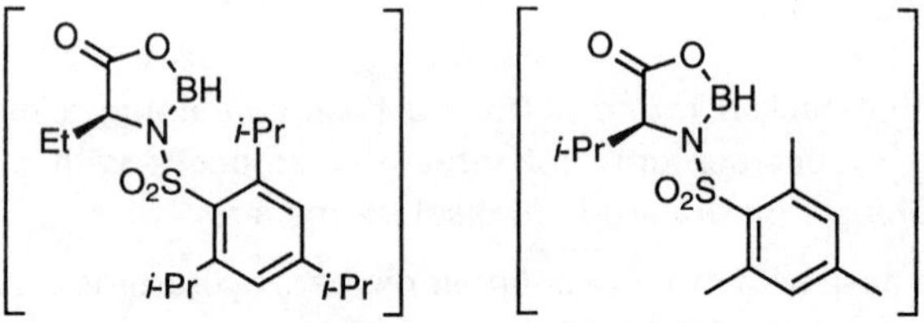

CAB **11** (Yamamoto *et al.*)   CAB **12** (Helmchen *et al.*)

Scheme 3.13

A similar effect was also published by Corey *et al.* on CAB **13**.[20–24] Especially efficient is the asymmetric catalysis of cycloaddition between 2-bromoacrolein and various dienes (>90–95% ee). The transition-state is believed to be as represented in Scheme 3.14.[21] Attractive interactions between the indolyl moiety and the π-acidic dienophile protect one face of the dienophile. CAB **14** derived from *N*-tosyl (α*S*, β*R*)-β-methyltryptophan catalyses the cycloaddition of 2-bromoacrolein and furan with 92% ee.[24]

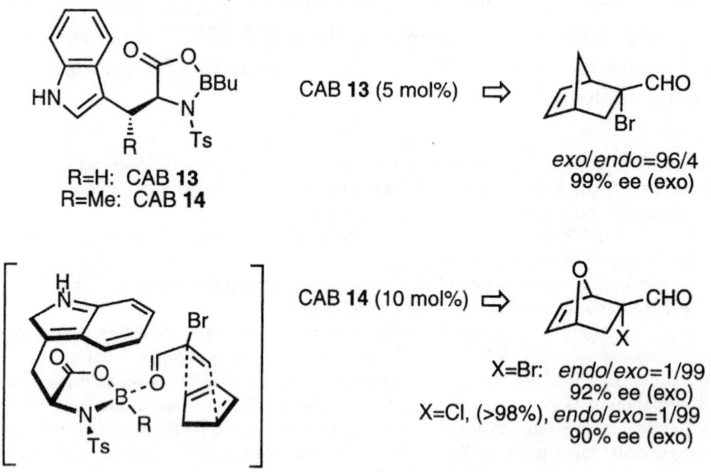

R=H: CAB **13**
R=Me: CAB **14**

CAB **13** (5 mol%) ⇒

exo/endo=96/4
99% ee (exo)

CAB **14** (10 mol%) ⇒

X=Br: endo/exo=1/99
92% ee (exo)
X=Cl, (>98%), endo/exo=1/99
90% ee (exo)

Scheme 3.14

Protocol 5.
Enantioselective Diels–Alder reaction catalysed by CAB 14[24] (Scheme 3.15)

Caution! Carry out all procedures in a well-ventilated hood, and wear disposable vinyl or latex gloves and chemical-resistant safety goggles.

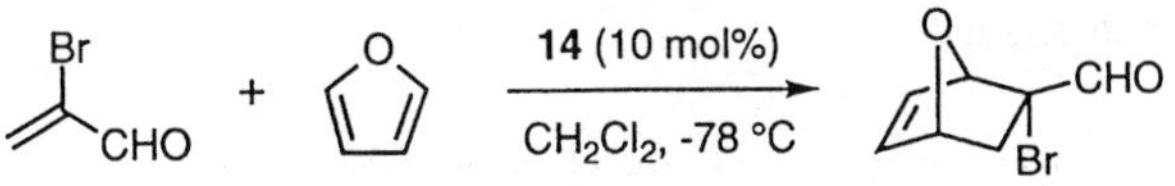

Scheme 3.15

Equipment

- Three-necked round-bottomed flask
- Separating funnel
- Magnetic stirrer bar
- Sintered glass filter funnel
- Magnetic stirrer
- Erlenmeyer flasks

- Column for flash chromatography
- Three-way stopcock
- Soxhlet extractor
- Reflux condenser
- Rubber septum
- Vacuum/nitrogen source

Materials

- N-Tosyl ($\alpha S,\beta R$)-methyltryptophan[21] (FW 358.4), 2.06 g, 5.6 mmol — irritant
- Butylboronic acid (FW 101.9), 0.60 g, 6.7 mmol — hygroscopic
- 2-Bromoacrolein (FW 135.0)[20] — flammable liquid, corrosive
- Furan (FW 68.1) — highly toxic cancer suspect agent
- Dry toluene[a] — flammable liquid, toxic
- Dry THF[b] — flammable liquid, irritant
- Dry dichloromethane[a] — toxic, irritant
- CaH$_2$ — flammable solid, moisture sensitive
- Sand

1. Add N-tosyl ($\alpha S,\beta R$)-methyltryptophan and butylboronic acid in a round-bottomed flask equipped with magnetic stirrer and a Soxhlet extractor with reflux condenser with a rubber stopper at the top, connect to a vacuum manifold and place under nitrogen.

2. A thimble in the extractor contains layers of sand (3 cm, bottom) and CaH$_2$ (3 cm, top).

3. After the addition of 20 mL of dry THF and 40 mL of dry toluene, heat the mixture at rapid reflux for 20 h using an oil bath heated to >165 °C.

4. Quickly disconnect the reaction flask and attach to a vacuum line, and remove the solvent *in vacuo* leaving catalyst **14** as a colourless viscous oil. Store **14** in a tightly sealed flask as a 0.1 M solution in 9:1 toluene–THF.

5. Remove the solvent *in vacuo* and exchange with CH$_2$Cl$_2$ toluene or other solvent just prior to the Diels–Alder reaction. THF serves to stabilize toluene solutions of catalyst **14**.

Protocol 5. *Continued*

6. The reaction of 5 equiv. of furan with 2-bromoacrolein in the presence of 10 mol% of **14** in dichloromethane at −78°C is complete in 5 h and gives the Diels–Alder adduct in 98% yield and 96:4 enantioselectivity, as determined by 500 MHz analysis of the α-methoxy-α-(trifluoromethyl)phenylacetic ester of the corresponding primary alcohol (from $NaBH_4$ reduction of the alde-hyde). The *N*-tosyl carboxylic acid precursor of **14** is efficiently recovered for reuse in each case.

[a] Distil toluene and dichloromethane from calcium hydride under nitrogen.
[b] Distil THF from sodium and benzophenone under nitrogen.

Itsuno *et al.* explored the possibility of using polymer-supported chiral Lewis acids in the cycloaddition of methacrolein with cyclopentadiene.[25–28] Using insoluble polymer-supported CAB, high enantioselectivity (up to 95% ee) has been achieved.[28]

Hawkins *et al.*[29,30] developed a simple and efficient catalyst for Diels–Alder reaction based on a chiral alkyldichloroborane **15** (Scheme 3.16). They have predicted the approach of the diene on one of the faces of the methyl croto-nate because the other face is protected by π–π donor-acceptor interactions on the basis of the crystal structure study of a molecular complex between methyl crotonate and **15**.

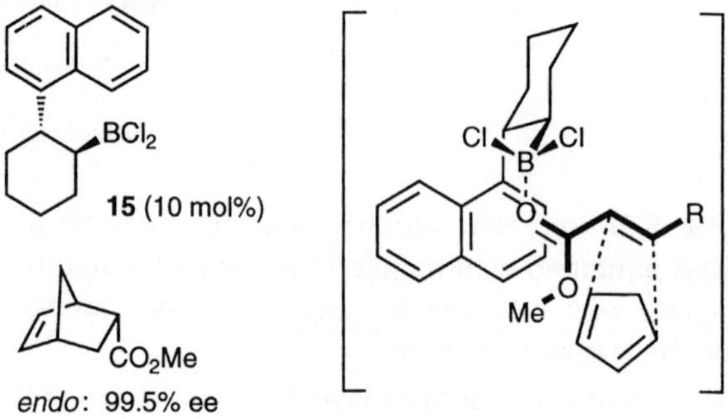

15 (10 mol%)

endo: 99.5% ee

Scheme 3.16

Protocol 6.
Enantioselective Diels–Alder reaction catalysed by chiral
alkyldichloroborane 15[30] (Scheme 3.17)

Caution! Carry out all procedures in a well-ventilated hood, and wear disposable vinyl or latex gloves and chemical-resistant safety goggles.

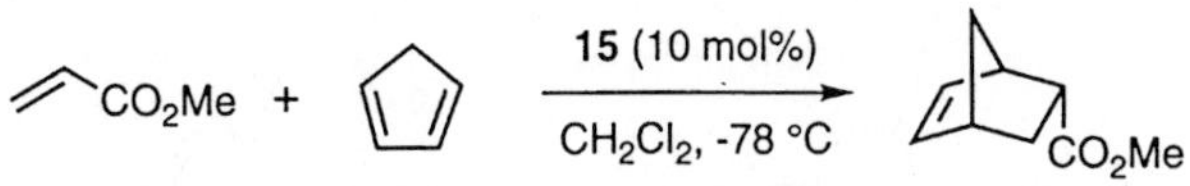

Scheme 3.17

Equipment
- Schlenk flask (100 mL)
- Separating funnel (100 mL)
- Magnetic stirrer bar
- Sintered glass filter funnel

- Magnetic stirrer
- Erlenmeyer flasks (100 mL)
- Column for flash chromatography
- Vacuum/nitrogen source

Materials
- (1*R*,2*R*)-15[29,30] (FW 291.0), 352.1 mg, 1.21 mmol
- Cyclopentadiene (FW 66.1), 4.66 g, 70.5 mmol
- Methyl acrylate (FW 86.1), 1.21 g, 14.1 mmol
- Dry dichloromethane

moisture sensitive
flammable liquid, toxic
flammable liquid, lachrymator
toxic, irritant

1. Add (1*R*,2*R*)-15 in a Schlenk flask, cool the flask to −78°C, and add 14 mL of freshly distilled CH_2Cl_2 slowly. Add enough Ch_2Cl_2 to establish a solvent to diene ratio of 3:1 (v/v).
2. Then, add cyclopentadiene and methyl acrylate successively.
3. Seal the reaction flask and allow to warm to the reaction temperature.
4. Quench the reaction with 10% $NaHCO_3$ (aq) and isolate the products (95% yield) by flash chromatography. Determine enantiomeric excess with chiral GC (97% ee).[a]

[a] Chiral GC analyses are performed with J&W Cyclodex-B β-cyclodextrin column.

Mukaiyama *et al.* have found that prolinol derivatives combined with BBr_3 are good chiral catalysts for some Diels–Alder reactions.[31,32] For example, methacrolein and cyclopentadiene afford the exo adduct (exo:endo = >99:1) in 97% ee (20 mol% of catalyst). The chiral catalyst is believed to be the HBr adduct salt of the amino boron derivative (Scheme 3.18).

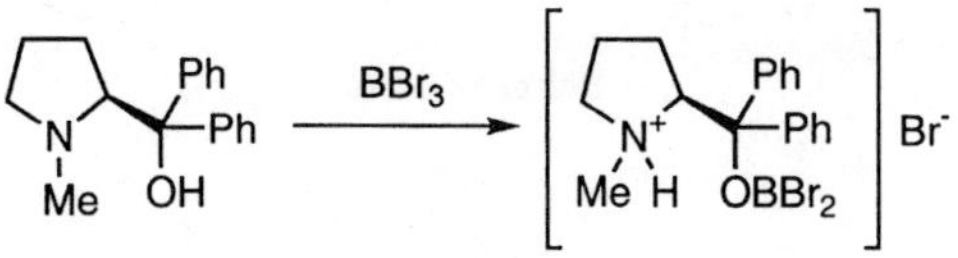

Scheme 3.18

Recently, a new chiral Lewis acid **16** was developed, and its utility was demonstrated in cycloadditions with both reactive and unreactive 1,3-dienes (Scheme 3.19).[33] In the hypothetical transition-state model **17**, one of the *N*-CH$_2$Ar substituents serves to block attack on the lower face of the *s-trans*-co-ordinated dienophile whereas the other screens another region in space and limits the rotational position of dienophile and *N*-CH$_2$Ar moieties.

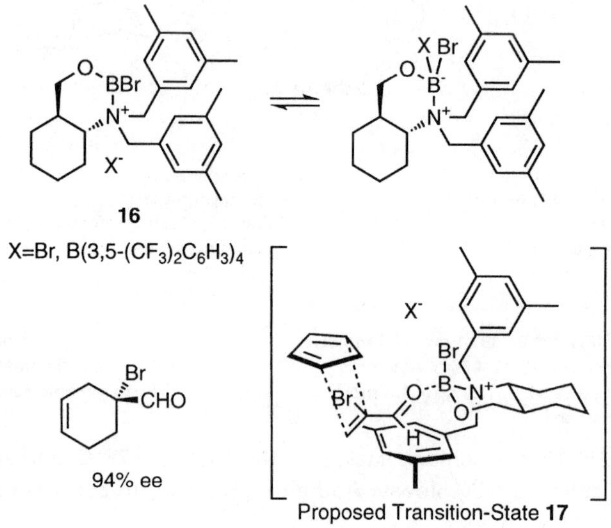

16

X=Br, B(3,5-(CF$_3$)$_2$C$_6$H$_3$)$_4$

94% ee

Proposed Transition-State **17**

Scheme 3.19

Protocol 7.
Enantioselective Diels–Alder reaction catalysed by chiral cationic Lewis acid 16[33] (Scheme 3.20)

Caution! Carry out all procedures in a well-ventilated hood, and wear disposable vinyl or latex gloves and chemical-resistant safety goggles.

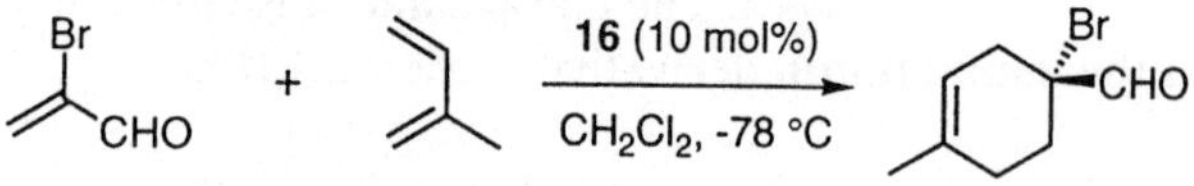

16 (10 mol%)

CH$_2$Cl$_2$, -78 °C

Scheme 3.20

Equipment
• Vacuum/argon source
• There is no information about apparatus.[33]

3: Boron reagents

Materials

- The silyl ether of chiral ligand (FW 423.7),[33] 40.3 mg, 0.095 mmol
- Boron tribromide (FW 250.5), 0.46 M in CH_2Cl_2 187 μL, 0.086 mmol   **corrosive moisture-sensitive**
- $Ag^+B[C_6H_3$-3,5-$(CF_3)_2]_4^-$ (FW 971.1),[a] 83.5 mg, 0.086 mmol   **toxic**
- 2-Bromoacrolein (FW 135.0), 76 μL, 0.94 mmol   **flammable liquid, irritant**
- Isoprene (FW 68.1), 500 μL 5 mmol   **cancer suspect agent, flammable liquid**
- Dry dichloromethane   **toxic, irritant**

1. Add a solution of BBr_3 in CH_2Cl_2 to a solution of the silyl ether of chiral ligand in 1 mL of CH_2Cl_2 at −94 °C (hexane–liquid N_2 bath) under dry argon, and after 5 min, replace the cooling bath with dry ice–acetone to bring the temperature to −78 °C.

2. After adding a freshly prepared dry solution of $Ag^+B[C_6H_3$-3,5-$(CF_3)_2]_4^-$ in 1 mL of CH_2Cl_2, stir the reaction mixture for 20 min at −78 °C and then cool to −94 °C.

3. Add 2-bromoacrolein and isoprene successively (each sropwise), and stir the reaction mixture for 1 h at −94 °C and then quench with 150 μL of triethylamine.

4. After warming the reaction mixture to room temperature and removing the inorganic salts by filtration, evaporate the solvents, and purify the residue by chromatography on silica gel to give 191 mg (99%) of the isoprene Diels–Alder adduct, $[\alpha]^{23}_D$ + 82.2° (c 0.8, CH_2Cl_2) and also recover ligand (28.6 mg, 83%).

---

[a] Prepare the silver salt as follows. Shake an ethereal solution of $NAB[C_6H_{13}$-3,5-$(CF_3)_2]_4$ (Broolhart, M.; Grant, B.; Volpe, A. F., Jr Organometallics **1992**, 11, 3920) with 2 equiv. of aqueous $AgNO_3$ in a separating funnel for 5 min, and separate the layers. Evaporation of the ether layer affords a quantitative yield of the colourless silver salt which is dissolved in ether to give a clear 0.1 M solution. Store this at −78 °C in a flask wrapped with aluminium foil to exclude light. To prepare the catalyst, concentrate a measured amount of this ethereal solution in vacuo, dissolve in dry CH_2Cl_2, and dry over activated molecular sieves 4 Å for 1 h at room temperature (with constant protection from light).

We found that Brønsted acid-assisted chiral Lewis acid (BLA) **18** achieved high selectivity through the double effect of intramolecular hydrogen binding interaction and attractive π-π donor–acceptor interaction in the transition-state (Scheme 3.21).[34] Extremely high enantioseletivity (>99–92% ee) and exo-selectivity (>99–97% exo) are obtained for cycloadditions of α-substituted α,β-enals with dienes. The absolute stereoference in the reaction can be easily understood in terms of the most favourable transition-state assembly **19**. The co-ordination of a proton of 2-hydroxyphenyl group with an oxygen of the adjacent B–O bond in complex **19** plays an important role in asymmetric induction; this hydrogen binding interaction via Brønsted acid causes Lewis acidity of boron and π-basicity of phenoxy moiety to increase.

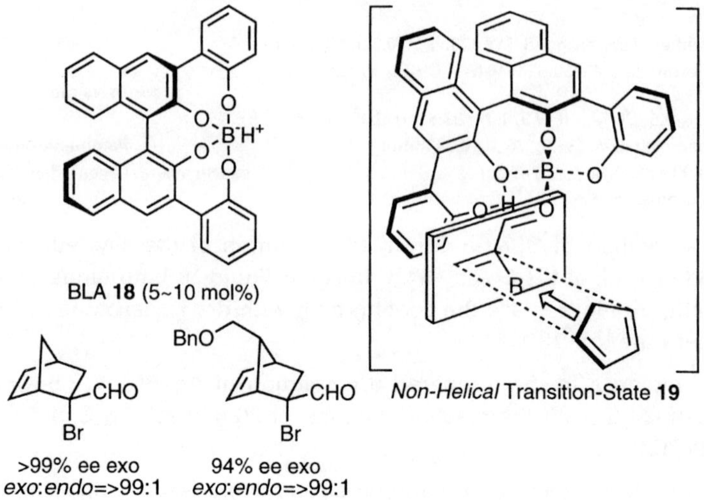
BLA 18 (5~10 mol%)
BnO
CHO
Br
CHO
Br
>99% ee exo
exo:endo=>99:1
94% ee exo
exo:endo=>99:1
O
O
B
H+
O
O
O
O
B
O
O
H
O
B
Non-Helical Transition-State 19

Scheme 3.21

Protocol 8.
Enantioselective Diels–Alder reaction catalysed by chiral BLA 18[34] (Scheme 3.22)

Caution! Carry out all procedures in a well-ventilated hood, and wear disposable vinyl or latex gloves and chemical-resistant safety goggles.

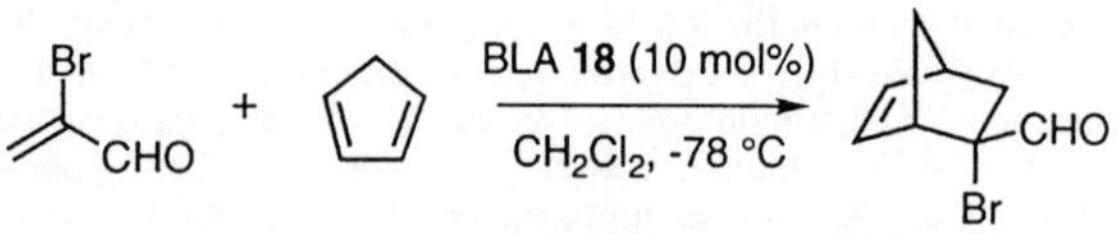
Br
CHO
+
BLA 18 (10 mol%)
CH2Cl2, -78 °C
CHO
Br

Scheme 3.22

Equipment

- Round-bottomed flask (10 mL)
- Separating funnel (50 mL)
- Magnetic stirrer bar
- Sintered glass filter funnel
- Magnetic stirrer
- Erlenmeyer flasks (50 mL)
- Column for flash chromatography
- Vacuum/argon source
- Pressure-equalized addition funnel
- Three-way stopcock

Materials

- (*R*)=3,3′-bis(2-hydroxyphenyl)-2,2′-dihydroxy-1,1′-binaphthyl (FW 472.5),[34] 23.5 mg, 0.05 mmol — **white solid, irritant**
- Trimethyl borate (FW 103.9), 0.1 M in CH_2Cl_2 0.5 mL, 0.05 mmol — **flammable liquid, moisture-sensitive**
- 2-bromoacrolein (FW 135.0), 80.8 μL, 1 mmol — **flammable liquid, irritant**
- Cyclopentadiene (FW 66.1), 332 μL, 4 mmol — **flammable liquid, toxic**

- Dry dichloromethane[a]   **toxic, irritant**
- Dry THF[b]   **flammable irritant**
- 4-Å molecular sieves   **pellets**

1. Add (*R*)-3,3'-bis(2-hydroxyphenyl)-2,2'-dihydroxy-1,1'-binaphthyl, trimethyl borate, and 3 mL of dichloromethane in a dry round-bottomed flask fitted with a stirrer bar and a pressure-equalized addition funnel (containing a cotton plug and ca. 4 g of 4-Å molecular sieves (pellets) and functioning as a Soxhlet extractor) surmounted by a reflux condenser.

2. Secure an argon atmosphere, and bring the solution to reflux (bath temperature 50–60 °C).

3. After 2 h, cool the reaction mixture to 25 °C and quickly remove the addition funnel and condenser and replace with a septum.

4. Add dry THF to the white precipitate in dichloromethane at 25 °C, and after 2 h the precipitate is completely dissolved.

5. After cooling a colourless solution of the catalyst (*R*)-**18** to −78 °C, add dropwise 2-bromoacrolein and cyclopentadiene.

6. After 4 h, add 50 mL of H_2O and warm the mixture to 25 °C, dry over $MgSO_4$, filter, and purify by eluting with hexane/ethyl acetate (10:1) to afford 201 mg of Diels–Alder adduct (1*S*, 2*S*, 4*S*)-bromo aldehyde as a white solid (1.0 mmol, >99% yield, exo:endo = >99:1, >99% ee) and quantitative recovery of pure chiral ligand. Determine enantioselectivity by reduction with $NaBH_4$, conversion to the Mosher ester, and ^{1}H NMR and HPLC analysis (Daicel AD).

[a] Distil dichloromethane from calcium hydride under argon.
[b] Distil THF from sodium and benzophenone under argon.

Diels–Alder reactions of α-unsubstituted α,β-enals with BLA **18** as well as most chiral Lewis acids exhibit low enantioselectivity and/or reactivity. We have developed a new type of BLA, **20** which was prepared from a chiral triol and **5** (Scheme 3.23).[35] **20** is extremely effective in enantioselective cyclo-addition of both α-substituted and α-unsubstituted α,β-enals with various dienes. The Brønsted acid in the BLA clearly accelerates the cycloaddition. The high enantioselectivity and stereochemical results attained in this reaction can be understood in terms of the transition-state model **21**.

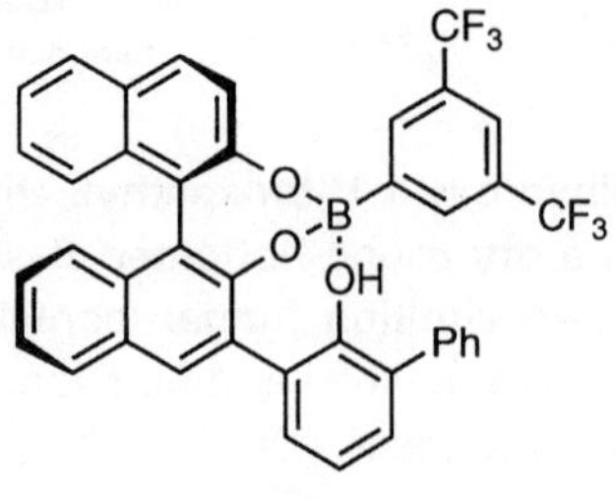

BLA 20

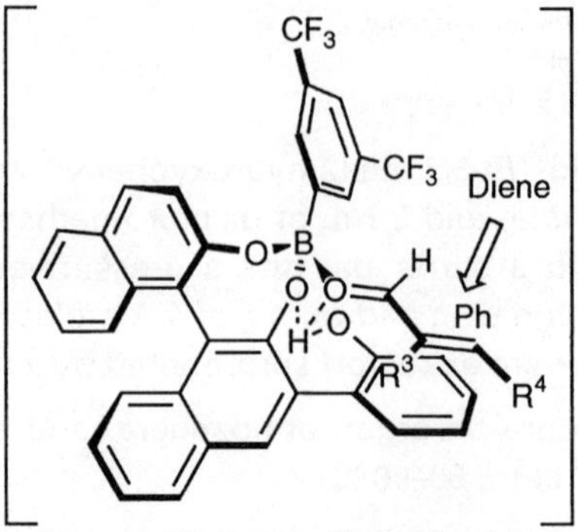

Proposed Transition State Model 21

Examples

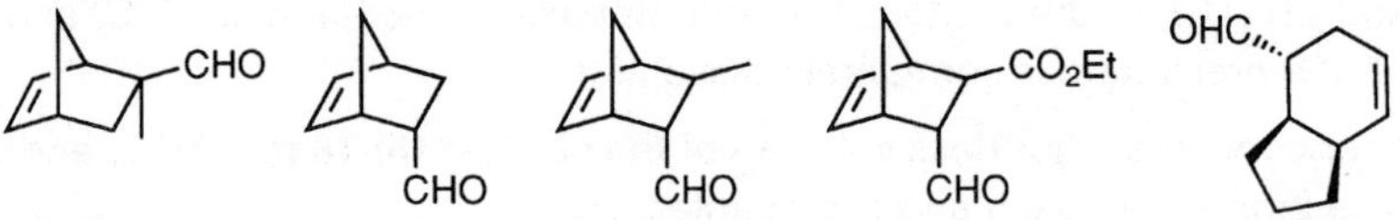

99% ee (*S*)   95% ee (*S*)   95% ee (*S*)   95% ee (*R*)   80% ee (*R*)

Scheme 3.23

Protocol 9.
Enantioselective Diels–Alder reaction catalysed by BLA 20[35] (Scheme 3.24)

Caution! Carry out all procedures in a well-ventilated hood, and wear disposable vinyl or latex gloves and chemical-resistant safety goggles.

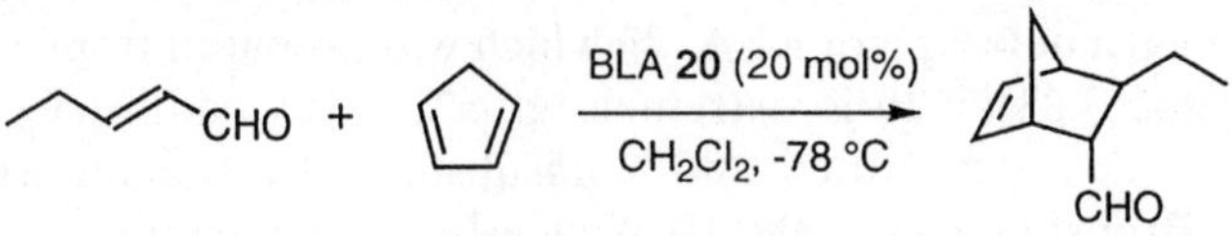

Scheme 3.24

Equipment

- Schlenk flasks (10 mL)
- Separating funnel (50 mL)
- Magnetic stirrer bar
- Sintered glass filter funnel
- Magnetic stirrer
- Erlenmeyer flasks (50 mL)

- Column for flash chromatography
- Vacuum/argon source
- Pressure-equalized addition funnel
- Three-way stopcock
- Oil bath

50

Materials

- The chiral ligand (FW 454.5),[35] 54.5 mg, 0.12 mmol — white solid
- Monomeric 5 (0.043 M) in CH_2Cl_2-THF-H_2O (20:3:0.054), 23.3 μL, 0.1 mmol — hygroscopic
- (E)-2-Pentenal (FW 84.12), 42.1 mg, 0.5 mmol — flammable liquid, irritant
- Cyclopentadiene (FW 66.1), 166 μL, 2 mmol — flammable liquid, toxic
- Dry THF[a] — flammable irritant
- Dry dichloromethane[b] — toxic, irritant
- 4-Å molecular sieves — powder

1. Stir a mixture of the chiral ligand and a solution of monomeric **5** in CH_2Cl_2–THF–H_2O (20:3:0.054) at ambient temperature for 2 h.

2. Transfer the resulting colourless solution into a Schlenk tube containing c. 0.5 mL of anhydrous dichloromethane and MS 4Å (powder, 250 mg, activated by heating at 200 °C under vacuum [c. 3 Torr] for 12 h), and use a further c. 0.5 mL of dichloromethane to transfer the residue into the Schlenk.

3. Stir the mixture at ambient temperature for another 12 h.

4. Then, evaporate the solvents and heat the resulting solid to 100 °C (oil bath) for 2 h under vacuum (c. 3 Torr) to dry catalyst.

5. After cooling to ambient temperature, purge the flask with argon and add dichloromethane (2 mL).

6. Cool the mixture to −78 °C, add dropwise (E)-pentenal, and 1 min later add freshly distilled cyclopentadiene slowly along the wall of the flask.

7. After stirring the reaction mixture at −78 °C for 72 h, quench the reaction with pyridine (20 μL, 0.25 mmol), warm to ambient temperature, and filter to remove molecular sieves. Wash the filtrate with ether, dry over $MgSO_4$, and concentrate to afford the crude products. Purify by silica gel chromatography eluting with pentane–ether to provide the pure Diels–Alder adduct (73% yield, endo/exo = 91:9, 98% ee for endo-isomer). Determine the exo/endo ratio by ^{1}H NMR analysis (500 MHz): δ 9.38 (d, J = 3.3 Hz, 1H, CHO (endo)), 9.79 (d, J = 2.9 Hz, 1H, CHO (exo)). Determine the ee by acetalization with (−)-(2R,4R)-2,4-pentanediol and GC analysis (90 °C, PEG-HT Bonded (25 m × 0.25 mm)): t_R = 29.0 min (major endo-isomer), 36.0 min (minor endo-isomer), 37.8 min (minor exo-isomer), 38.5 min (major exo-isomer). The absolute configuration is not established.

[a] Distil YHF from sodium and benzophenone under argon.
[b] Distil dichloromethane from calcium hydride under argon.

3.2 Catalytic enantioselective aldol and allylation reactions

Asymmetric aldol and allylation reactions are now of great interest because of their utility for introduction of asymmetric centres and functional groups.

We have reported that CAB **10**, R^2 = H, is an excellent catalyst (20 mol%)

for the enantioselective and diastereoselective Mukaiyama condensation of simple enol silyl ethers with various aldehydes.[36,37] The reaction is accelerated without reducing the enantioselectivity by using 10–20 mol% of **10**, R^2 = 3,5-$(CF_3)_2C_6H_3$. The enantioselectivity is increased without reducing the chemical yield by using 20 mol% of **10**, R^2 = *o*-$PhOC_6H_4$. Another aldol-type reaction of ketene silyl acetals derived from phenyl esters with achiral aldehydes also proceeds smoothly with **10** and can furnish *syn* β-hydroxy esters with high optical purity.[37,38] Regardless of the stereochemistry of enol silyl ethers, *syn* aldols are highly selectively obtained *via* the acyclic extended transition-state mechanism (Scheme 3.25).

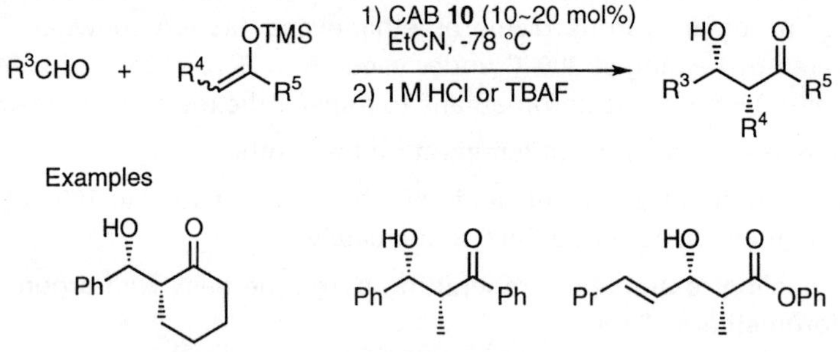

(83%), 97% ee *syn* *syn*:*anti*=>95:5 (**10**, R^2=3,5-$(CF_3)_2C_6H_3$)

(92%), 96% ee *syn* *syn*:*anti*=99:1 (**10**, R^2=3,5-$(CF_3)_2C_6H_3$)

97% ee *syn* *syn*:*anti*=96:4 (**10**, R^2=H)

Scheme 3.25

Protocol 10.
Enantioselective Mukaiyama Aldol reaction catalysed by CAB 10[37] (Scheme 3.26)

Caution! Carry out all procedures in a well-ventilated hood, and wear disposable vinyl or latex gloves and chemical-resistant safety goggles.

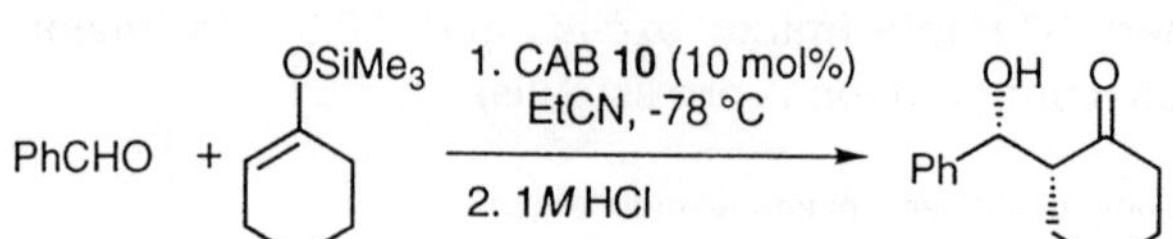

Scheme 3.26

Equipment

- Schlenk flask (10 mL)
- Separating funnel (50 mL)
- Magnetic stirrer bar
- Sintered glass filter funnel
- Magnetic stirrer
- Erlenmeyer flasks (50 mL)
- Column for flash chromatography
- Vacuum/argon source
- Three-way stopcock
- Rubber septum

Materials

- Mono(2,6-diisopropoxybenzoyl)tartaric acid[13]
  (FW 370.4), 74.1 mg, 0.2 mmol — **irritant**
- **5** (FW 257.9), 51.6 mg, 0.2 mmol — **hygroscopic**
- 1-Trimethylsiloxy-1-cyclohexene (FW 170.3),
  204.4 mg, 1.2 mmol — **flammable liquid**
- Benzaldehyde (FW 106.12), 106.1 mg, 1.0 mmol — **highly toxic cancer suspect agent**
- Dry propionitrile[a] — **highly toxic flammable liquid**

1. Equip a Schlenk flask containing a magnetic stirrer bar with a rubber septum and three-way stopcock with an argon inlet. Repeat flushing with dry argon to displace the air.

2. Add mono(2,6-di-isopropoxybenzoyl)tartaric acid and **5** in the flask, dissolve in 1 mL of dry propionitrile, stir the resulting solution at 25 °C for 30 min, and cool the reaction flask to −78 °C.

3. Add 1-trimethylsiloxy-1-cyclohexene and benzaldehyde successively and stir the reaction mixture for 12 h at low temperature.

4. Pour this cold solution into water and extract the product with ether repeatedly. Dry the combined ether layers over $MgSO_4$, concentrate *in vacuo* and treat the residue with 1 M HCl (aq)–THF solution (2 mL, 1/1 in vol.). After usual work-up, purify the crude product by column chromatography on silica gel to give aldol adducts (83% yield, *syn:anti* = >95:5, 97% ee for syn diastereomer). The product is >98% pure by ¹H NMR, IR analysis, and may be characterized further by elemental analysis. Determine the *syn/anti* ratio by ¹H NMR analysis (500 MHz), and determine the ee by ¹H NMR analysis fo (+)-MTPA ester.

[a] Distil propionitrile from calcium hydride under argon.

We have found that CAB **10** has a powerful activity for the Sakurai–Hosomi allylation reaction of aldehydes to furnish homoallylic alcohols in excellent enantiomeric excess (Scheme 3.27).[39] γ-Alkylated allysilanes exhibit excellent diastereo- and enantioselectivities affording *syn* homoallylic alcohols of higher optical purity. Regardless of the geometry of starting allylsilanes, the predominant isomer in this reaction has *syn* configuration. The observed preference for relative and absolute configurations for the adducts is predicted on the basis of an extended transition-state model similar to that for the CAB **10** catalysed aldol reaction.[36–38] The 3,5-bis(trifluoromethyl)phenyl group is the most effective *B*-substituent of **10**.[40]

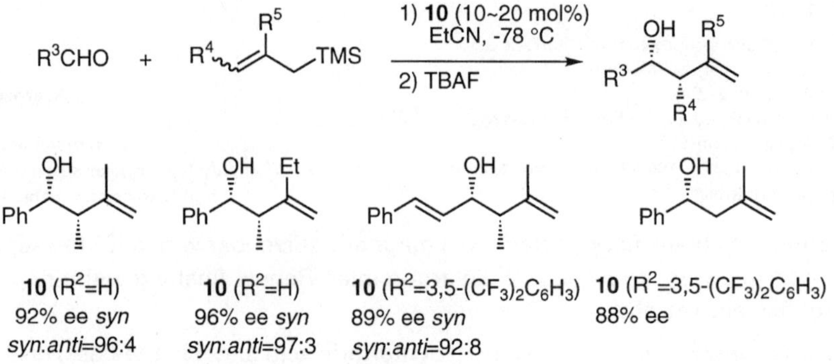

Scheme 3.27

Protocol 11.
Enantioselective Sakurai–Hosomi allylation reaction catalysed by CAB 10[40] (Scheme 3.28)

Caution! Carry out all procedures in a well-ventilated hood, and wear disposable vinyl or latex gloves and chemical-resistant safety goggles.

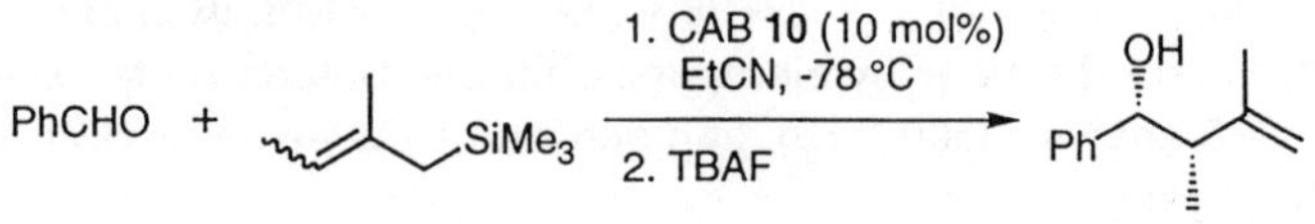

Scheme 3.28

Equipment

- Schlenk flasks (10 mL)
- Separating funnel (50 mL)
- Magnetic stirrer bar
- Sintered glass filter funnel
- Magnetic stirrer

- Erlenmeyer flasks (50 mL)
- Column for flash chromatography
- Vacuum/argon source
- Three-way stopcock
- Rubber septum

Materials

- Mono(2,6-diisopropoxybenzoyl)tartaric acid[13]
  (FW 370.4) 74.1 mg, 0.2 mmol — **irritant**
- **5** (FW 257.9), 51.6 mg, 0.2 mmol — **hygroscopic**
- 1-(Trimethylsilyl)-2-methyl-2-butene (FW 142.3),
  170.8 mg, 1.2 mmol — **flammable liquid**
- Benzaldehyde (FW 106.12), 106.1 mg, 1.0 mmol — **highly toxic cancer suspect agent**
- Dry propionitrile[a] — **highly toxic flammable liquid**

1. Equip a Schlenk flask containing a magnetic stirrer bar with a rubber septum and three-way stopcock with an argon inlet. Repeat flushing with dry argon to displace the air.

2. Add mono(2,6-di-isopropoxybenzoyl)tartaric acid and **5** in the flask, dissolve in 1 mL of dry propionitrile, stir the resulting solution at 25°C for 30 min, and cool the reaction flask to −78°C.

3. Add 1-(trimethylsilyl)-2-methyl-2-butene and benzaldehyde successively and stir the reaction mixture for 12 h at low temperature.

4. Pour this cold solution into brine and extract with ether repeatedly, dry the combined ether layers over $MgSO_4$, and concentrate *in vacuo*.

5 Treat the residue with tetrabutylammonium fluoride (1.5 mL of 1 M solution in THF, 1.5 mmol) in THF (3 mL).

6. Usual work-up followed by chromatographic separation gives products (82% yield, *syn:anti* = 94.6, 91% ee for *syn* diastereomer). The product is >98% pure by ^{1}H NMR, IR analysis, and may be characterized further by elemental analysis. Determine the *syn/anti* ratio and the ee by ^{1}H NMR analysis of (+)-MTPA ester: ^{1}H NMR δ 5.80 (minor enantiomer of *anti* diastereomer), 5.87 (major enantiomer of *anti* diastereomer), 5.88 (minor enantiomer of *syn* diasetereomer), 5.93 (major enantiomer of *syn* diastereomer).

a Distil propionitrile from calcium hydride under argon.

Marshall *et al.*[41] reported that more reactive allyltin analogues can be used in place of allylsilane nucleophiles in our CAB **9** catalyst system, and that trifluoroacetic anhydride is an efficient promoter (Scheme 3.29).

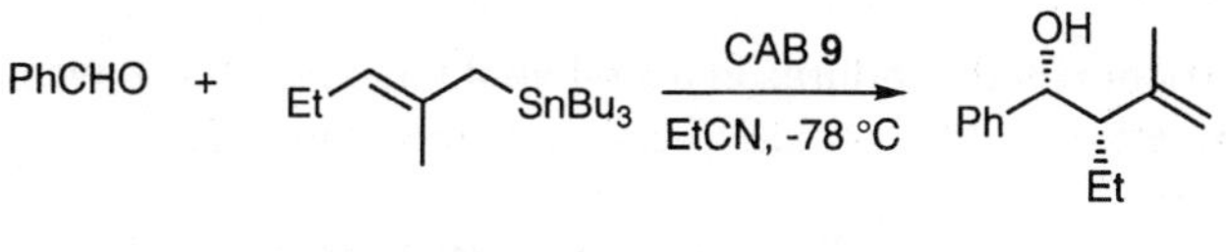

9 (20 mol%) + $(CF_3)_2CO$ (40 mol%)   : (88%),*syn:anti*=85:15, 74% ee syn
9 (100 mol%) + $(CF_3CO)_2O$ (200 mol%) : (99%),*syn:anti*=90:10, 85% ee syn

Scheme 3.29

After the enantioselective aldol reaction using chiral oxaborolidines (another type of CAB) under a stoichiometric condition was reported by Kiyooka *et al.*[42] Masamune *et al.*,[43,44] Kiyooka *et al.*,[45] and Corey *et al.*[46] independently developed CAB-catalysed-aldol reactions (Scheme 3.30). Masamune *et al.* suggested that the initial aldol adduct must undergo ring closure as indicated by the arrows in **26** to release the final product **25** and to regenerate the catalyst **22** or **23**.[43,44] In many cases, slow addition of the aldehyde to the reaction mixture has proved beneficial (permitting enough time for **26** to undergo ring closure) in improving the enantioselectivity of the reaction. Kiyooka *et al.*[45] have reported straightforward improvement of this reaction to its catalytic version by employing nitroethane instead of dichloramethane as solvent.

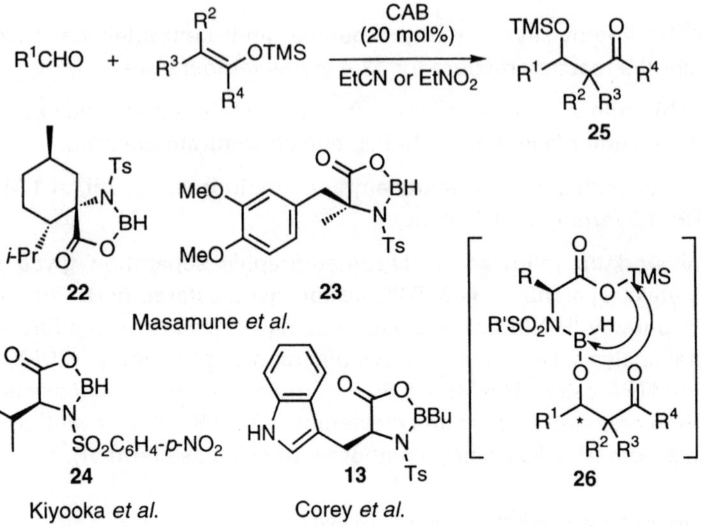

Scheme 3.30

Protocol 12.
Enantioselective Mukaiyama aldol reaction catalysed by CAB 24[45] (Scheme 3.31)

Caution! Carry out all procedures in a well-ventilated hood, and wear disposable vinyl or latex gloves and chemical-resistant safety goggles.

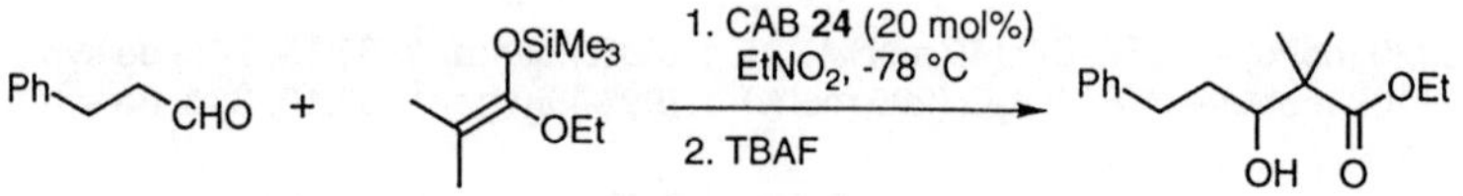

Scheme 3.31

Equipment

• Vacuum/argon source

Materials

• The sulfonamide ligand[40] (FW 302.3), 66.5 mg, 0.22 mmol — **irritant**
• Borane-THF (1.0 M solution in THF), 0.2 mL, 0.2 mmol — **flammable liquid, moisture-sensitive**
• Hydrocinnamaldehyde (FW 134.18), 134.2 mg, 1 mmol — **irritant**
• 1-Ethoxy-2-methyl-1-trimethylsiloxypropene (FW 188.3), 207.2 mg, 1.1 mmol — **flammable liquid**
• Nitroethane — **flammable liquid irritant**

1. Add BH₃–THF complex in dropwise fashion to a solution of the sulfonamide in nitroethane at 0 °C under Ar.

2. Stir the mixture for 0.5 h, allow to warm to room temperature, and stir for another 0.5 h.

3. Add a solution of hydrocinnamaldehyde in THF and a solution of the ketene silyl acetal in THF to the resulting mixture successively at −78 °C.

4. Stir the mixture at −78 °C for 4 h, whereupon quench the reaction by the introduction of buffer solution (pH 6.8; 5 mL). Extract the mixture with Et₂O (30 mL) twice. After the usual work-up, isolate the pure products of the silylated aldol, the corresponding aldol, and the sulfonamide by silica-gel chromatography. After desilylation with Bu₄NF (a 1.0 M solution in THF), the aldol product is obtained in 97% yield with 95% ee. Determine the ee by HPLC analysis with a chiral Daicel OD column.

Harada *et al.* reported that CAB **14** serves as an excellent catalyst for enantioselective ring-cleavage of 2-substituted 1,3-dioxolanes with silyl enol ethers.[47–49]

3.3 Asymmetric reactions of imines mediated by chiral boron reagents

Imines are important starting materials for nitrogen-containing compounds such as alkaloids, amino acids, β-lactams, amino sugars, and terpenes. We developed chiral Lewis acids **27** and **28** for asymmetric reactions of imines.

(*R*)-**27**   (*R*)-**28**

3.3.1 Enantioselective Diels–Alder reaction of imines mediated by chiral boron reagent 27[50,51]

Asymmetric aza Diels–Alder reaction of imines is promoted by an *in situ* generated boron complex **27**. (*R*)-**27** is prepared *in situ* simply by mixing a 1:1 molar ratio of optically active binaphthol and triphenyl borate in dichloromethane. The aza Diels–Alder reaction of imine **29** with Danishefsky diene **30** is promoted by (*R*)-**27** at −78 °C to give the adduct **31** in 75% yield with 82% ee (Scheme 3.32). The reaction is applicable to aromatic and aliphatic imines and Danishefsky-type dienes.

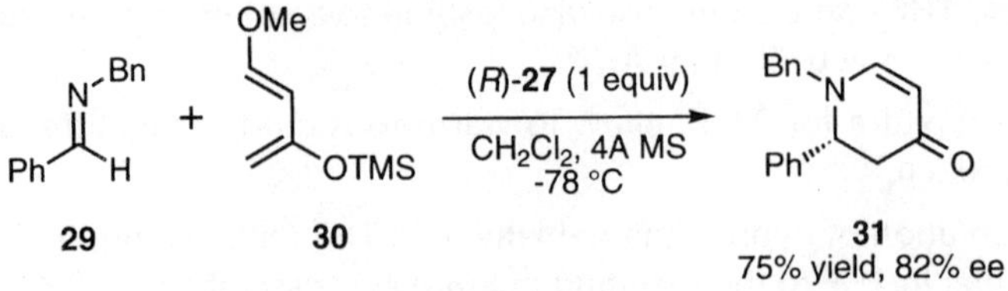

Scheme 3.32

3.3.2 Double asymmetric induction of Diels–Alder reaction of imines mediated by chiral boron reagents[50–53]

There have been a number of investigations into reactions exhibiting diastereofacial selectivity with imines containing a chiral auxiliary. We succeeded in the double asymmetric induction of the aza Diels–Alder reaction of chiral imines derived from α-methylbenzylamine mediated by **27**, and thereafter found that new chiral boron reagent **28** is a more efficient promoter than **27**.[54–56] The X-ray analysis of **28** demonstrates that it exists as a Brønsted acid-assisted chiral Lewis acid (BLA). **28** is prepared by the reaction of B(OMe)$_3$ (1 equiv) with (R)-binaphthol (2 equiv) in dichloromethane at reflux by removing methanol. Almost complete diastereoselectivities are obtained for a variety of imines under an optimum condition with a matching pair (Scheme 3.33).

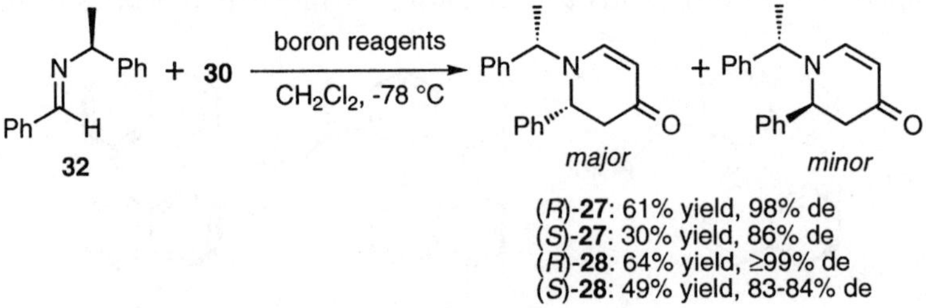

(R)-**27**: 61% yield, 98% de
(S)-**27**: 30% yield, 86% de
(R)-**28**: 64% yield, ≥99% de
(S)-**28**: 49% yield, 83-84% de

Scheme 3.33

3.3.3 Double asymmetric induction of aldol-type reaction of imines mediated by chiral boron reagents[52–56]

Largely stimulated by the synthesis of β-lactam antibiotics, there have been a great number of investigations into stereochemical aspects of imine condensations. We developed highly stereoselective aldol-type reaction of chiral imines derived from α-methylbenzylamine and aldehydes with ketene silyl acetals in the presence of chiral Lewis acid **27**[54–56] or BLA **28**.[52,53] Some examples are listed in Scheme 3.34. The matching and mismatching pairs in double stereodifferentiation depend on the structure of the ketene silyl acetals.

Examples

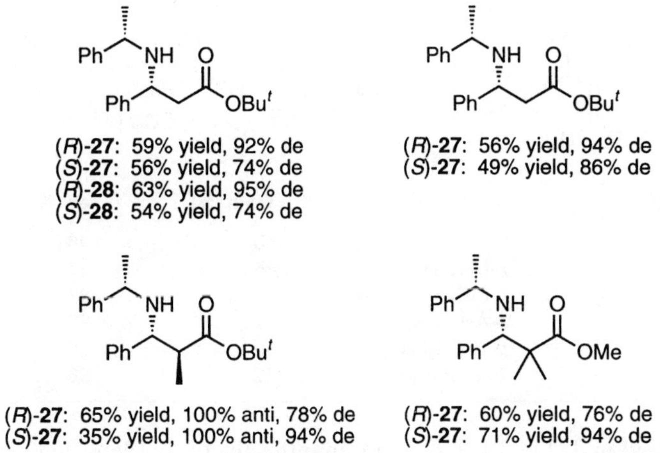

Ph—NH O
Ph—⋮—CH₂—C—OBuᵗ

(*R*)-**27**: 59% yield, 92% de
(*S*)-**27**: 56% yield, 74% de
(*R*)-**28**: 63% yield, 95% de
(*S*)-**28**: 54% yield, 74% de

Ph—NH O
Ph—⋮—CH₂—C—OBuᵗ

(*R*)-**27**: 56% yield, 94% de
(*S*)-**27**: 49% yield, 86% de

Ph—NH O
Ph—⋮—CH—C—OBuᵗ

(*R*)-**27**: 65% yield, 100% anti, 78% de
(*S*)-**27**: 35% yield, 100% anti, 94% de

Ph—NH O
Ph—⋮—C(CH₃)₂—C—OMe

(*R*)-**27**: 60% yield, 76% de
(*S*)-**27**: 71% yield, 94% de

Scheme 3.34

The absolute configuration of the adducts can be understood in terms of a rational model **33** involving an intramolecular hydrogen binding interaction *via* Brønsted acid (Scheme 3.35).

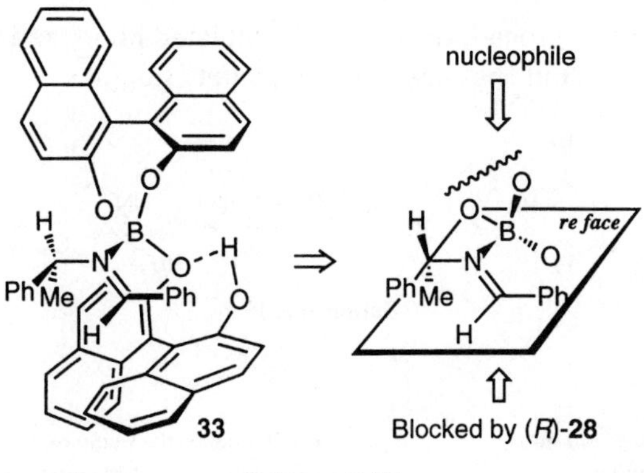

Scheme 3.35

3.3.4 Enantioselective aldol-type reaction of imines mediated by BLA 28[52,53]

We developed the enantioselective synthesis of chiral β-amino acid esters from *achiral* imines and ketene silyl acetals using chiral BLA **28**. The enantioselectivity of the aldol-type reaction is dramatically increased by substituting sterically bulky *N*-substituents. The enantioselective aldol-type reaction of a variety of *N*-benzhydryl imines with **30** by (*R*)-**28** under optimum conditions is

Table 3.1 Enantioselective Aldo-Type Reaction of *N*-Benzhydrylimines

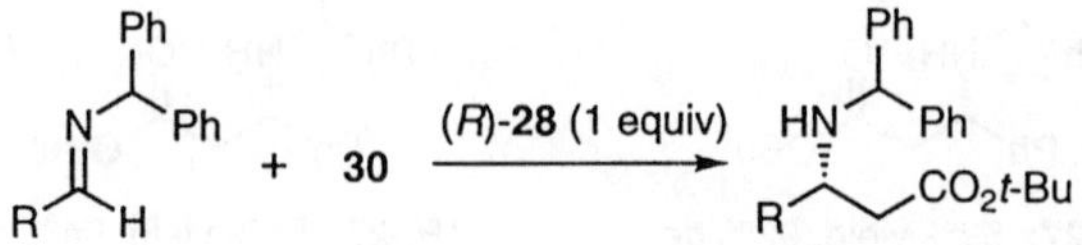

Entry	R	Yield (%)	ee (%)
1	C$_6$H$_5$	58	96 (*R*)
2	*p*-MeC$_6$H$_4$	35	97
3	*p*-ClC$_6$H$_4$	45	98
4	*p*-AcOC$_6$H$_4$	52	98
5	*p*-2,4,Cl$_2$C$_6$H$_3$	49	95
6	2-Naphthyl	43	96

summarized in Table 3.1. Excellent enantioselectivity (95% ee or better) has been achieved in the reactions of aromatic aldehyde-derived imines. The removal of *N*-benzhydryl protecting group from β-aryl β-amino acid esters is easily advanced by catalytic hydrogenation (10% Pd/C, H$_2$, MeOH).

Protocol 13.
Enantioselective aldol-type reaction of imines[52,53] (Scheme 3.36)

Caution! Carry out all procedures in a well-ventilated hood, and wear disposable vinyl or latex gloves and chemical-resistant safety goggles.

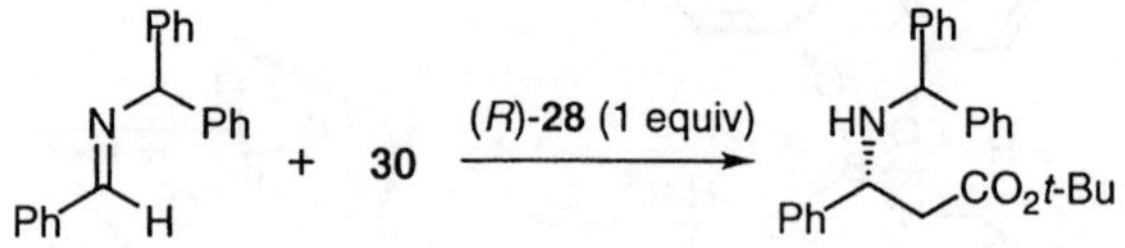

Scheme 3.36

Equipment
- Round-bottomed flask (10 mL)
- Separating funnel (50 mL)
- Magnetic stirrer bar
- Sintered glass filter funnel
- Magnetic stirrer
- Erlenmeyer flasks (50 mL)
- Column for flash chromatography
- Vacuum/argon source
- Pressure-equalized addition funnel
- Three-way stopcock

Materials
- (*R*)-(+)-1,1′-Bi-2-naphthol (FW 286.3), 200.4 mg, 0.70 mmol — **irritant**
- Trimethyl borate (FW 103.9), 0 1 M in CH$_2$Cl$_2$ 3.5 mL, 0.35 mmol — **flammable liquid, moisture-sensitive**
- 4Å molecular sieves — **pellets**
- *N*-Benzylidenebenzhydrylamine[52,53] (FW 271.4), 95.0 mg — **colourless crystal**

3: Boron reagents

- **3 0** (FW 188.3) 131.8 mg, 0.7 mmol — flammable
- Dry dichloromethane[a] — toxic, irritant
- Dry toluene[a] — flammable liquid, toxic

1. Add (R)-binaphthol, trimethyl borate, and dichloromethane in a dry round-bottomed flask fitted with stirrer bar and a pressure-equalized addition funnel (containing a cotton plug and c. 4 g of 4Å molecular sieves and functioning as a Soxhlet extractor) surmounted by a reflux condenser.

2. Secure an argon atmosphere, and bring the solution to reflux (bath temperature 50–60 °C).

3. After 2–3 h, cool the reaction mixture to 25 °C, quickly remove the solution funnel and condenser and replace with a septum.

4. After adding 2 mL of dichloromethane, 5 mL of toluene, and N-benzylidenebenzhydrylamine to the white precipitate of (R)-**28** in dichloromethane at 0 °C, stir the yellow suspension at 0 °C for 10 min.

5. After cooling the suspension to −78 °C, add **3 0** dropwise.

6. After stirring for 20 h, wash the solution with water and saturated NaHCO₃ and then dry over MgSO₄. Evaporation of the solvent and purification by column chromatography on silica gel gives tert-butyl 3-(benzhydrylamino)-3-phenylpropionate in 58% yield with 96% ee (R). The product is >98% pure by ¹H NMR, IR analysis, and may be characterized by elemental analysis. Determine the ee by HPLC analysis with a chiral Daicel OD-H column (hexane: i-PrOH = 100:1, 0.5 mL/min): t_R = 10.1 min for (R)-enantiomer, t_R = 12.1 min for (S)-enantiomer.

[a] Distil dichloromethane and toluene from calcium hydride under argon.

References

1. Messey, A. G.; Park, A. J. *J. Organomet. Chem.* **1964**, *2*, 245.
2. Messey, A. G.; Park, A. J. *J. Organomet. Chem.* **1966**, *5*, 218.
3. Ishihara, K.; Hanaki, N.; Yamamoto, H. *Synlett* **1993**, 577.
4. Ishihara, K.; Funahashi, M.; Hanaki, N.; Miyata, M.; Yamamoto, H. *Synlett* **1994**, 963.
5. Ishihara, K.; Hanaki, N.; Funahashi, M.; Miyata, M.; Yamamoto, H. *Bull. Chem. Soc. Jpn.* **1995**, *68*, 1721.
6. Ishihara, K.; Hanaki, N.; Yamamoto, H. *Synlett* **1996**, 721.
7. Parks, D. J.; Piers, W. E. *J. Am. Chem. Soc.* **1996**, *118*, 9440.
8. Ishihara, K.; Kurihara, H.; Yamamoto, H. *Synlett* **1997**, 597.
9. Ishihara, K.; Ohara, S.; Yamamoto, H. *J. Org. Chem.* **1996**, *61*, 4196.
10. Kaufmann, D.; Boese, R. *Angew, Chem., Int. Ed. Engl.* **1990**, *29*, 545.
11. Furuta, K.; Miwa, Y.; Iwanaga, K.; Yamamoto, H. *J. Am. Chem. Soc.* **1988**, *110*, 6254.
12. Furuta, K.; Shimizu, S.; Miwa, Y.; Yamamoto, H. *J. Org. Chem.* **1989**, *54*, 1481.

13. Gao, Q.; Yamamoto, H. *Org. Synth.* **1995**, *72*, 86.
14. Furuta, K.; Kanematsu, A.; Yamamoto, H.; Takaoka, S. *Tetrahedron Lett.* **1989**, *30*, 7231.
15. Ishihara, K.; Gao, Q.; Yamamoto, H. *J. Org. Chem.* **1993**, *58*, 6917.
16. Ishihara, K.; Gao, Q.; Yamamoto, H. *J. Am. Chem. Soc.* **1993**, *115*, 10412.
17. Sartor, D.; Saffrich, J.; Helmchen, G. *Synlett* **1990**, 197.
18. Takasu, M.; Yamamoto, H. *Synlett* **1990**, 194.
19. Sartor, D.; Saffrich, J.; Helmchen, G.; Richards, C. J.; Lambert, H. *Tetrahedron: Asymmetry* **1991**, *2*, 639.
20. Corey, E. J.; Loh, T.-P. *J. Am. Chem. Soc.* **1991**, *113*, 8966.
21. Corey, E. J.; Loh, T.-P.; Roper, T. D.; Azimioara, M. D.; Noe, M. C. *J. Am. Chem. Soc.* **1992**, *114*, 8290.
22. Marshall, J. A.; Xie, S. *J. Org. Chem.* **1992**, *57*, 2987.
23. Corey, E. J.; Guzman-perez, A. Loh, T.-P. *J. Am. Chem. Soc.* **1994**, *116*, 3611.
24. Corey, E. J.; Loh, T.-P. *Tetrahedron Lett.* **1993**, *34*, 3979.
25. Itsuno, S.; Kamahori, K.; Watanabe, K.; Koizumi, T.; Ito, K. *Tetrahedron: Asymmetry* **1994**, *5*, 523.
26. Kamahori, K.; Tada, S.; Ito, K.; Itsuno, S. *Tetrahedron: Asymmetry* **1995**, *6*, 2547.
27. Itsuno, S.; Watanabe, K.; Koizumi, T.; Ito, K. *React. Polym.* **1995**, *24*, 219.
28. Kamahori, K.; Ito, K.; Itsuno, S. *J. Org. Chem.* **1996**, *61*, 8321.
29. Hawkins, J. M.; Loren, S. *J. Am. Chem. Soc.* **1991**, *113*, 7794.
30. Hawkins, J. M.; Loren, S.; Nambu, M. *J. Am. Chem. Soc.* **1994**, *116*, 1657.
31. Kobayashi, S.; Murakami, M.; Harada, T.; Mukaiyama, T. *Chem. Lett.* **1991**, 1341.
32. Aggarwal, V. K.; Anderson, E.; Giles, R.; Zaparucha, A. *Tetrahedron: Asymmetry*, **1995**, *6*, 1301.
33. Hayashi, Y.; Rohde, J.; Corey, E. J. *J. Am. Chem. Soc.* **1996**, *118*, 5502.
34. Ishihara, K.; Yamamoto, H. *J. Am. Chem. Soc.* **1994**, *116*, 1561.
35. Ishihara, K.; Kurihara, H.; Yamamoto, H. *J. Am. Chem. Soc.* **1996**, *118*, 3049.
36. Furuta, K.; Maruyama, T.; Yamamoto, H. *J. Am. Chem. Soc.* **1991**, *113*, 1041.
37. Ishihara, K.; Maruyama, T.; Mouri, M.; Gao, Q.; Furuta, K.; Yamamoto, H. *Bull. Chem. Soc. Jpn* **1993**, *66*, 3483.
38. Furuta, K.; Maruyama, T.; Yamamoto, H. *Synlett* **1991**, 439.
39. Furuta, K.; Mouri, M.; Yamamoto, H. *Synlett* **1991**, 561.
40. Ishihara, K.; Mouri, M.; Gao, Q.; Maruyama, T.; Furuta, K.; Yamamoto, H. *J. Am. Chem. Soc.* **1993**, *115*, 11490.
41. Marshall, J. A.; Tang, Y. *Synlett* **1992**, 653.
42. Kiyooka, S.; Kaneko, Y.; Komura, M.; Matsuo, H.; Nakano, M. *J. Org. Chem.* **1991**, *56*, 2276.
43. Parmee, E. R.; Tempkin, O.; Masamune, S. *J. Am. Chem. Soc.* **1991**, *113*, 9365.
44. Parmee, E. R.; Hong, Y. H.; Tempkin, O.; Masamune, S. *Tetrahedron Lett.* **1992**, *33*, 1729.
45. Kiyooka, S.; Kaneko, TY.; Kume, K. *Tetrahedron Lett.* **1992**, *33*, 4927.
46. Corey, E. J.; Cywin, C. L.; Roper, T. D. *Tetrahedron Lett.* **1992**, *33*, 6907.
47. Kinugasa, M.; Harada, T.; Fujita, K.; Oku, A. *Synlett* **1996**, 43.
48. Kinugasa, M.; Harada, T.; Egusa, T.; Fujita, K.; Oku, A. *Bull. Chem. Soc. Jpn* **1996**, *69*, 3639.
49. Kinugasa, M.; Harada, T.; Oku, A. *J. Org. Chem.*, **1996**, *61*, 6772.
50. Hattori, K.; Yamamoto, H. *J. Org. Chem.* **1992**, *57*, 3264.

51. Hattori, K.; Yamamoto, H. *Tetrahedron* **1993**, *49*, 1749.
52. Ishihara, K.; Miyata, M.; Hattori, K.; Yamamoto, H.; Tada, T. *J. Am. Chem. Soc.* **1994**, *116*, 10520.
53. Ishihara, K.; Kuroki, Y.; Yamamoto, H. *Synlett* **1995**, 41.
54. Hattori, K.; Miyata, M.; Yamamoto, H. *J. Am. Chem. Soc.* **1993**, *115*, 1151.
55. Hattori, K.; Yamamoto, H. *Synlett* **1993**, 239.
56. Hattori, K.; Yamamoto, H. *BioMed. Chem. Lett.* **1993**, *3*, 2337.

Magnesium(II) and other alkali and alkaline earth metals

AKIRA YANAGISAWA

1. Introduction

Alkali and alkaline earth metal salts have mild Lewis acidities and have been utilized as a promoter or catalyst for organic reactions. Among the alkali metal salts, $LiClO_4$,[1–3] usually used as a concentrated solution in diethyl ether, is the most popular reagent to induce various transformations, including Diels–Alder reactions,[4,5] 1,3-sigmatropic rearrangements,[6] Mukaiyama aldol additions,[7,8] and Michael additions.[9] A catalytic amount of the lithium compound has also been found effective to promote these reactions.[7,8,10] Additionally, $LiBF_4$ and $LiNTf_2$ have been used as lithium Lewis acid alternatives for Diels–Alder reactions.[11,12] The alkaline earth metal salt $MgBr_2$[13] is commonly used as a Lewis acid capable of forming strong bidentate chelates with α-alkoxy carbonyl compounds. It has thus been successfully applied to diastereoselective hetero-Diels–Alder reactions,[14,15] Mukaiyama aldol additions,[16,17] and allylation reactions.[18–21] Chiral bis(oxazoline)-Mg(II) complexes have been developed and proven to be excellent chiral Lewis acid catalysts for enantioselective Diels–Alder reactions[22,23] and 1,3-dipolar cycloaddition reactions.[24] The moderate Lewis acidity and reactivity of the chiral magnesium reagents are useful in obtaining high enantioselectivity and designing new catalysts for asymmetric synthesis.

2. Cycloaddition reactions

In 1986 Braun and Sauer found the enhanced endo selectivity for the Diels–Alder reaction of methyl acrylate and cyclopentadiene in concentrated solutions of lithium perchlorate in diethyl ether, THF, and DME.[4] Four years later Grieco et al. described such a solvent system, a 5.0 M solution of $LiClO_4$ in diethyl ether, which had a greater accelerating effect on the [4+2]-cycloaddition.[5] For example, the reaction of trans-piperylene with 2,6-dimethylbenzoquinone in a 5.0 M ether solution of $LiClO_4$ at ambient temperature and pressure is complete within 15 min, affording an 80% yield of cycloadduct

(Scheme 4.1), whereas the same reaction in benzene requires a longer reaction time and higher reaction temperature for completion of the reaction. The observed rate acceleration, which has been believed to come from a high internal solvent pressure, is due to the Lewis acidity of the lithium salt.[25] In 1993 Reetz *et al.* found that a catalytic amount of $LiClO_4$ (3–25 mol%) suspended in dichloromethane is effective in promoting Diels–Alder reactions[10] in addition to Mukaiyama aldol reactions,[8] Michael additions,[8] and 1,3-Claisen rearrangements.[10] An example is provided by the reaction of methyl acrylate with cyclopentadiene in the presence of 25 mol% of $LiClO_4$ (Scheme 4.2 in Protocol 1).[10] The conversion in 87% at $-15°C$ after a reaction time of 24 h and the *endo:exo* ratio is shown to be 7.3:1.

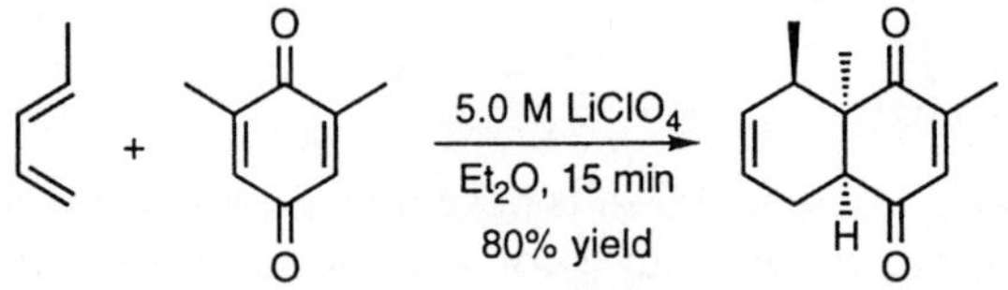

Scheme 4.1

Protocol 1.
Diels–Alder reaction of methyl acrylate with cyclopentadiene catalysed by $LiClO_4$[10] (Scheme 4.2)

Caution! Carry out all procedures in a well-ventilated hood, and wear disposable vinyl or latex gloves and chemical-resistant safety goggles.

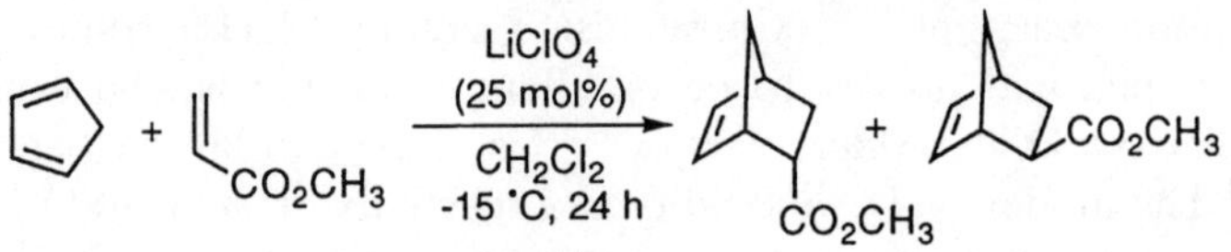

Scheme 4.2

Equipment
- Magnetic stirrer
- Syringes
- Cooling bath with dry ice–acetone
- Schlenk flask (10 mL)
- Vacuum/inert gas source (argon)

Materials
- Cyclopentadiene, 330 mg, 5 mmol — flammable, toxic
- Methyl acrylate, 86 mg, 1 mmol — flammable, lachrymator
- Lithium perchlorate, 25 mg, 0.25 mmol — oxidizer, irritant
- Dichloromethane, 16 mL — toxic, irritant
- Water
- Magnesium sulfate — hygroscopic

66

1. Ensure that the Schlenk flask and syringes have been dried and flushed with argon before use.
2. Dry dichloromethane by distillation over P_2O_5 and then CaH_2 prior to use.
3. Prepare a solution of methyl acrylate (86 mg, 1 mmol) and cyclopentadiene (330 mg, 5 mmol) in dry CH_2Cl_2 (1 mL) under argon atmosphere.
4. After cooling the solution to $-15°C$, add $LiClO_4$ (25 mg, 0.25 mmol).
5. After stirring for 24 h, dilute the reaction mixture with CH_2Cl_2 (15 mL).
6. Wash the dichloromethane solution with water twice (15 mL each). Dry the organic layer and then remove the organic solvent using a rotary evaporator to obtain the crude product (87% conversion, an *endo:exo* ratio of 7.3:1).
7. Determine the conversion based on methyl acrylate and the *endo:exo* ratio of the crude product by GC analysis.

Magnesium halides have been utilized as Lewis acids for [4+2]-cycloaddition reactions and shown to increase the regioselectivity, having an opposite effect to that of $BF_3•OEt_2$.[14,15,26] For example, the $MgBr_2$-promoted hetero-Diels–Alder reaction of a siloxy diene with α-alkoxy aldehydes gives a single diastereomer (Scheme 4.3 in Protocol 2),[15] whereas under Lewis acid catalysis by either $BF_3•OEt_2$ or $ZnCl_2$, another diastereomer is selectively obtained with a 1.4:1 ratio. The stereochemical outcome of the $MgBr_2$-promoted reaction reveals that the cycloaddition occurs from the less hindered face of a *syn* conformer of the aldehyde in which a chelation of the magnesium exists between the two oxygens.

Protocol 2.
Hetero-Diels–Alder reaction of 2,4-Bis(trimethylsiloxy)-1,3-pentadiene with 2-(benzyloxy)butyraldehyde promoted by magnesium bromide. synthesis of (2S*,1'S*)-2-[(1'-benzyloxy)propyl-6-methyl-2,3-dihydro-4H-pyran-4-one[15] (Scheme 4.3)

Caution! Carry out all procedures in a well-ventilated hood, and wear disposable vinyl or latex gloves and chemical-resistant safety goggles.

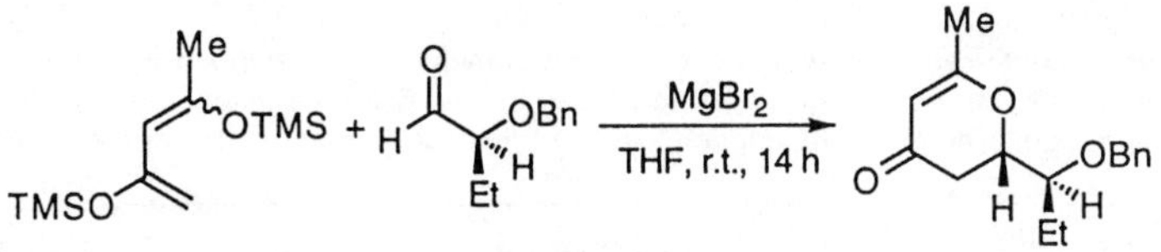

Scheme 4.3

Equipment
- Magnetic stirrer
- Syringes
- Cooling bath with ice water
- Round-bottomed flask (200 mL)
- Vacuum/inert gas source (argon)

Akira Yanagisawa

Protocol 2. Continued

Materials

• 2,4-Bis(trimethylsiloxy)-1,3-pentadiene, 9.10 g, 42.1 mmol	flammable, moisture-sensitive
• 2-(Benzyloxy)butyraldehyde, 3.00 g, 16.9 mmol	irritant
• Magnesium bromide (2.95 M solution in 10% benzene/ether),[a] 5.7 mL, 16.9 mmol	irritant, hygroscopic
• Dry tetrahydrofuran (THF), 80 mL	flammable, irritant
• Sodium bicarbonate	moisture-sensitive
• Ethyl ether	flammable, irritant
• Magnesium sulfate	hygroscopic
• Dichloromethane, 100 mL	toxic, irritant
• Trifluoroacetic acid, 4 mL	corrosive, toxic
• Sodium chloride	hygroscopic, irritant

1. Ensure that all glass equipment has been dried in an oven before use.

2. Place 2-(benzyloxy)butyraldehyde (3.00 g, 16.9 mmol) in a 200 mL flask with a magnetic stirring bar under argon atmosphere and dissolve the aldehyde with dry THF (80 mL).

3. Cool the solution to 0°C and add a solution of magnesium bromide (2.95 M, 5.7 mL, 16.9 mmol) in 10% benzene/ether over 2 min.

4. After stirring for 10 min, add 2,4-bis(trimethylsilyloxy)-1,3-pentadiene (9.10 g, 42.1 mmol) and warm the solution slowly to room temperature.

5. After stirring for 14 h, pour the solution into a saturated aqueous sodium bicarbonate solution and extract with ether. Dry the organic layer over anhydrous MgSO$_4$ and concentrate in vacuo.

6. Dissolve the resultant oil in 100 mL of CH$_2$Cl$_2$ and add trifluoroacetic acid (4 mL). After stirring for 3 h, wash the solution with water, saturated aqueous sodium bicarbonate solution, and brine, then dry the organic layer over anhydrous MgSO$_4$ and concentrate in vacuo.

7. Purify the crude product by flash column chromatography on silica gel (40% ethyl acetate/hexane) to give 3.41 g (78%) of the title compound as a clear colourless oil and characterize by IR, ^{1}H NMR, ^{13}C NMR, and elemental analysis.

8. Examine the crude and purified products by GLC (column: 4 ft, 3% OV-17, 210°C, 30 mL/min flow rate, t_R = 10.1 min) and 500 MHz ^{1}H NMR experiments to confirm there is no detectable amount of the diastereomer.

[a] Magnesium bromide can be prepared from 1,2-dibromoethane and magnesium turnings in ether. Add benzene (c. 10%) to homogenize the two-phase mixture. Calculate the molarity by weighing the residual magnesium. The resultant solution can be stored for several months at room temperature.

In 1992 the first example of a chiral magnesium reagent as a catalyst for an asymmetric reaction was reported by Corey and Ishihara.[22] The chiral Mg catalyst is prepared by treatment of a chiral bis(oxazoline) ligand with MgI$_2$ and I$_2$ (co-catalyst) in CH$_2$Cl$_2$. When 3-acryloyloxazolidine-2-one is reacted with cyclopentadiene in the presence of the catalyst (10 mol%) at −80°C, the Diels–Alder product was obtained in 82% yield with 91% ee and an *end/exo* ratio of 97/3 (Scheme 4.4).[22] A 1:1:1 complex of the chiral ligand, Mg^{2+}, and the dienophile is believed to be the reactive species in which the *s-cis* form of oxazolidinone co-ordinates to the metal. The diene approaches from the front side of the complex avoiding repulsion by the phenyl group (Scheme 4.4).

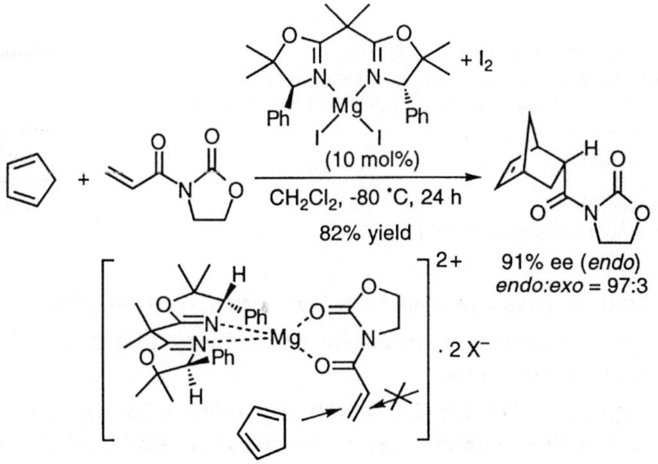

91% ee (*endo*)
endo:exo = 97:3

Scheme 4.4

This chiral bis(oxazoline)-magnesium catalyst has been successfully applied to the enantioselective 1,3-dipolar cycloaddition reaction of alkenes with nitrones to lead high *endo*-selectivity with up to 82% ee (Scheme 4.5 in Protocol 3).[24]

Protocol 3.

Asymmetric 1,3-dipolar cycloaddition reaction of 3-((*E*)-2'-butenoyl)-1,3-oxazolidin-2-one with benzylidenephenylamine *N*-oxide catalysed by chiral Mg(II)-bisoxazoline complex[24] (Scheme 4.5)

Caution! Carry out all procedures in a well-ventilated hood, and wear disposable vinyl or latex gloves and chemical-resistant safety goggles.

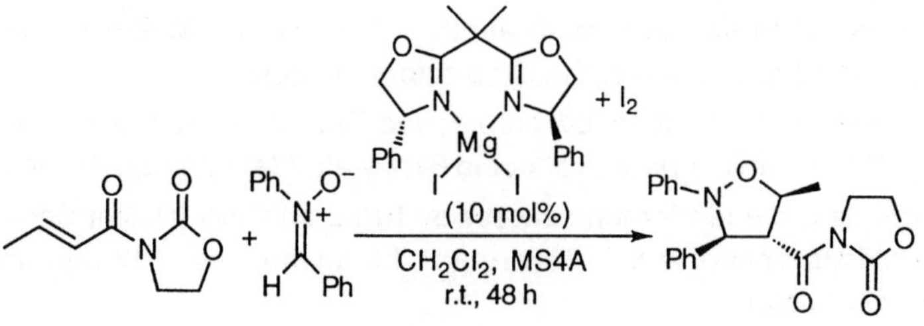

Scheme 4.5

Protocol 3.   *Continued*

Equipment

- Magnetic stirrer
- Syringes

- Schlenk flask (10 mL, 5mL)
- Vacuum/inert gas source (argon)

Materials

- 3-((*E*)-2'-butenoyl)-1,3-oxazolidin-2-one, 78 mg, 0.5 mmol   **flammable, toxic**
- Benzylidenephenylamine *N*-oxide, 123 mg, 0.625 mmol
- (*R*)-2,2'-isopropylidenebis(4-phenyl-2-oxazoline), 410 mg, 1.25 mmol
- Magnesium, 48 mg, 2.0 mmol   **moisture-sensitive**
- Iodine, 506 mg, 2.0 mmol   **toxic, corrosive**
- Ethyl ether, 5 mL   **flammable, irritant**
- Dichloromethane, 15 mL   **toxic, irritant**
- 4Å powdered molecular sieves, 50–100 mg
- MeOH

1. Ensure that all glass equipment has been dried in an oven before use.
2. Activate the 4Å powdered molecular sieves by heating to 250°C for 3 h in high vacuum prior to use.
3. Place magnesium (48 mg, 2.0 mmol) and iodine (253 mg, 1.0 mmol) in a 10 mL flask with a magnetic stirring bar under N_2. Add diethyl ether (5 mL) and stir the mixture at room temperature until the iodine colour disappears (2–3 h). Filter off the unreacted Mg and remove the solvent of the filtrate under reduced pressure at room temperature.
4. Dissolve the white MgI_2 in CH_2Cl_2 and add (*R*)-2,2'-isopropylidenebis(4-phenyl-2-oxazoline) (410 mg, 1.25 mmol) dissolved in CH_2Cl_2 (5 mL) to the resulting solution.
5. After stirring the milky suspension for 2 h, add I_2 (253 mg, 1.0 mmol) and stir the deep-red suspension for 2 h before use.
6. Add 3-((*E*)-2'-butenoyl)-1,3-oxazolidin-2-one (78 mg, 0.5 mmol) and benzylidenephenylamine *N*-oxide (123 mg, 0.625 mmol) to a suspension of 50–100 mg of 4Å powdered molecular sieves in CH_2Cl_2 (2 mL) in a 5 mL flask.
7. After stirring the mixture for 15 min, add the catalyst (0.05 mmol) with a glass pipette.
8. After stirring for 48 h, treat the reaction mixture with 5% MeOH in CH_2Cl_2 (1 mL) and filter the mixture through a 20 mm layer of silica gel.
9. Wash the silica gel layer with another 2 mL of 5% MeOH in CH_2Cl_2 and remove the organic solvent using a rotary evaporator.
10. Purify the crude product by preparative TLC (silica gel, ether–petroleum ether (3:1)) to give a pure product in 81% yield (84% *endo*, 75% ee).
11. Determine the enantiomeric excess by HPLC analysis (Daicel Chiralcel OD, hexane/isopropanol = 9:1, flow rate = 1.0 mL/min, t_R = 42 min (minor), t_R = 58 min (major).

3. Aldol additions, allylations, and other reactions

Lithium and magnesium Lewis acid reagents are also effective promoters for nucleophilic carbon–carbon bond forming reactions. The Mukaiyama aldol reaction of ketene silyl acetals with aldehydes occurs at room temperature under the influence of a stoichiometric or catalytic amount of $LiClO_4$.[7,8] For example, the silyl ether derived from methyl acetate reacts with benzaldehyde in a 5.0 M ether solution of $LiClO_4$ to provide the silylated aldol adduct in 98% yield (Scheme 4.6).[7] The use of a catalytic amount (3 mol%) of $LiClO_4$ suspended in CH_2Cl_2 results in complete conversion within 15 min.[8]

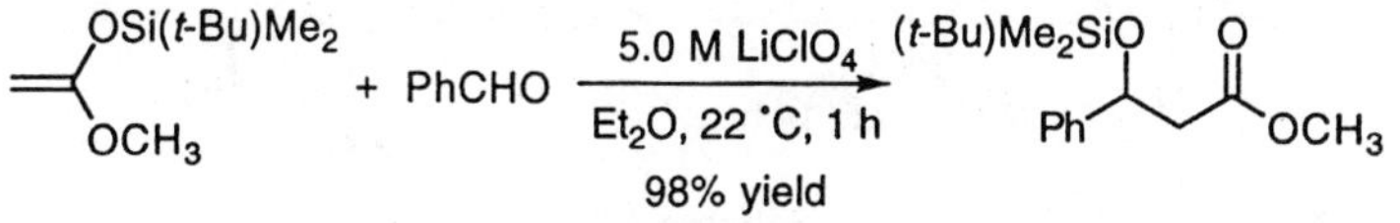

Scheme 4.6

The addition of allylstannanes to α-alkoxy aldehydes has been shown to proceed with high diastereoselectivity under chelation control in the presence of $LiClO_4$[27] or $MgBr_2$.[18–21] When a mixture of α-benzyloxy aldehyde and allyltributylstannane is treated with $MgBr_2$ in CH_2Cl_2, the *syn*-product is obtained with >250:1 stereoselectivity.[18] In marked contrast, non-chelation-controlled stereochemistry is observed with $BF_3\bullet OEt_2$ (Scheme 4.7).[18] The high *syn*-selectivity provided by $MgBr_2$ is due to the formation of a discrete bidentate chelate with the α-alkoxy aldehyde.

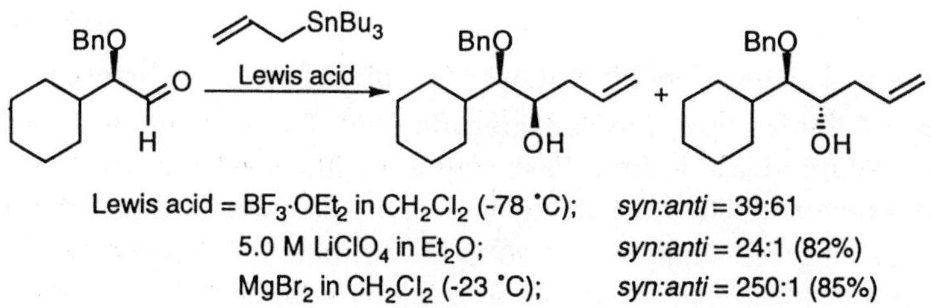

Lewis acid = $BF_3 \cdot OEt_2$ in CH_2Cl_2 (-78 °C); *syn:anti* = 39:61
5.0 M $LiClO_4$ in Et_2O; *syn:anti* = 24:1 (82%)
$MgBr_2$ in CH_2Cl_2 (-23 °C); *syn:anti* = 250:1 (85%)

Scheme 4.7

The high levels of diastereofacial selectivity observed for the allylations are preserved for the case of crotylation where three contiguous chiral centres are defined during the reaction.[19,20] With α-benzyloxy aldehyde, *syn,syn*-selectivity is observed regardless of the geometry of the crotyltin. However, higher diastereoselectivities are given with the *E*-stannane, for instance, a 12.3:1 ratio of *syn,syn/syn,anti* products is obtained with a 90:10 mixture of the *E*- and *Z*-stannanes (Scheme 4.8 in Protocol 4).[21]

Akira Yanagisawa

Protocol 4.
Addition of crotyltributylstannane to (R)-2-(benzyloxy)propionaldehyde promoted by magnesium bromide etherate. Preparation of (2R,3R,4S)-2-benzyloxy-4-methyl-5-hexen-3-ol[21] (Scheme 4.8)

Caution! Carry out all procedures in a well-ventilated hood, and wear disposable vinyl or latex gloves and chemical-resistant safety goggles.

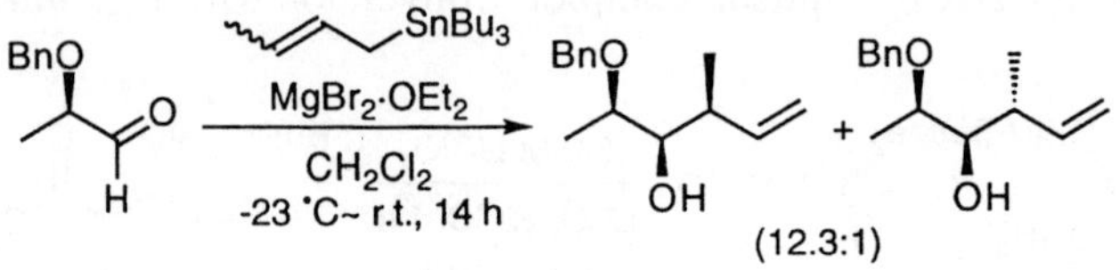

Scheme 4.8

Equipment
- Magnetic stirrer
- Syringes
- Cooling bath with dry ice/acetone
- Round-bottomed flask (100 mL)
- Vacuum/inert gas source (argon)

Materials
- (R)-2-(Benzyloxy)propionaldehyde, 450 mg, 2.74 mmol — irritant
- Crotyltributylstannane (E:Z = 90:10),[a] 1.42 g, 4.11 mmol — irritant
- Magnesium bromide etherate, 1.14 g, 5.48 mmol — flammable, moisture-sensitive
- Dry dichloromethane, 30 mL — toxic, irritant
- Sodium bicarbonate — moisture-sensitive
- Dichloromethane for extraction (80 mL) — toxic, irritant
- Sodium sulfate — irritant, hygroscopic

1. Ensure that all glass equipment has been dried in an oven before use.

2. Place (R)-2-(benzyloxy)propionaldehyde (450 mg, 2.74 mmol) in an oven-dried 100 mL round bottom flask with a magnetic stirring bar under argon atmosphere and dilute the aldehyde with dry dichloromethane (30 mL).

3. After cooling the solution to −23°C for 10 min, add magnesium bromide etherate (1.14 g, 5.48 mmol) all at once as a white powder. The mixture will become cloudy, and then over the next 10 min the solid material will partially dissolve and the solution will become slightly yellow.

4. After stirring for 15 min, add crotyltributylstannane (1.42 g, 4.11 mmol) dropwise via a syringe down the side of the flask to allow for ample cooling. Stir the mixture for an additional 2 h at −23°C before the bath is allowed to expire.

5. After stirring for 12 h at room temperature, add saturated aqueous sodium bicarbonate solution (30 mL) and continue stirring for 25 min. Extract the mixture with CH$_2$Cl$_2$ (4 × 20 mL) and dry the organic layer over anhydrous Na$_2$SO$_4$. Filter the organic layer through a plug of Celite® (1 cm) and silica gel (2 cm).

6. Remove the solvent *in vacuo* and purify the crude product via radial plate liquid chromatography. Load the material onto a 4 mm plate with hexane (3 mL) and elute with 150 mL of hexane, 100 mL of 5% EtOAc/hexane, 100 mL of 10% EtOAc/hexane, and 150 mL of 15% EtOAc/hexane to give the title compound (509 mg, 84%, (2*R*,3*R*,4*S*)-isomer:(2*R*,3*R*,4*R*)-isomer = 12.3:1) as a clear colourless liquid.
7. Determine the diastereomeric ratio by GLC analysis (GC DX-4 15 m column 120–180 °C at 3.5°C/min, major, 13.72 min, minor, 14.00 min).

[a] Crotyltributylstannane can be obtained by tin anion displacement of the corresponding crotyl chloride.

The oxophilic nature of MgBr$_2$ is valid for mediating numerous rearrangements of oxygen-containing compounds. A typical example of the stereospecific conversion of a β-lactone to a butyrolactone including methyl group migration is shown in Scheme 4.9 (Protocol 5).[28] Use of other conventional Lewis acids than MgBr$_2$•OEt$_2$ results in recovery of the unreacted β-lactone or the occurrence of undesired side reactions. A stepwise mechanism involving a rate-determining ionization step is proposed for the rearrangement rather than a concerted mechanism.[29]

Protocol 5.
Rearrangement of a β-lactone to a butyrolactone promoted by magnesium bromide etherate. Preparation of *trans*-4,5-dihydro-3-phenyl-4,5,5-trimethyl-2(3*H*)-furanone[29] (Scheme 4.9)

Caution! Carry out all procedures in a well-ventilated hood, and wear disposable vinyl or latex gloves and chemical-resistant safety goggles.

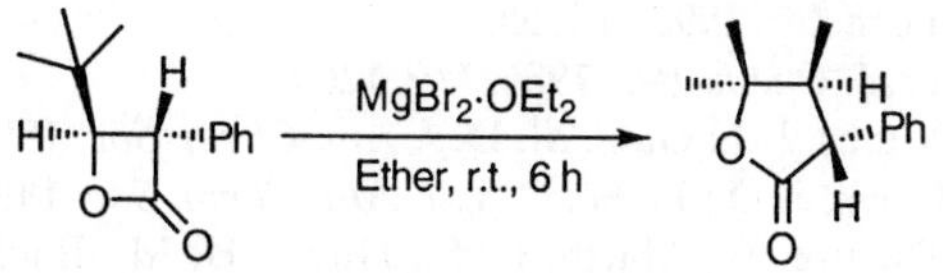

Scheme 4.9

Equipment
- Magnetic stirrer
- Syringes
- Three-necked flask (25 mL)
- Vacuum/inert gas source (nitrogen)

Materials
- *cis*-4-*tert*-Butyl-3-phenyloxetan-2-one,[a] 2.04 g, 10 mmol — irritant
- Magnesium bromide etherate, 2.58 g, 10 mmol — flammable, moisture-sensitive
- Dry ether,[b] 10 mL — flammable, irritant
- Water, 10 mL
- Magnesium sulfate — hygroscopic
- Hexane for recrystallization — flammable, irritant

Akira Yanagisawa

Protocol 5. *Continued*

1. Ensure that all glass equipment has been dried in an oven at 120°C for a minimum of 4 h and then assembled under a nitrogen stream before use.
2. Equip an oven-dried 25-mL three-necked flask with a nitrogen inlet and stirring bar and charge the flask with 10 mL of a 1.0 M solution of *cis-4-tert*-butyl-3-phenyloxetan-2-one (2.04 g, 10 mmol) in anhydrous ether.
3. Begin stirring and add magnesium bromide etherate (2.58 g, 10 mmol) in a single portion. Stir the light-yellow mixture under nitrogen for 6 h at room temperature, then terminate the reaction by the gradual addition of 10 mL of water.
4. Separate the layers and dry the ether layer over anhydrous MgSO₄. After filtering the organic layer, remove the solvent under reduced pressure. Purify the crude product by recrystallization from hexane to afford the title compound (1.75 g, 86%) which displays the appropriate ^{1}H NMR (in CDCl$_3$) and IR (KBr).

^{a} *cis-4-tert*-Butyl-3-phenyloxetan-2-one can be conveniently prepared from phenylacetic acid by a two-step sequence: (1) deprotonation of the acid by 2 equiv of lithium di-isopropylamide in THF followed by treatment with 2,2-dimethylpropanal to afford the resulting β-hydroxy acid (74%), (2) cyclization of the acid with benzenesulfonyl chloride in pyridine at 0°C (94%).
^{b} Dry ether is distilled from sodium benzophenone ketyl prior to use.

References

1. Grieco, P. A. *Aldrichimica Acta* **1991**, *24*, 59.
2. Waldmann, H. *Angew. Chem. Int. Ed. Engl.* **1991**, *30*, 1306.
3. Charette, A. B. In *Encyclopedia of reagents for organic synthesis* (ed. L. A. Paquette), John Wiley, Chichester, **1995**, p 3155.
4. Braun, R.; Sauer, J. *Chem. Ber.* **1986**, *119*, 1269.
5. Grieco, P. A.; Nunes, J. J.; Gaul, M. D. *J. Am. Chem. Soc.* **1990**, *112*, 4595.
6. Grieco, P. A.; Clark, J. D.; Jagoe, C. T. *J. Am. Chem. Soc.* **1991**, *113*, 5488.
7. Reetz, M. T.; Raguse, B.; Marth, C. F.; Hügel, H. M.; Bach, T.; Fox, D. N. A. *Tetrahedron* **1992**, *48*, 5731.
8. Reetz, M. T.; Fox, D. N. A. *Tetrahedron Lett.* **1993**, *34*, 1119.
9. Grieco, P. A.; Cooke, R. J.; Henry, K. J.; VanderRoest, J. M. *Tetrahedron Lett.* **1991**, *32*, 4665.
10. Reetz, M. T.; Gansäuer, A. *Tetrahedron* **1993**, *49*, 6025.
11. Smith, D. A.; Houk, K. N. *Tetrahedron Lett.* **1991**, *32*, 1549.
12. Handy, S. T.; Grieco, P. A.; Mineur, C.; Ghosez, L. *Synlett* **1995**, 565.
13. Black, T. H. In *Encyclopedia of reagents for organic synthesis* (ed. L. A. Paquette). John Wiley, Chichester, **1995**, p 3197.
14. Danishefsky, S. J.; Pearson, W. H.; Harvey, D. F. *J. Am. Chem. Soc.* **1984**, *106*, 2455, 2456.
15. Danishefsky, S. J.; Pearson, W. H.; Harvey, D. F.; Maring, C. J.; Springer, J. P. *J. Am. Chem. Soc.* **1985**, *107*, 1256.

16. Bernardi, A.; Cardani, S.; Colombo, L.; Poli, G.; Schimperna, G.; Scolastico, C. *J. Org. Chem.* **1987**, *52*, 888.
17. Annunziata, R.; Cinquini, M.; Cozzi, F.; Cozzi, P. G.; Consolandi, E. *J. Org. Chem.* **1992**, *57*, 456.
18. Keck, G. E.; Boden, E. P. *Tetrahedron Lett.* **1984**, *25*, 265.
19. Keck, G. E.; Boden, E. P. *Tetrahedron Lett.* **1984**, *25*, 1879.
20. Keck, G. E.; Abbott, D. E. *Tetrahedron Lett.* **1984**, *25*, 265, 1883.
21. Keck, G. E.; Savin, K. A.; Cressman, E. N. K.; Abbott, D. E. *J. Org. Chem.* **1994**, *59*, 7889.
22. Corey, E. J.; Ishihara, K. *Tetrahedron Lett.* **1992**, *33*, 6807.
23. Desimoni, G.; Faita, G.; Righetti, P. P. *Tetrahedron Lett.* **1996**, *37*, 3027.
24. Gothelf, K. V.; Hazell, R. G.; Jørgensen, K. A. *J. Org. Chem.* **1996**, *61*, 346.
25. Forman, M. A.; Dailey, W. P. *J. Am. Chem. Soc.* **1991**, *113*, 2761.
26. Kelly, T. R.; Montury, M. *Tetrahedron Lett.* **1978**, 4311.
27. Henry, K. J., Jr.; Grieco, P. A.; Jagoe, C. T. *Tetrahedron Lett.* **1992**, *33*, 1817.
28. Black, T. H.; Hall, J. A.; Sheu, R. G. *J. Org. Chem.* **1988**, *53*, 2371.
29. Black, T. H.; Eisenbeis, S. A.; McDermott, T. S.; Maluleka, S. L. *Tetrahedron* **1990**, *46*, 2307.

Zinc(II) Reagents

NOBUKI OGUNI

1. Introduction

There are two kinds of organozinc compounds: Grignard-type reagents and dialkylzincs. Organozinc reagents have been used for the carbon–carbon bond-forming reactions, for example, Reformatsky and Simmons-Smith reactions. Nevertheless dialkylzinc compounds have been little used for organic reactions except organometallic component of Ziegler catalyst. After our first report of the discovery in 1883 on the enantioselective addition of diethylzinc to benzaldehyde catalyzed by chiral β-aminoalcohols, this field of asymmetric carbon–carbon bond forming catalytic reaction was developed largely by many researchers. Also the asymmetric amplification, that is extreme amplification in enantioselectivity was found in this reaction and gave the extensive effect for all enantioselective catalytic reactions.[1-4]

2. Scope and limitations

Dialkylzincs can be prepared easily by transmetallation from alkylaluminiums and alkylboranes. Particularly, diethylzinc and dimethylzinc among many dialkylzincs are commercially available in most countries. Therefore the synthetic method of non-commercially available organozinc compounds using diethylzinc will be convenient and advantageous in comparison with other methods.[3,4]

$$R(CH_2)nI + Et_2Zn \xrightarrow[\text{r.t. 1--12 h}]{\text{neat}} [R(CH_2)_n]_2Zn + EtI$$
$$(3\text{--}5 \text{ equiv})$$

$$RCH=CH_2 + Et_2BH \longrightarrow RCH_2CH_2BEt_2 \xrightarrow[\text{0\,°C}]{\text{neat}} (RCH_2CH_2)_2Zr$$

Dialkylzincs have been known to be unreactive to aldehydes, esters, imines, and so on, in polar solvents at room temperature. This characteristic of organozincs is quite different from alkyl-lithium and dialkylmagnesium which react easily on mixing with aldehydes, ketones, and other functionalized compounds even at low temperature. Dialkylzinc only reacts with aldehydes to give addition products in the presence of basic catalysts such as trialkyl amines. The reaction products are monoalkylzinc alkoxides which easily associate to afford cubic tetramers co-ordinated with oxygen atoms of alkoxides. The alkyl group of alkylzinc alkoxide does not react with aldehydes.

$$R_2Zn + R'CHO \xrightarrow{\text{catalyst}} \overset{\displaystyle R'}{\underset{\displaystyle |}{R}}CHOZnR$$

Dialkylzincs also react with β-aminoalcohols or β-thioalcohols to give 2-valence 3-co-ordinated alkylzinc compounds which also associate to form dimers. When aldehydes and excess dialkylzincs are present in the media, dimer easily dissociate to catalyse the addition reaction of dialkylzincs to aldehydes. There have been reports of the use of chiral β-aminoalcohols[5-50] containing β-diamines, β-aminothiols,[68-72] hydroxy-pyridines[51-57] and amino-alcohol derivatives of ferrocene[58-63] and benzene–chrom complexes[64-72] in the reaction media of dialkylzincs with aldehydes to afford optically active products. Generally the chirality of the carbon bonded with a hydroxy group affects predominantly the chirality of the product rather than one with an amino group. In general aromatic and α,β-unsaturated aldehydes give high enantioselective products in reactions using β-difunctionalized chiral auxiliaries. In enantioselective addition of dialkylzincs to aldehydes, asymmetric amplification was found, even if a small amount of β-aminoalcohols bearing low optical purity was used as a catalyst, very high enantiomeric products were obtained. Investigations of the mechanism of this curious reaction estimated that the monomeric zincaminoalkoxide is the real active catalyst, and its racemic dimer associate cannot dissociate in reaction media.[73-76] When the chemical constitution and absolute configuration of the product are exactly the same with chiral auxiliary used, the product itself can also catalyse the reaction to produce the same compound. This reaction was realized in iso-propylation of pyridinecarbaldehydes with di-isopropylzinc, by a reaction called autocatalysis.[77-82]

Chiral Lewis acid-catalysed alkylation of aldehydes with dialkylzincs gave highly enantioselective products. Two outstanding catalysts are the titanium–chiral TADDOL and titanium–chiral disulfonamido systems.[83-105] Both catalysts give the very high enantiomeric products from all kinds of aldehydes, both aliphatic and alicyclic. However, in the former catalyst system an equimolar amount of titanium alkoxides to aldehydes is needed to obtain addition products in high yield.

Highly enantioselective 1,4-addition of diethylzinc to α,β-unsaturated ketones was attained using nickel or cupper complexes of β-aminoalcohols.[106-110] Also the catalytic enantioselective cyclopropanation of allylic alcohols with diethylzinc-CH_2I_2 was found using chiral titanium-disulfonamido complexes as catalyst.[111-124]

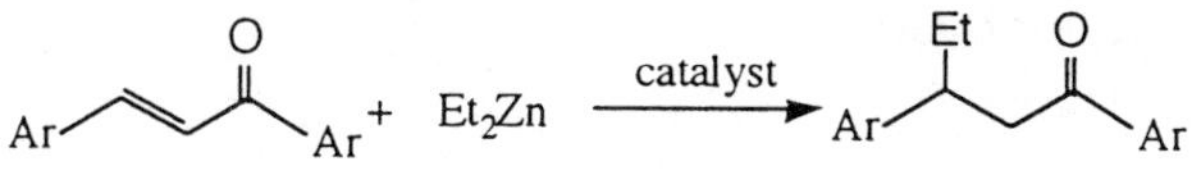

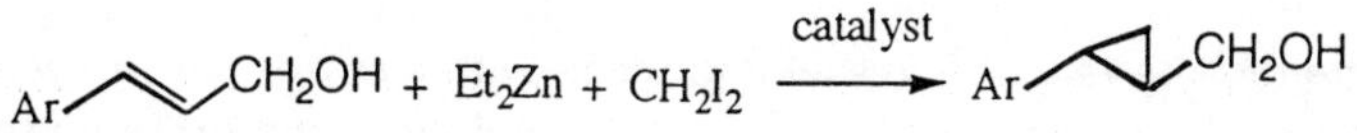

Dialylzinc can add to imino groups by catalysis of chiral titanium complexes to give chiral amine derivatives, followed by hydrolysis to afford optical active primary amines.[125-129]

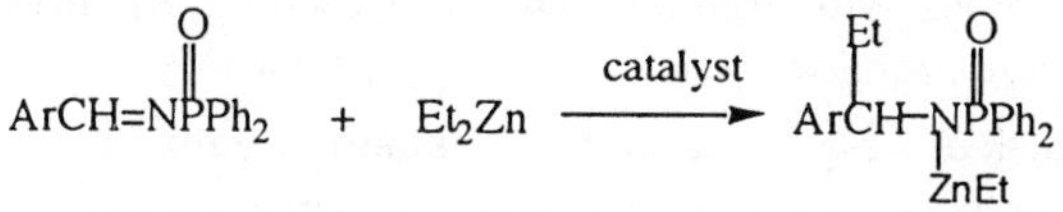

Cautions: Dialkylzincs are highly flammable liquids, which induce spontaneous ignition in air, and react violently with water and alcohols. Therefore, dialkylzincs should be kept under an inert atmosphere.

Protocol 1.
Chiral β-aminoalcohol-mediated asymmetric addition of diethylzinc to benzaldehyde[8] (Scheme 5.1)

Alkylation, particularly ethylation of aromatic compounds has been investigated with many kinds of β-aminoalcohols and highly enantioselective reactions also reported.

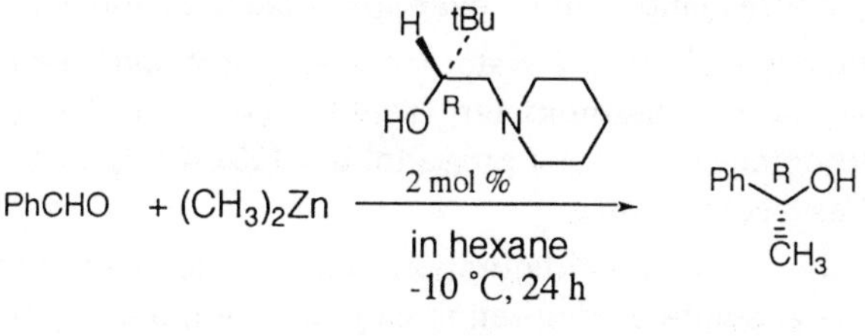

Scheme 5.1

Equipment

- Two-necked, round-bottomed flask (500 mL) fitted with rubber septum and a Tefon-coated magnetic stirring bar
- Three-way stopcock fitted to the top of flask and connected to a vacuum/argon (or dry nitrogen) source
- Magnetic stirrer
- All-glass syringe with a needle-lock Luer and medium-gauge needle
- Water-jacketed short-path distillation apparatus
- Separating funnel (500 mL)
- Rotary evaporator

Protocol 1. *Continued*

Materials

- Dry toluene
- Benzaldehyde
- Diethylzinc   **explosive on contact to air**
- (*R*)-1-*t*-Butyl-2-piperidinoethanol
- Technical ether for extraction   **flammable**

1. Flame dry the reaction vessel and place with a stirrer bar *in vacuo*. Fill the flask with argon and maintain a slightly positive argon pressure throughout this reaction.

2. Assemble the syringe and needle while hot and flush the syringe with argon.

3. Support the flask using clamp and a stand with a heavy base.

4. Charge the flask with toluene (200 mL) using a syringe by piercing the septum on the reaction flask.

5. Keep the flask at $-40\,°C$ with a dry ice/acetone bath.

6. Add benzaldehyde (10 g, 94.2 mmol) and (*R*)-1-*t*-butyl-2-piperidinoethanol (0.37 g, 2.0 mmol) using syringes under stirring.

7. Add diethylzinc (10.5 mL, 102 mmol) dropwise from the syringe through the septum on the reaction flask to the mixture in the flask with vigorous stirring.

8. Allow the temperature of the flask to rise from $-40\,°C$ to $-10\,°C$ with an ice/NaCl bath, and keep the mixture at $-10\,°C$ for 24 h.

9. Quench the mixture by dropping 2M HCl (100 mL) with stirring at $0\,°C$.

10. Transfer the mixture to a separating funnel and separate the two layers. Extract the water layer with ether (2×50 mL).

11. Transfer the organic layer to a 500 mL flask. Dry the organic layer over Na_2SO_4, and filter through a filter paper. Concentrate the filtrate under reduced pressure using a rotary evaporator ($30\,°C$/30 mm Hg).

12. Transfer the oily residue to a water-jacketed, short path distillation apparatus equipped with a thermometer. Distil the crude product under reduced pressure to obtain 1-phenyl-1-propanol(b.p.103–104 $°C$ 14 mm Hg; 12.6 g, 98% yield) as a colourless oil.

13. Determine the absolute configuration and enantiomeric excess(ee) of the product. The absolute configuration can be determined by the sign of $[\alpha]_D$ $[\alpha]_D$ $-45.4.(c\ 2.0,\ C_2H_5OH)$ indicated a (*S*)-configuration. Ee (>98%) was determined by HPLC analysis (column, Daicel CHIRALCEL OB; 100/0.2 hexane/2-propanol; t_R of (*S*)-isomer, 12 min; t_R of (*R*)-isomer, 13 min.

(*R*)-1-Dialkylamino-3-3-dimethylbutan-2-ol was prepared by the reaction of (*R*)-*t*-butyl-ethylene oxide[a] with bromomagnesium dialkylamide as follows.

To a solution of ethylmagnesium bromide (5.0 mmol) in tetrahydrofuran (THF)(10 mL) was added a solution of a secondary amine (5.0 mmol) in THF(5 mL). After the mixture has been stirred at 35°C for 1 h, (R)-t-butylethylene oxide (500 mg, 5 mmol) was added to the solution, which was then stirred for 6 h at room temperature before being poured into saturated aq.NH$_4$Cl. THe aqueous solution was acidified with 1 mol dm^{-3} HCl (20 ml), extracted with ethyl acetate (30 mL × 2), and then made alkaline by 10% aq. NaOH and extracted with ethyl acetate (20 mL × 2). The combined organic layer was washed with brine (20 mL × 2), dried over Na$_2$SO$_4$, and distilled or recrystallized to give the corresponding (R)-1-dialkylamino-3,3-dimethylbutan-2-ol in good yield. The enantiomeric excess of the β-aminoalcohol thus obtained was determined as over 99% by HPLC analysis of the corresponding 3,5-dinitrophenyl carbamate, using a chiral stationary phase(column, Sumitomo Chemical Co. Sumipax OA 4000).

(R)-3,3-Dimethyl-1-piperidinobutan-2-ol (1) was obtained from the reaction of piperidine (425 mg, 5 mmol) and (R)-t-butylethylene oxide (500 mg, 5 mmol) in yield of 743 mg (80%), b.p. 59–61°C/1 mmHg). [α]$_D^{25}$ −72.4 (C 1.8, CHCl$_3$).

(R)-1-(3-Azabicyco[3.2.2]nonan-3-yl)-3-dimethylbutan-2-ol was obtained from the reaction of 3-azabicyclo[3.3.3]nonane (630 mg, 5 mmol) and (R)-t-butyl-ethylene oxide (500 mg, 5 mmol) in yield of 1.12 g (88%), m.p.105–108°C). [α]$_D^{22}$ −64.2° (C 1.0, CHCl$_3$).

[a] Leven, P. A.; Walti, A. Organic Synthesis 1943, Coll. Vol. 2, 5–9. Hurst, S. J.; Bruce, J. M. J. Chem. Soc. 1963, 147.

Protocol 2.
Chiral β-aminoalcohol mediated asymmetric addition of dimethylzinc to methylacrylaldehyde[29] (Scheme 5.2)

Methylation of aldehydes has been investigated less frequently, because dimethylzinc has a very low boiling point like diethyl ether. However, it can be handled easily as a toluene solution.

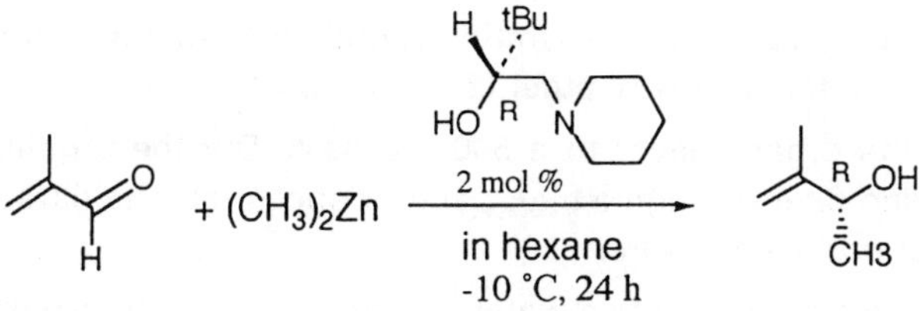

Scheme 5.2

Equipment

- Two-necked, round-bottomed flask (500 mL) fitted with rubber septum and a Teflon-coated magnetic stirring bar
- Three-way stopcock fitted to the top of the flask and connected to a vacuum/argon (or dry nitrogen) source
- Magnetic stirrer
- All-glass syringe with a needle-lock Luer and medium-gauge needle
- Water-jacketed short-path distillation apparatus
- Separating funnel (500 mL).

Nobuki Oguni

Protocol 2. *Continued*

Materials

- Dry ether and hexane
- Methylacrylaldehyde
- Dimethylzinc   **explosive on contact with air**
- (*R*)-1-*t*-Butyl-2-piperidinoethanol
- Technical ether for extraction   **flammable**

1. Flame dry the reaction vessel and place with a stirrer bar *in vacuo*. Fill the flask with argon and maintain a slightly positive argon pressure throughout the reaction.

2. Assemble the syringe and needle while hot and flush the syringe with argon.

3. Support the flask using clamp and a stand with a heavy base.

4. Charge the flask with hexane (100 mL) and there (100 mL) using a syringe by piercing the septum on reaction flask.

5. Keep the flask at −40 °C with a dry ice/acetone bath.

6. Charge methylacrylaldehyde (6.6 g, 94.2 mmol) and (*R*)-1-*t*-butyl-2-piperidinoethanol (0.37 g, 2.0 mmol) with syringes under stirring.

7. Remove dimethylzinc in a syringe from the cylinder of dimethylzinc stored at −10 °C under argon pressure.

8. Add dimethylzinc (10.5 mL, 102 mmol) dropwise from the syringe through the septum on the reaction flask to the mixture in the flask with vigorous stirring.

9. Allow the temperature of the flask to rise from −40 °C to 15 °C and keep the mixture at this temperature for 96 h.

10. Quench the mixture by adding 2 M HCl (100 mL) dropwise with stirring at 0 °C.

11. Transfer the mixture to a separating funnel and separate into two layers. Extract the water layer with ether (2 × 50 mL).

12. Transfer the organic layer to a 500 mL flask. Dry the organic layer over Na_2SO_4, and filter through a filter paper. Concentrate the filtrate by distillation under ordinal pressure.

13. Transfer the oily residue to a water-jacketed, short path distillation apparatus equipped with a thermometer. Distill the crude product under reduced pressure to obtain 3-methylbut-3-en-2-ol (b.p. 106–108 °C/760 mm Hg; 6.8 g, 70% yield) as a colourless oil. The product is .99% pure by ^{1}H-NMR; d($CDCl_3$) 1.29 (3H,d, *J* 6.7Hz), 1.60(1H,s), 1.75(3H,s), 4.25(1H,q, *J* 6.7Hz), 4.80(1H,s), and 4.96(1H,s).

14. Determine the absolute configuration of the product from the sign of [α]$_D$. [α]$_D$ −5.6 (*c* 8.0, $CHCl_3$) showed it to be (*S*)-configurational product.

15. Enantiomeric excess(ee) (>95%) was determined by HPLC analysis of the 3,5-dinitrophenyl carbamate[a] (column, Sumitomo Chemical Co. SUMIPAX OA 4000) in hexane/ethanol 100:1.5; t_R of (*R*)-isomer, 41 min; t_R of (*S*)-isomer, 36 min.

[a] React 3-methylbut-3-en-2-ol (8.6 mg, 1.0 mmol) with 3,5-dinitrophenylisocyanide (21 mg, 1.0 mmol) in toluene (2.0 mL) at room temperature for 1 h with stirring. This mixture is used directly for HPLC.

Protocol 3.
Chiral Taddolated catalysed asymmetric addition of dimethylzinc to trimethylsilylpropynal[105] (Scheme 5.3)

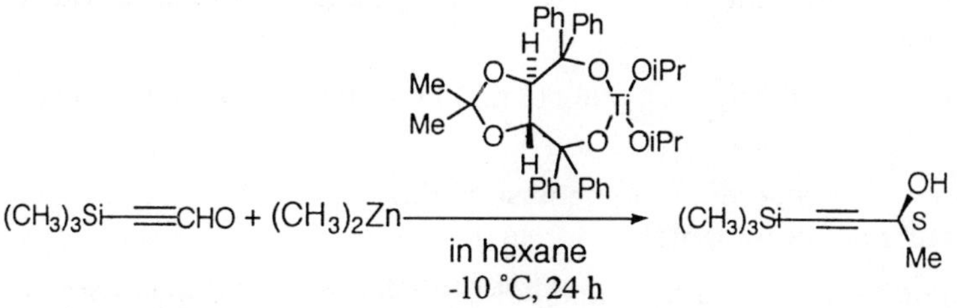

$$(CH_3)_3Si\text{---}\!\!\equiv\!\!\text{CHO} + (CH_3)_2Zn \xrightarrow[\substack{\text{in hexane} \\ -10\,°C,\,24\,h}]{} (CH_3)_3Si\text{---}\!\!\equiv\!\!\text{---}\langle^{OH}_{S,Me}$$

Scheme 5.3

Equipment

- Two-necked, round-bottomed flask (500 mL) fitted with rubber septum and a teflon-coated magnetic stirring bar
- Three-way stopcock fitted to the top of the flask and connected to a vacuum/argon (or dry nitrogen) source
- Magnetic stirrer
- All-glass syringe with a needle-lock Luer and medium-gauge needle
- Water-jacketed short-path distillation apparatus
- Separatory funnel (500 mL)

Materials

- Dry toluene
- 3-Trimethylsilylpropynal
- Dimethylzinc   **explosive on contact with air**
- (4*R*,5*R*)-2,2'-dimetrhyl-α,α,α',α'-Tetraphenyl-1,3-dioxolane-4,5-dimethanol(TADDOL)
- Technical ether for extraction   **flammable**

1. Flame dry the reaction vessel and place with a stirrer bar *in vacuo*. Fill the flask with argon and maintain a slightly positive argon pressure throughout the reaction.

2. Assemble the syringe and needle while hot and flush the syringe with argon.

3. Support the flask using clamp and a stand with a heavy base.

Protocol 3. *Continued*

4. Charge the flask with TADDOL (15.1 g, 20 mmol) and toluene (100 mL) using a syringe by piercing the septum on reaction flask.

5. Add titanium tetraisopropoxide (6 mL, 20 mmol) to the above mixture at room temperature and stir the resulting solution for 10 h.

6. Remove toluene and isopropanol produced under reduced pressure with stirring.

7. Dissolve the residue in *t*-butylmethyl ether (100 mL) solution containing titanium tetraisopropoxide (35 mL, 120 mmol).

8 Transfer dimethylzinc in a syringe from the cylinder of dimethylzinc stored at $-10\,°C$ under an argon pressure.

9. Add dimethylzinc (12 mL, 180 mmol) dropwise from the syringe through the septum on the reaction flask to the mixture in the flask with vigorous stirring.

10. Add 3-trimethylsilylpropynal (16 g, 100 mmol) with a syringe with stirring at $-20\,°C$.

11. Allow the temperature of the flask to rise from $-20\,°C$ to $0\,°C$ and keep the mixture at this temperature for 48 h.

12. Quench the mixture by adding 2M HCl(100 mL) dropwise with stirring at $0\,°C$.

13. Transfer the mixture to a separating funnel and separate into two layers. Extract the water layer with ether (2 × 50 mL).

14. Transfer the organic layer to a 500 mL flask. Dry the organic layer over Na_2SO_4, and filter through a filter paper. Concentrate the filtrate under reduced pressure using a rotary evaporator (25 °C/50 mm Hg).

15. Transfer the oily residue to a water-jacketed, short-path distillation apparatus equipped with a thermometer. Distil the crude product under reduced pressure to obtain 4-trimethylsilyl-3-butyn-2-ol (b.p. 77°C/20 mm Hg; 14.7 g, 98% yield) as a colourless oil. The product is 99% pure by ^1H NMR; d(CDCl$_3$) 0.15(9H,s), 1.42 (3H,d,*J* 4.6 Hz), 1.85 (1H, d, *J* 3.8 Hz), and 4.50 (1H, m).

16. Determine the absolute configuration of the product from the sign of $[\alpha]_D$. $[\alpha]_D$ -24.2 (*c* 2.0, CHCl$_3$) showed it to be (*S*)-configurational product.

17. Enantiomeric excess(ee) (96%) was determined by HPLC analysis of the 3,5-dinitrobenzoate[a] (column, Daicel, CHIRAPAK AD in volume ratio of hexane/2-propanol (100:1); t_R of (*R*)-isomer, 30 min; t_R of (*S*)-isomer, 40 min.

[a] React 3-trimethylsilylbut-3-yn-2-ol (14.2 mg, 0.1 mmol) with 3,5-dinitrobenzoyl chloride (23 mg, 0.1 mmol) in toluene (2.0 mL) in the presence of triethylamine (1 mL) at room temperature for 1 h with stirring. Remove Et$_3$NHCl by filtration and use the filtrate directly for HPLC.

Protocol 4.
Asymmetric 1,4-conjugate addition of diethylzinc to chalcone catalysed by chiral nickel complex[106] (Scheme 5.4)

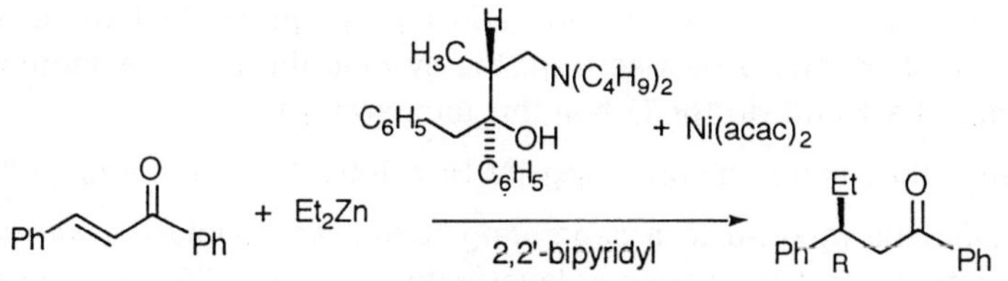

Scheme 5.4

Equipment

- Two-necked, round-bottomed flask (300 mL) fitted with a rubber septum and a Teflon-coated magnetic stirring bar
- Three-way stopcock fitted to the top of flask and connected to a vacuum/argon (or dry nitrogen) source
- Magnetic stirrer

- All-glass syringe with a needle-lock Luer and medium-gauge needle
- Water-jacketed short-path distillation apparatus
- Separating funnel (300 mL)
- Rotary evaporator
- Chiralcel OK for HPLC

Materials

- Dry toluene
- Bis(acetylacetonato)nickel(II)
- R-(−)-(1S,2R)-N,N-Dibutylnorephedrine(DBNE)
- 2,2′-Bipyridyl
- Chalcone
- Diethylzinc   **explosive on exposure to air**
- Acetonitrile
- Hexane
- Technical ether for extraction   **flammable**

1. Flame dry the reaction vessel and place with a stirrer bar *in vacuo*. Fill the flask with argon and maintain a slightly positive argon pressure throughout this reaction.

2. Assemble the syringe and needle while hot and flush the syringe with argon.

3. Support the flask using clamp and a stand with a heavy base.

4. Place bis(acetylacetonato)nickel(ii) (3.6 g, 14 mmol) in the flask.

5. Add (1S,2R)-(−)-DBNE (0.896 g, 34 mmol) to the flask in CH_3CN (20 mL) using a syringe by piercing the septum on reaction flask. Stir the mixture at 80°C for 1 h.

6. Remove the acetylacetone liberated and CH_3CN under reduced pressure.

7. Add 2,2′-bipyridyl (0.22 g, 14 mmol) in CH_3CN (20 mL) to the flask and stir at 80°C for 1 h.

Protocol 4. *Continued*

8. Cool the resulting green solution to room temperature.

9. Charge chalcone (4.16 g, 0.2 mol) inCH_3CN (100 mL) to above mixture followed by stirring for 20 min at room temperature.

10. Cool to −30°C this solution and add diethylzinc (1.0M toluene solution, 240 mL, 0.24 mol) dropwise from the syringe through the septum on the reaction flask and stir for 12 h at this temperature.

11. Quench the mixture by dropping 1M HCl (100 mL) with stirring at 0°C.

12. Transfer the mixture to a separating funnel and allow the two layers to separate. Extract the aqueous layer with ether (4 × 20 mL), and dry over anhydrous Na_2SO_4.

13. Transfer the organic layer to a 300 mL round-bottomed flask and concentrate under reduced pressure using a rotary evaporator (30°C/30 mm Hg).

14. Purify the residue by silica gel chromatography (eluent, hexane/chloroform, 1:1, v/v) to obtain (R)-(−)-1,3-diphenylpentan-1-one (2.26 g, 47% yield), which displays the appropriate ^1H NMR in $CDCl_3$.

15. Determine the absolute configuration and enantiomeric excess(ee) of the product. The absolute configuration is determined from the sign of $[\alpha]_D$. $[\alpha]_D$ −4.7 (c 2.5, EtOH) indicated (R)-configuration. Ee (90%) was determined by HPLC analysis (column, Daicel CHIRALCEL OD; hexane/2-propanol, 100:0.2, flow rate 0.5 mL/min; t_R of (S)-isomer, 40.1 min; t_R of (R)-isomer, 44.8 min.

Protocol 5.
Chiral disulfonamide-titanate catalysed asymmetric cyclopropanation of cinnamyl alcohol[124] (Scheme 5.5)

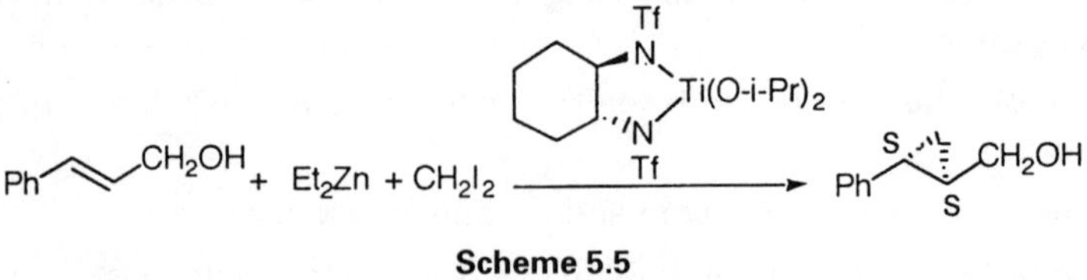

Scheme 5.5

Equipment

- Two-necked, round-bottomed flask (500 mL) fitted with a rubber septum and a Teflon-coated magnetic stirring bar
- Three-way stopcock fitted to the top of the flask and connected to a vacuum/argon (or dry nitrogen) source
- Magnetic stirrer
- All-glass syringe with a needle-lock Luer and medium-gauge needle
- Water-jacketed short-path distillation apparatus
- Separating funnel (500 mL)

5: Zinc(II) Reagents

Materials

- Dry dichloromethane
- Di-iodomethane
- Diethylzinc   **explosive on exposure to air**
- (4R,5R)-2,2′-Dimethyl-α,α,α′,α′-tetraphenyl-1,3-dioxolane-4,5-dimethanol(TADDOL)
- Ti (OiPr)$_4$
- Technical ether for extraction   **flammable**
- Molecular sieve 4A and Ti(OiPr)$_4$ **flammable**

1. Flame dry the reaction vessel and place with a stirrer bar *in vacuo*. Fill the flask with argon and maintain a slightly positive argon pressure throughout the section.

2. Assemble the syringe and needle while hot and flush the syringe with argon.

3. Support the flask using clamp and a stand with a heavy base.

4. Charge the flask with CH$_2$I$_2$ (1.6 mL, 20 mmol) and CH$_2$Cl$_2$ (50 mL) using a syringe by piercing the septum on reaction flask.

5. Remove diethylzinc in a syringe from the cylinder of diethylzinc stored under argon pressure.

6. Add diethylzinc (1 mL, 10 mmol) dropwise to the flask at 0 °C and stir the resulting solution for 15 min. A white precipitate is formed (solution **A**).

7. Mix (4R,5R)-TADDOL (1.4 g, 2.9 mmol) and molecular sieve 4A (10 g) in CH$_2$Cl$_2$ (5 mL), and add Ti(OiPr)$_4$ (0.74 mL, 2.5 mmol) with stirring. Stir for 1.5 h at room temperature.

8. Remove the solvent and isopropanol produced by the reaction under reduced pressure and leave the residue under high vacuum. Dissolve the residue into CH$_2$Cl$_2$ (50 mL) (solution **B**).

9. Mix solutions **A** and **B** in the flask at −40 °C and add immediately cinnamyl alcohol (1.4 g, 10.4 mmol) to the mixture with stirring.

10. Stir the mixture at 0 °C for 90 min and add TiCl$_4$ (0.16 mL, 1.5 mmol).

11. Cool the solution to −40 °C and quench the reaction by pouring the mixture into saturated aq. NH$_4$Cl solution (300 mL).

12. Transfer the mixture to a separating funnel and allow the two layers to separate. Extract the water layer with ethyl acetate (50 mL).

13. Wash the organic layer with saturated aqueous NH$_4$Cl, and brine, then dry over MgSO$_4$ and concentrate under reduced pressure (25 °C/30 mm Hg).

14. Purify the crude product by flash chromatography (EtOAc/hexane, 80:20) to obtain (1.2 g, 80% yield) as a colourless oil. The product is .99% pure by ^{1}H NMR.

15. Determine the absolute configuration of the product from the sign of [α]$_D$. [α]$_D$ + 84 (*c* 1.3, EtOH) indicated a (2S*S*,3*S*)- configurational product.

16. Enantiomeric excess(ee) (90%) was determined by GC analysis of the trifluoroacetate derivative on a chiral stationary phase: Cyclodex GT-A column, 0.32 mm × 30 m (25 psi, 110 °C), t_r 11.5 min (minor), 12.0 (major).

References

1. Noyori, R.; Kitamura, M. *Angew. Chem. Int. Ed.* **1991**, *30*, 49–69.
2. Soai, K.; Niwa, S. *Chem. Rev.* **1992**, *92*, 833–856.
3. Knochel, P.; Singer, R. D. *Chem. Rev.* **1993**, *93*, 2117–2188.
4. Knochel, P. *Synlett* **1995**, 393–403.
5. Oguni, N.; Omi, T. *Tetrahedron Lett.* **1984**, *25*, 2823–2824.
6. Kitamura, M.; Suga, S.; Kawai, K.; Noyori, R. *J. Am. Chem. Soc.* **1986**, *108*, 6071–6072.
7. Alberts, A. H.; Wynberg, H. *J. Am. Chem. Soc.* **1989**, *111*, 7265–7266.
8. Noyori, R.; Suga, S.; Kawai, K.; Okada, S.; Kitamura, M.; Oguni, N.; Hayashi, M.; Kaneko, K.; Matsuda, Y. *J. Organometal. Chem.* **1990**, *382*, 19–37.
9. Smaardijk, A. A.; Wynberg, H. *J. Org. Chem.* **1987**, *52*, 135–137.
10. Soai, K.; Ookawa, A.; Ogawa, K.; Kaba, T. *J. Chem. Soc. Chem. Commun.* **1987**, 467–468.
11. Soai, K.; Ookawa, A.; Kaba, K.; Ogawa, K. *J. Am. Chem. Soc.* **1987**, *109*, 7111–7115.
12. Chaloner, P. A.; Perera, S. A. R. *Tetrahedron Lett.* **1987**, *28*, 3013–3016.
13. Soai, K.; Niwa, S.; Yamada, Y.; Inoue, H. *Tetrahedron Lett.* **1987**, *28*, 4841–4842.
14. Soai, K.; Yokoyama, S.; Ebihara, K.; Hayasaka, T. *J. Chem. Soc. Chem. Commun.* **1987**, 1690–1691.
15. Soai, K.; Nishi, S.; Ito. Y. *Chem. Lett.* **1987**, 2405–2406.
16. Corey, E. J.; Hannon, F. J. *Tetrahedron Lett.* **1987**, *28*, 5233–5236; 5237–5240.
17. Muchow, G.; Vannonoorenberghe, Y.; Buono, G. *Tetrahedron Lett.* **1987**, *28*, 6163–6166.
18. Soai, K.; Niwa, S.; Watanabe, M. *J. Org. Chem.* **1988**, *55*, 927–928.
19. Soai, K.; Watanabe, M.; Koyano, M. *J. Chem. Soc. Chem. Commun.* **1989**, 534–536.
20. Corey, E. J.; Yuen, P.; Hannon, F. J.; Wierda, D. A. *J. Org. Chem.* **1990**, *55*. 784–746.
21. Oppolzer, W.; Radinov, R. N. *Tetrahedron Lett.* **1988**, *29*, 5645–5648.
22. Itsuno, S.; Frechet, J. M. *J. Org. Chem.* **1987**, *52*, 4140–4142.
23. Itsuno, S.; Sakurai, Y.; Ito, K.; Maruyama, T.; Nakahama, S.; Frechet, J. M. *J. Org. Chem.* **1990**, *55*, 304–310.
24. Tanaka, K.; Ushio, H.; Suzuki, H. *J. Chem. Soc. Chem. Commun.* **1989**, 1700–1701.
25. van Oeveren, A.; Menge, W.; Feringa, B. L. *Tetrahedron Lett.* **1989**, *30*, 6427–6430.
26. Soai, K.; Watanabe, M. *J. Chem. Soc., Chem. Commun.* **1990**, 43–44.
27. Chaloner, P. A.; Longadianon, E. *Tetrahedron Lett.* **1990**, *31*, 5185–5186.
28. Niwa, S.; Soai, K. *J. Chem. Soc., Perkin Trans I* **1990**, 937–943.
29. Hayashi, M.; Kaneko, T.; Oguni, N. *J. Chem. Soc. Perkin. Trans I* **1991**, 25–28.
30. Rosini, C.; Franzini, L.; Iuliano, A.; Pini, D.; Salvadori, P. *Tetrahedron Asymmetry* **1990**, *1*, 587–588; **1991**, *2*, 363–366.
31. Oppolzer, W.; Radinov, R. N. *Tetrahedron Lett.* **1991**, *32*, 5777–5780.
32. Hayashi, M.; Miwata, H.; Oguni, N. *Chem. Lett.* **1989**, 1969–1970.
33. Hayashi, M.; Miwata, H.; Oguni, N. *J. Chem. Soc. Perkin Trans I* **1991**, 1167–1171.
34. Soai, K.; Niwa, S.; Hatanaka, T. *J. Chem. Soc., Chem. Commun.* **1990**, 709–711.

35. Hatanaka, T.; Yamashita, T.; Soai, K. *J. Chem. Soc., Chem. Commun*, **1992**, 927–928.
36. Conti, S.; Falorni, M.; Giacomelli, G.; Soccolini, F. *Tetrahedron* **1992**, *48*, 8993–9000.
37. Matsumoto, Y.; Ohno, A.; Lu, S.; Hayashi, T.; Oguni, N.; Hayashi, M. *Tetrahedron Asymmetry* **1993**, *4*, 1763–1766.
38. Vries, E. F. J.; Brussee, J.; Kruse, C. G.; van der Gen, A. *Tetrahedron Asymmetry* **1993**, *4*, 1987–1990.
39. Watanabe, M.; Soai, K. *J. Chem. Soc., Perkin Trans I* **1994**, 837–842, 3125–3128.
40. Soai, K.; Shimada, C.; Takeuchi, M.; Itabashi, M. *J. Chem. Soc. Chem. Commun.* **1994**, 567–568.
41. Soai, K.; Hayase, T.; Takai, K.; Sugiyama, T. *J. Org. Chem.* **1994**, *59*, 7908–7909.
42. Shi, M.; Satoh, Y.; Makihara, T.; Masaki, Y. *Tetrahedron Asymmetry* **1995**, *6*, 2109–2112.
43. Dai, W. M.; Zhu, H. J.; Hao, X-J. *Tetrahedron Asymmetry* **1995**, *6*, 1857–1860.
44. Nakano, H.; Kumagai, H.; Kabuto, C.; Matsuzaki, H.; Hongo, H. *Tetrahedron Asymmetry* **1995**, *6*, 1233–1236.
45. Collomb, P.; von Zelewsky, A. *Tetrahedron Asymmetry* **1995**, *6*, 2903–2904.
46. Soai, K.; Inoue, Y.; Takahashi, T.; Shibata, T. *Tetrahedron* **1996**, *52*, 13355–13362.
47. Peper, V.; Martens, J. *Chem. Ber.* **1996**, *129*, 691–698.
48. Audres, J. M.; Barrio, R.; Martine, M. A.; Redrosa, R. *J. Org. Chem.* **1996**, *61*, 4210–4213.
49. Falorni, M.; Collu, C.; Conti, S.; Giacomelli, G. *Tetrahedron Asymmetry* **1996**, *7*, 293–299.
50. Kitajima, H.; Ito, K.; Katsuki, T. *Chem. Lett.* **1996**, 343–344.
51. Bolm, C.; Zehnder, M.; Bur, D. *Angew. Chem. Int. Ed.* **1990**, *29*, 205–207.
52. Bolm, C.; Schlingloff, G.; Harmo, K. *Chem. Ber.* **1992**, *125*, 1191.
53. Ishizuka, M.; Hoshino, O. *Chem. Lett.* **1994**, 1337–1338.
54. Ishizaki, M.; Hoshino, O. *Tetrahedron Asymmetry* **1994**, *5*, 1901–1904.
55. Ishizaki, M.; Fujita, K.; Shimamoto, M.; Hoshino, O. *Tetrahedron Asymmetry* **1994**, *5*, 411–412.
56. Macedo, E.; Moberg, C. *Tetrahedron Asymmetry* **1995**, *6*, 549–548.
57. Kotsuki, H.; Hayakawa, H.; Wako, M.; Shimanouchi, T.; Ochi, M. *Tetrahedron.*
58. Wally, H.; Widhalm, M.; Weisensteiner, W.; Schlogl, K. *Tetrahedron Asymmetry* **1993**, *4*, 285–288.
59. Watanabe, M.; Komota, M.; Nishimura, M.; Araki, S.; Butsugan, Y. *J. Chem. Soc. Perkin Trans I* **1993**, *18*, 2193–2196.
60. Nicolosi, G.; Patti, A.; Morrone, R.; Piattelli, M. *Tetrahedron Asymmetry* **1994**, *5*, 1639–1642.
61. Watanabe, M. *Synlett* **1995**, 1050–1052.
62. Watanabe, M. *Tetrahedron Lett.* **1995**, *36*, 8991–8994.
63. Fukuzawa, S.; Tsuduki, K.; *Tetrahedron Asymmetry* **1995**, *6*, 1039–1942.
64. Watanabe, M.; Araki, S.; Butsugan, Y.; Uemura, M. *J. Org. Chem.* **1991**, *56*, 2218–2224.
65. Uemura, M.; Miyake, R.; Nakayama, K.; Shiro, M.; Hayashi, Y. *J. Org. Chem.* **1993**, *58*, 1238–1244.
66. Heaton, S. B.; Jones, G. B. *Tetrahedron Lett.* **1992**, *33*, 1693–1696.
67. Jones, G. B.; Heaton, S. B. *Tetrahedron Asymmetry* **1993**, *4*, 261–262.

68. Rijinberg, E.; Jastrzebski, J. T. B. H.; Jannsen, M. D.; Boersma, J.; Koten, G. *Tetrahedron Lett.* **1994**, *35*, 6521–6524.
69. Kang, J.; Kim, D. S.; Kim, J. I. *Synlett* **1994**, 842–843.
70. Kang, J.; Lee, J. W.; Kim, J. I. *J. Chem. Soc., Chem. Commun.* **1994**, 2009–2010.
71. Fitzpatrick, K.; Hulst, R.; Kellogg, R. M. *Tetrahedron Asymmetry* **1995**, *6*, 1861–1864.
72. Hulst, R.; Heres, H.; Fitzpatrick, K.; Peper, N. C. M. W.; Kellogg, R. M. *Tetrahedron Asymmetry* **1996**, *7*, 2755–2760.
73. Oguni, N.; Matsuda, Y.; Kaneko, T. *J. Am. Chem. Soc.* **1988**, *110*, 7877–7878.
74. Kitamura, M.; Okada, S.; Suga, S.; Noyori, R. *J. Am. Chem. Soc.* **1989**, *111*, 4028–4036.
75. Kitamura, M.; Suga, S.; Niwa, M.; Noyori, R. *J. Am. Chem. Soc.* **1995**, *117*, 4832–4842.
76. Kitamura, M.; Yamakawa, M.; Oka, da, S.; Suga, S.; Noyori, R. *Chem. Eur. J.* **1996**, *2*, 1173–1181.
77. Wynberg, H. *Chimia* **1989**, *43*, 150–152.
78. Soai, K.; Hayase, T.; Takai, K.; Sugiyama, T. *J. Org. Chem.* **1994**, *59*, 7908–7909.
79. Soai, K.; Hayase, T.; Shimada, C.; Isobe, K. *Tetrahedron Asymmetry* **1994**, *5*, 789–792.
80. Soai, K.; Hayase, T.; Takai, K. *Tetrahedron Asymmetry* **1995**, *6*, 637–638.
81. Soai, K.; Shibata, T.; Morioka, H.; Choji, K. *Nature* **1995**, *378*, 767–768.
82. Shibata, T.; Morioka, H.; Hayase, T.; Choji, K.; Soai, K. *J. Am. Chem. Soc.* **1996**, *118*, 471–472.
83. Oguni, N.; Omi, T.; Yamamoto, Y.; Nakamura, A. *Chem. Lett.* **1983**, 841–842.
84. Yoshioka, M.; Kawakita, T.; Ohno, M. *Tetrahedron Lett.* **1989**, *30*, 1657–1660.
85. Takahashi, H.; Kawakita, T.; Yoshioka, M.; Kobayashi, S.; Ohno, M. *Tetrahedron Lett.* **1989**, *30*, 7095–7098.
86. Joshi, N. N.; Screbnik, M.; Brown, H. C. *Tetrahedron Lett.* **1989**, *30*, 5551–5554.
87. Schmidt, B.; Seebach, D. *Angew. Chem.* **1991**, *103*, 100–101.
88. Schmidt, B.; Seebach, D. *Angew. Chem.* **1991**, *30*, 99–101.
89. Seebach, D.; Behrendt, L.; Felix, D. *Angew, Chem.* **1991**, *30*, 1008–1009.
90. Schmidt, B.; Seebach, D. *Angew. Chem.* **1991**, *30*, 1231–1233.
91. Bussche-Hunnefeld, J. L.; Seebach, D. *Tetrahedron* **1992**, *48*, 5719–5730.
92. Itoh, K.; Kimura, Y.; Okamura, H.; Katsuki, T. *Synlett* **1992**, 573–574.
93. Mori, A.; Yu, D.; Inoue, S. *Synlett* **1992**, 427–428.
94. Knochel, P.; Brieden, W.; Rozema, M. J.; Eisenberg, C. *Tetrahedron Lett.* **1993**, *34*, 5881–5884.
95. Soai, K.; Hirose, Y.; Ohno, Y. *Tetrahedron Asymmetry* **1993**, *4*, 1473–1474.
96. Klement, I.; Knochel, P. *Tetrahedron Lett.* **1994**, *35*, 1177–1180.
97. Schwink, L.; Knochel, P. *Tetrahedron Lett.* **1994**, *35*, 9007–9010.
98. Vettel, S.; Knochel, P. *Tetrahedron Lett.* **1994**, *34*, 5849–5852.
99. Ostwald, R.; Chavant, P-Y.; Stadtmuller, H.; Knochel, P. *J. Org. Chem.* **1994**, *59*, 4143–4153.
100. Lutjens, H.; Knochel, P. *Tetrahedron Asymmetry* **1994**, *5*, 1161–1162.
101. Nowotny, S.; Vettel, S.; Knochel, P. *Tetrahedron Lett.* **1994**, *35*, 4539–4640.
102. Seebach, D.; Beck, A. K.; Schmidt, B.; Wang, Y. M. *Tetrahedron* **1994**, *50*, 4363–4384.

5: Zinc(II) Reagents

103. Lutjens, H.; Nowotny, S.; Knochel, P. *Tetrahedron Asymmetry* **1995**, *6*, 2675–2678.
104. Vettel, S.; Vaupel, A.; Knochel, P. *Tetrahedron Lett.* **1995**, *36*, 1023–1026.
105. Oguni, N.; Satoh, N.; Fujii, H. *Synlett* **1995**, 1043–1044.
106. Soai, K.; Hayasaka, T.; Ukaji, S. *J. Chem. Soc. Chem. Commun.* **1989**, 516–517.
107. Bolm, C. *Tetrahedron Asymmetry* **1991**, *2*, 701–704.
108. Uemura, M. *Tetrahedron Asymmetry* **1992**, *3*, 713–714.
109. Felder, M.; Muller, J.; Bolm, C. *Synlett* **1992**, 439–441.
110. Asami, M.; Usui, K.; Higuchi, S.; Inoue, S. *Chem. Lett.* **1994**, 297–298.
111. Kobayashi, S. *Tetrahedron Lett.* **1992**, *33*, 2575–2578.
112. Denmark, S. E. *Synlett* **1992**, 229–230.
113. Ukaji, Y. *Chem. Lett.* **1992**, 61–64.
114. Ukaji, Y.; Sada, K.; Inomata, K. *Chem. Lett.* **1993**, 1227–1230.
115. Imai, N.; Sakamoto, K.; Takahashi, H.; Kobayashi, S. *Tetrahedron Lett.* **1994**, *35*, 7045–7048.
116. Imai, N.; Takahashi, H.; Kobayashi, S. *Chem. Lett.* **1994**, 177–180.
117. Charette, A. B.; Brochu, C. *J. Am. Chem. Soc.* **1995**, *117*, 11367–11368.
118. Charette, A. B.; Brochu, C. *J. Org. Chem.* **1995**, *60*, 181–183.
119. Audres, J. M.; Charette, B.; Marcoux, J. F. *Synlett* **1955**, 1197–1207.
120. Takahashi, H.; Yoshioka, Y.; Shibasaki, M.; Ohno, M.; Imai, N.; Kobayashi, S. *Tetrahedron* **1995**, *51*, 12013–12026.
121. Kitajima, H.; Aoki, Y.; Ito, K.; Katsuki, T. *Chem. Lett.* **1955**, 1113–1114.
122. Charette, A. B.; Prescott, S.; Brochu, C. *J. Org. Chem.* **1995**, 1081–1083.
123. Denmark, S. E.; Christensen, B. L.; Coe, D. M.; O'Cornor, S. P. *Tetrahedron Lett.* **1995**, *36*, 2215–2218; 2219–2122.
124. Charette, A. B.; Brochu, C. *J. Am. Chem. Soc.* **1995**, *117*, 11367–11368.
125. Ukaji, Y.; Sada, K.; Inomata, K. *Chem. Lett.* **1993**, 1847–1850.
126. Anderson, P. G.; Guijarro, D.; Tanner, D. *Synlett* **1996**, 727–728.
127. Hayase, T.; Inoue, Y.; Shibata, T.; Soai, K. *Tetrahedron Asymmetry* **1996**, *7*, 2509–2510.
128. Suzuki, T.; Narisada, N.; Shibata, T.; Soai, K. *Tetrahedron Asymmetry* **1996**, *7*, 2519–2522.
129. Nakamura, N.; Hirai, A.; Nakamura, E. *J. Am. Chem. Soc.* **1996**, *118*, 8489–8490.

6

Chiral titanium complexes for enantioselective catalysis

KOICHI MIKAMI and MASAHIRO TERADA

1. Introduction

Enantioselective catalysis is an economical and environmentally benign process, since it achieves 'multiplication of chirality'[1] thereby affording a large amount of the enantio-enriched product, while producing a small amount of waste material, due to the very small amount of chiral catalyst used. Thus, the development of enantioselective catalysts is a most challenging and formidable endeavour for synthetic organic chemists.[2,3] Highly promising candidates for such enantioselective catalysts are metal complexes bearing chiral organic ligands. Thus, the choice of the central metal and the molecular design of the chiral organic ligand are most crucial for the development of enantioselective catalysts. The degree of enantioselectivity should be critically influenced by metal–ligand bond lengths, particularly metal–oxygen and –nitrogen bond lengths in case of metal alkoxide and amide complexes (Table 6.1),[4,5] as well as the steric demand of the organic ligands. The shorter the bond length, in principle, the higher the enantioselectivity, because the asymmetric environment constructed by the chiral ligand is closer to the reaction centre. Therefore, boron and aluminium are the main group elements of choice, and

Table 6.1 Metal-oxygen (M-O) and metal-nitrogen (M-N) bond length

Metal	M-O bond length (Å)	M-N bond length (Å)
Li	1.92–2.00	2.12
Mg	2.01–2.13	2.22
Zn	1.92–2.16	2.16
Sn	2.70	2.25
Al	1.92	1.95
B	1.36–1.47	1.40
Ti	1.62–1.73	2.30
Zr	2.15	2.30

titanium is one of the best early transition metals with hexa- and penta-co-ordination (Scheme 6.1).

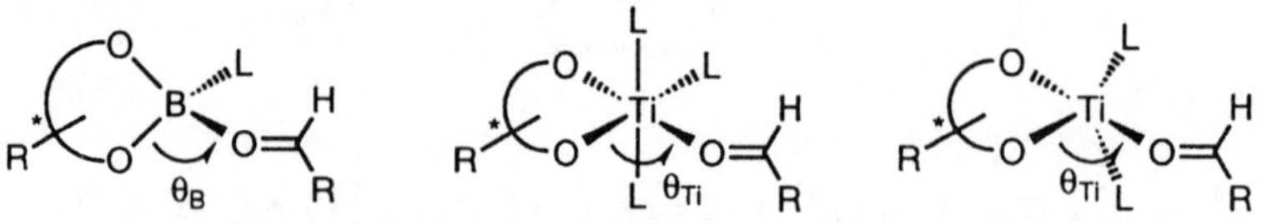

Scheme 6.1

The Lewis acidity of metal complexes is generally proportional to the value of (charge density) × (ionic radius)$^{-3}$.[6] Electron donating and sterically demanding ligands decrease the Lewis acidity but increase the configurational stability of titanium complexes in the order: Cp (cyclopentadienyl) > NR$_2$ > OR > X (halides).[4,7,8] Therefore, the Lewis acidity of titanium complexes decreases on going from titanium halides to titanium alkoxides to titanium amides. The Lewis acidity can be fine-tuned by mixing ligands such as in alkoxytitaniumhalide complexes. By contrast, bond strengths with titanium decrease in the order: M–O > M–Cl > M–N > M–C.[4] The M–C bond strengths in Ti(IV) compounds are comparable with those of other metal–carbon bonds. However, the Ti–O bond is exceptionally strong.

General preparative procedures for chiral titanium complexes are classified in Scheme 6.2.[9-13] (1) A halide can be replaced with metalated ligands by transmetalation (eq. 1). (2) A halide can be replaced by metathesis of a silylated ligand with accompanying generation of silylhalide (eq. 2). (3) HCl is evolved with protic ligands, and hence must be evaporated or neutralized with a base (eq. 3). (4) Ligand redistribution results in disproportionation (eq. 4) (Protocol 1). (5) A chiral titanate ester is prepared using an alkoxy exchange reaction (*trans*-esterification) with a free chiral alcohol (eq. 5). The equilibrium is shifted towards the chiral titanium complex by azeotropic removal of

$$\text{TiCl}_4 \quad + \quad n\,\text{NaOR} \quad \longrightarrow \quad (\text{RO})_n\text{TiCl}_{4-n} \quad + \quad n\,\text{NaCl} \qquad (1)$$

$$+ \quad \text{SiCl(OR)}_3 \quad \longrightarrow \quad (\text{RO})\text{TiCl}_3 \quad + \quad \text{SiCl}_2(\text{OR})_2 \qquad (2)$$

$$\text{TiCl}_4 \quad + \quad 4\,\text{ROH} \quad \longrightarrow \quad (\text{RO})_2\text{TiCl}_2 \quad + \quad 2\,\text{HCl} \qquad (3)$$

$$+ \quad n\,\text{ROH} + n\,\text{NR}_3 \quad \longrightarrow \quad (\text{RO})_n\text{TiCl}_{4-n} \quad + \quad n\,\text{NR}_3\text{HCl}$$

$$4\text{-}n\,\text{TiCl}_4 \quad + \quad n\,\text{Ti(OR)}_4 \quad \longrightarrow \quad 4\,(\text{RO})_n\text{TiCl}_{4-n} \qquad (4)$$

$$(\text{RO})_3\text{TiCl} \quad + \quad 3\,\text{R'OH} \quad \longrightarrow \quad (\text{R'O})_3\text{TiCl} \quad + \quad 3\,\text{ROH} \qquad (5)$$

$$\text{CH}_3\text{TiCl}_3 \quad + \quad \text{ROH} \quad \longrightarrow \quad \text{ROTiCl}_3 \quad + \quad \text{CH}_4 \qquad (6)$$

Scheme 6.2

the volatile achiral alcohol. (6) Alkyl (methyl, in particular) titanium complexes can be used for deprotonation of the chiral ligands along with generation of alkane (methane) (eq. 6). Chiral titanium alkoxide complexes thus obtained usually form bridged dimers, or trimers in extreme cases. Such aggregates are the favoured form even in solution. As shown above, there are many ways to prepare chiral titanium complexes for the enantioselective catalysis of carbon–carbon bond forming reactions or asymmetric functional group transformations. However, reactions of the latter type such as the Sharpless oxidation[14-17] or reduction[18-23] are beyond the scope of this manuscript due to limited space.

Protocol 1.
Preparation of di-isopropoxytitanium(IV) dibromide *via* ligand redistribution[24,25]

Caution! Carry out all procedures in a well-ventilated hood, and wear disposable vinyl or latex gloves and chemical-resistant safety goggles.

$$TiBr_4 + Ti(OPr^i)_4 \xrightarrow[\text{hexane}]{} 2\ Br_2Ti(OPr^i)_4$$

Equipment
- A pre-dried, pre-weighed, two-necked, round-bottomed flask (50 mL) equipped with a three-way stopcock, a magnetic stirring bar, and a septum cap. The three way stopcock is connected to a vacuum/argon source
- Dry gas-tight syringe
- Thermometer for low temperatures

Materials
- Titanium(IV) bromide (FW 367.5), 7.35 g, 20 mmol — corrosive, lachrymator, moisture sensitive
- Titanium(IV) isopropoxide (FW 284.3), 5.94 mL, 20 mmol — flammable liquid, irritant, moisture sensitive
- Dry hexane, total 40 mL — flammable liquid, irritant
- Dry toluene, 87.6 mL — flammable liquid, toxic

1. Flame dry reaction vessel and accessories under dry argon. After cooling to room temperature, add the titanium (IV) bromide (7.35 g, 20 mmol) to the pre-weighed, two-necked, round-bottomed flask under a flow of argon.

2. Introduce 20 mL of hexane to this flask to form a red–brown suspension and then add the titanium(IV) isopropoxide (5.94 mL, 20 mmol) carefully over ~7 min at ambient temperature with the syringe. The addition of titanium (IV) isopropoxide causes the mixture to warm to about 40°C.

3. Stir for 10 min at that temperature to give a yellow solution. Then allow to stand for 6 h at room temperature. Isolate the pale-yellow precipitate that has formed by removal of the supernatant solvent with a syringe.

Protocol 1. *Continued*

4. Wash with hexane (5 mL × 2) and recrystallize from hexane (10 mL). The recrystallization is carried out in the same flask by heating to reflux and then leaving the solution at room temperature overnight.

5. Remove the supernatant solvent again with the syringe. Vacuum dry the pale-yellow crystalline solid to give di-isopropoxytitanium(IV) dibromide (5.71 g, 44%), a highly moisture sensitive product.

6. Dissolve the crystalline in dry toluene (87.6 mL) to give a 0.2 M solution. Store the solution in a refrigerator.

Additionally, the crystalline solid may be stored at $0 \sim -20\,^{\circ}\mathrm{C}$.

---

2. Carbonyl addition reaction

Alkyltitanium complexes can be obtained from alkaline metal carbanions via titanation. Introduction of chirality at the titanium centre or on the ligand (or a combination of both) (Scheme 6.3) allows for the possibility of asymmetric induction in the carbonyl addition reaction.

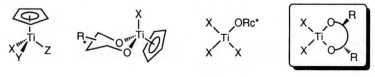

Scheme 6.3

However, the use of complexes which are chiral at the titanium centre, which is the closest to the reacting carbonyl group, generally affords only low enantioselectivity, because of the configurational lability of chiral titanium centre.[26] The use of a C_2 symmetric 1,1'-bi-2-naphthol (BINOL)[27] derived titanium complex has been unsuccessful so far in allylation or methylation reactions.[4] In an exceptional case, high enantioselectivity was obtained with a BINOL-modified phenyltitanium reagent. In that reaction, chiral titanium 'ate' complexes formed from BINOL-Ti(OPr^i)$_2$[28] and arylmagnesium halides could also be used.[29,30] Allylation of aromatic and aliphatic aldehydes by $(4R,5R)$-$\alpha,\alpha,\alpha',\alpha'$-tetra-aryl-1,3-dioxolane-4,5-dimethanol (TADDOL)[31–34]-derived cyclopentadienyltitanium complexes was found to give homoallyl alcohols with high enantioselectivity.[35] Exploitation of the Schlenk equilibrium of a mixture of a Grignard reagent, RMgX and 0.5 equiv. of ZnCl$_2$ with dioxane allowed the *in situ* generation of a functionalized alkyl zinc reagent along with MgX$_2$–dioxane complex (Protocol 2).[36,37] Such R$_2$Zn can be used as nucleophile for (TADDOL)$_2$–Ti complex-catalysed carbonyl addition reaction with high enantioselectivity. Chiral titanium bis-sulfonylamide complexes have also been used to accelerate the addition reaction of dialkylzinc.[38–43]

Protocol 2.
TADDOL–Ti complex-catalysed asymmetric carbonyl addition reaction with functionalized dialkylzinc compounds, generated *in situ* from Grignard reagents[36,37] (Scheme 6.4)

Caution! Carry out all procedures in a well-ventilated hood, and wear disposable vinyl or latex gloves and chemical-resistant safety goggles.

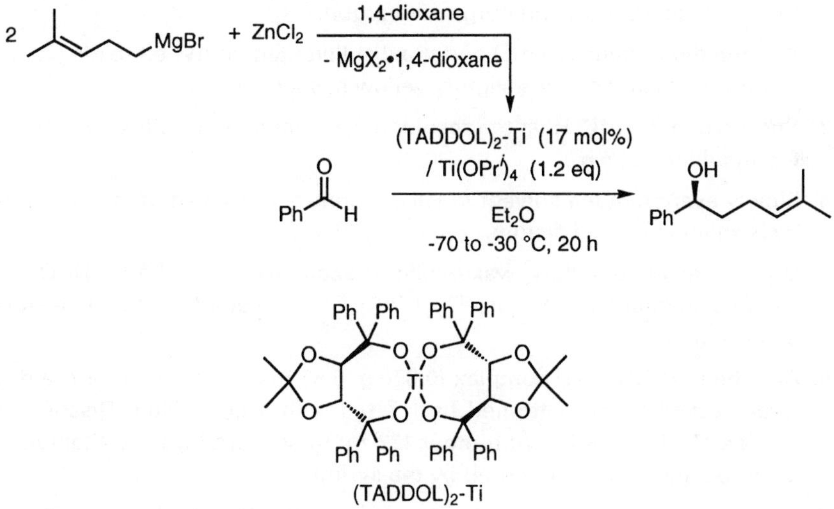

(TADDOL)$_2$-Ti

Scheme 6.4

Equipment

• A pre-dried, three-necked round-bottomed flask (100 mL) equipped with a condenser, a three-way stopcock, a magnetic stirring bar, and a septum cap. The three-way stopcock is connected to a vacuum/argon source
• A pre-dried, three-necked, round-bottomed flask (100 mL) equipped with a magnetic stirring bar, a glass filter, a three-way stopcock, and a septum cap. The three-way stopcock is connected to a vacuum/argon source

• A pre-dried, two-necked, round-bottomed flask (100 mL) equipped with a three-way stopcock, a magnetic stirring bar, and a septum cap. The three-way stopcock is connected to a vacuum/argon source
• Dry gas-tight syringe
• Thermometers for low temperature

Materials

• (4*R*,5*R*)-2,2-dimethyl-α,α,α′,α′-tetraphenyl-1,3-dioxolane-4,5-dimethanol (FW 466.6), 9.33 g, 20 mmol
• Distilled titanium(IV) ethoxide (FW 228.2), 2.3 mL, 11 mmol   **flammable liquid, moisture sensitive**
• Dry toluene, 20 mL   **flammable liquid, toxic**

Protocol 2. *Continued*

- Titanium(IV) isopropoxide (FW 284.3), 1.80 mL,
  6 mmol   **flammable liquid, irritant, moisture sensitive**
- 1.0 M solution of zinc chloride in ether, 10 mL,
  10 mmol   **flammable liquid, moisture sensitive**
- 1.05 M solution of 4-methyl-3-penten-1-ylmagnesium
  bromide in ether, 19 mL, 20 mmol   **flammable liquid, moisture sensitive**
- Dry ether   **flammable liquid, toxic**
- Dry 1,4-dioxane (FW 88.1), 6 mL, 70 mmol   **cancer suspect agent, flammable liquid**
- Benzaldehyde (FW 106.1) 0.51 mL, 5 mmol   **highly toxic, cancer suspect agent**

1. Add the (4R,5R)-2,2-dimethyl-$\alpha,\alpha,\alpha',\alpha'$-tetraphenyl-1,3-dioxolane-4,5-dimethanol (9.33 g, 20 mmol) to the pre-dried three-necked round-bottomed flask with condenser, and purge with argon.

2. Add the dry toluene (20 mL) and the distilled titanium(IV) ethoxide (2.3 mL, 11 mmol), giving to rise a slightly yellow suspension.

3. Stir for 12 h at 40°C. Heat the resulting clear solution to reflux temperature for an additional 5 h.

4. Slowly evaporate the solvent *in vacuo*. Control the speed of the distillation by regulation of the vacuum.

5. Dry the resulting yellow, waxy solid *in vacuo* to obtain the (TADDOL)$_2$–Ti complex in quantitative yield. The (TADDOL)$_2$–Ti complex should be stored under argon.

6. Add the (TADDOL)$_2$–Ti complex (0.856 g, 0.85 mmol) to the other pre-dried three-necked round-bottomed flask fitted with a glass filter. Dissolve the (TADDOL)$_2$–Ti complex with ether (12 mL) and introduce the titanium(IV) isopropoxide (1.8 mL, 6 mmol) by the syringe.

7. Introduce the 1.0 M ether solution of zinc chloride (10 mL, 10 mmol) into the pre-dried two-necked round bottomed flask with a syringe and dilute with ether (5 mL).

8. Add 1.05 M Grignard reagents (19 mL, 20 mmol), 4-methyl-3-penten-1-ylmagnesium bromide,[a] and stir at room temperature for 2 h.

9. Add 1,4-dioxane (6 mL, 70 mmol) to the suspension and stir for an additional 45 min. To obtain a clear solution of the dialkyl zinc reagent, filter under an argon atmosphere. Add the filtrate directly to the TADDOL–Ti solution, obtained above, at −78°C.

10. After stirring at −78°C for 1 h, add benzaldehyde (0.51 mL, 5 mmol) and then raise the reaction temperature to −30°C.

11. Stir at −30°C for 20 h and then quench with saturated NH$_4$Cl solution (10 mL) and add ether (25 mL) at −30°C. Filter through a pad of Celite and rinse the filter cake with ether. Wash the organic phase with H$_2$O and brine, dry over Na$_2$SO$_4$, filter the solution, and concentrate *in vacuo*.

12. To crystallize and separate from the TADDOL ligand, add pentane to the

resulting crude product. Again concentrate the supernatant solution and isolate by bulb to bulb distillation to collect 0.847 g (89%) of 1-phenyl-5-methylhex-4-en-1-ol.

13. The enantiomeric purity is determined by the capillary GC analysis using a heptakis(2,3,6-tri-*O*-ethyl)-β-cyclodextrin in OV 1701 as a chiral stationary phase column (96% ee).

[a] Prepared as usual from 5-bromo-2-methylpent-2-ene and magnesium turnings and titrated shortly before use.

Further pursuing our research project on the asymmetric catalysis of the carbonyl-ene reaction, we have found that the BINOL–Ti complexes (**1**),[44] which are prepared *in situ* in the presence of MS 4A from di-isopropoxytitanium dihalides ($X_2Ti(OPr^i)_2$: X = Br or Cl) and optically pure BINOL (see below), catalyse rather than promote stoichiometrically the carbonyl addition reaction of allylic silanes and stannanes.[45] The addition reactions to glyoxylate of (*E*)-2-butenylsilane and -stannane proceed smoothly to afford the *syn*-product in high enantiomeric excess (Scheme 6.5). The *syn*-product thus obtained could readily be converted into the lactone portion of verrucaline A.[46,47]

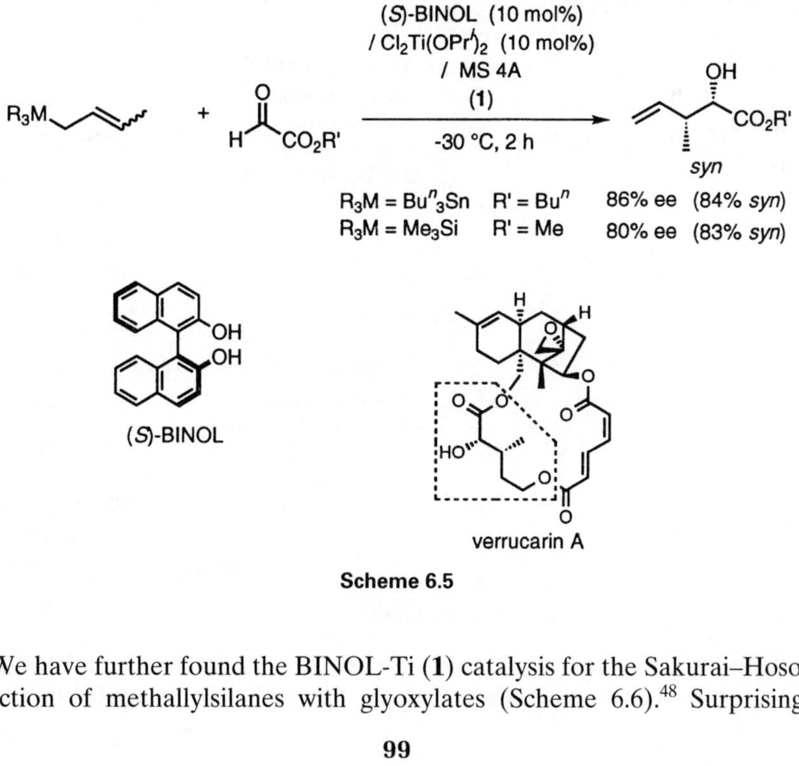
(S)-BINOL (10 mol%)
/ Cl2Ti(OPri)2 (10 mol%)
/ MS 4A
(1)
-30 °C, 2 h
R3M
H
CO2R'
OH
syn
R3M = Bun3Sn R' = Bun 86% ee (84% syn)
R3M = Me3Si R' = Me 80% ee (83% syn)
OH
OH
(S)-BINOL
H
HO
verrucarin A

Scheme 6.5

We have further found the BINOL-Ti (**1**) catalysis for the Sakurai–Hosomi reaction of methallylsilanes with glyoxylates (Scheme 6.6).[48] Surprisingly, however, the products were obtained in the allylic silane (ene product) form (see below), with high enantioselectivity.

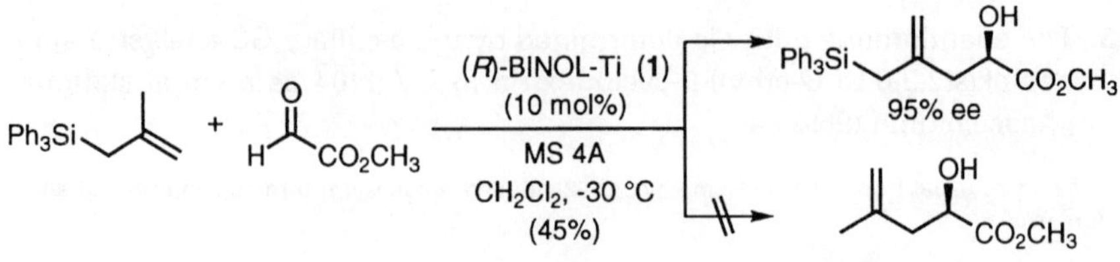

Scheme 6.6

Asymmetric catalysis by BINOL–Ti complexes of the reaction of aliphatic and aromatic aldehydes with an allylstannane has been reported independently by Tagliavini, Umani-Ronchi, and Keck.[49–52] In Tagliavini's case,[49] a new complex generated by reaction of the BINOL–Ti complex with allylstannane has been suggested to be the catalytic species which provides the remarkably high enantioselectivity (Scheme 6,7). Interestingly enough, no reaction occurs if MS 4A is not present during the preparation stage of the chiral catalyst, and MS 4A affects the subsequent allylation reaction. MS 4A dried for 12 h at 250 °C and 0.1 Torr was recommended. Keck reported that addition of CF_3CO_2H or CF_3SO_3H strongly accelerates the reactions catalysed by BINOL–Ti(OPr^i)$_2$ complex (**2**).[50]

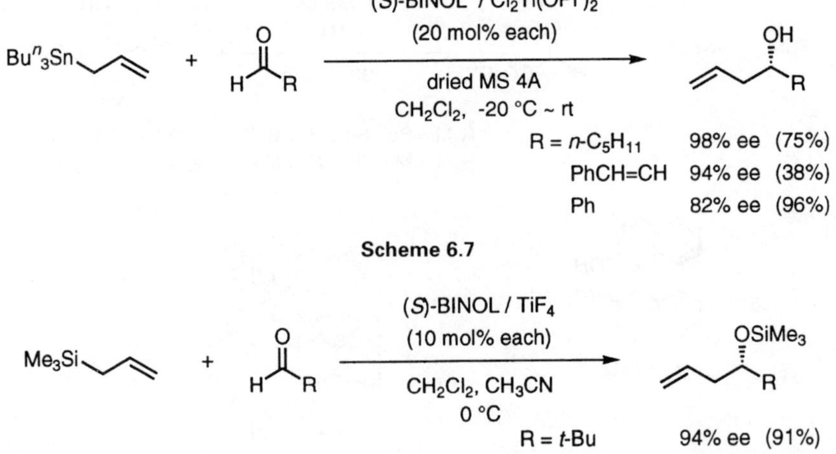

R = n-C$_5$H$_{11}$	98% ee	(75%)
PhCH=CH	94% ee	(38%)
Ph	82% ee	(96%)

Scheme 6.7

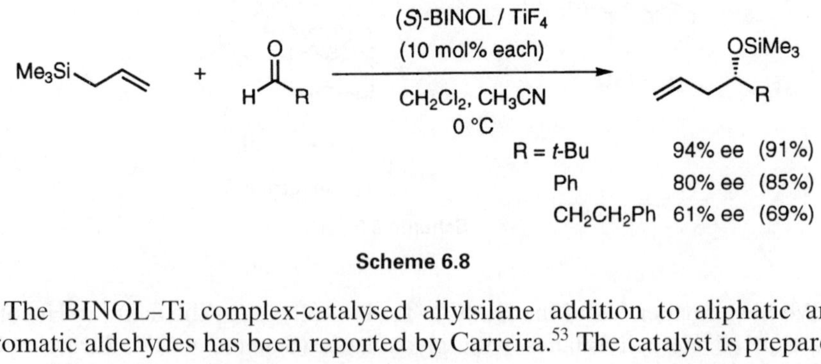

R = t-Bu	94% ee	(91%)
Ph	80% ee	(85%)
CH$_2$CH$_2$Ph	61% ee	(69%)

Scheme 6.8

The BINOL–Ti complex-catalysed allylsilane addition to aliphatic and aromatic aldehydes has been reported by Carreira.[53] The catalyst is prepared from BINOL and polymeric TiF_4 (Scheme 6.8). The presence of a small amount of CH_3CN is crucial to attain not only high catalytic activity but also high enantioselectivity.

3. Carbonyl-ene reaction

The class of ene reactions which involves a carbonyl compound as the enophile, which we refer to as the 'carbonyl-ene reaction',[54–57] constitutes a useful synthetic method for the construction of carbon skeleton (Scheme 6.9).

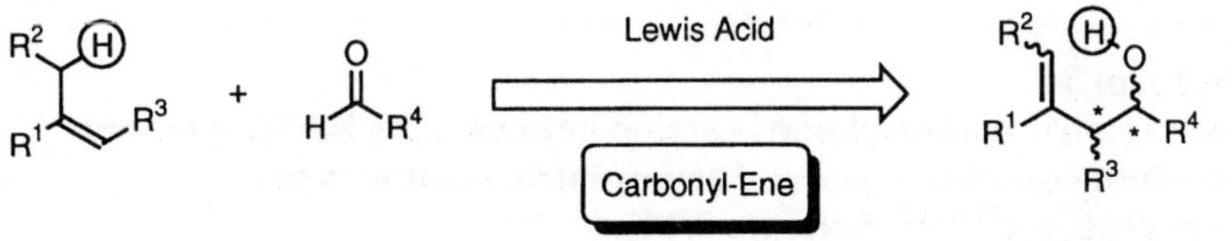

Scheme 6.9

We have been investigating the possibility of stereocontrol in carbonyl-ene reactions promoted by a stoichiometric or catalytic amount of various Lewis acids.[58–60] In particular, we have developed a chiral titanium catalyst for the glyoxylate-ene reaction which provides the α-hydroxy esters of biological and synthetic importance[61–64] in an enantioselective fashion (Scheme 6.10).[65–70] Various chiral titanium catalysts were screened.[71] The best result was obtained with the titanium catalyst (**1**) prepared *in situ* in the presence of MS

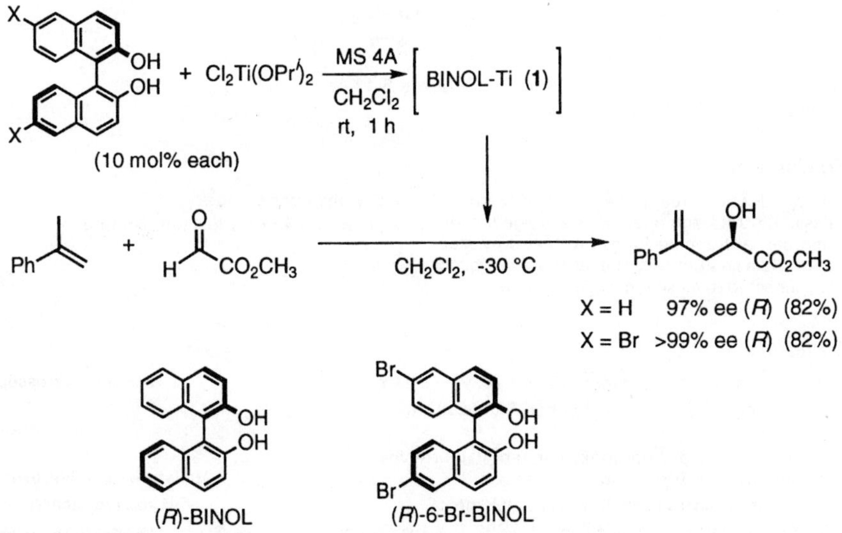

X = H   97% ee (*R*) (82%)
X = Br   >99% ee (*R*) (82%)

(*R*)-BINOL   (*R*)-6-Br-BINOL

Scheme 6.10

4A from di-isopropoxytitanium dihalides $(X_2Ti(OPr^i)_2 : X = Br^{24}$ or $Cl^{25})$ and optically pure BINOL or 6-Br-BINOL[72-74] (Protocol 3) (These two ligands are now commercially available in (R)- and (S)-forms.) The remarkable levels of enantioselectivity and rate acceleration observed with these BINOL–Ti catalysts (**1**) stem from the favourable influence of the inherent C_2 symmetry and the higher acidity of BINOLs compared to those of aliphatic diols. The reaction is applicable to a variety of 1,1-disubstituted olefins to provide the ene products in extremely high enantiomeric excess (ee) (Table 6.2). In the reactions with mono- and 1,2-disubstituted olefins, however, no ene product was obtained.

Protocol 3.
Asymmetric carbonyl-ene reaction catalysed by BINOL-derived titanium complex. Catalytic asymmetric synthesis of α-hydroxyester[24,68,69] (Scheme 6.11)

Caution! Carry out all procedures in a well-ventilated hood, and wear disposable vinyl or latex gloves and chemical-resistant safety goggles.

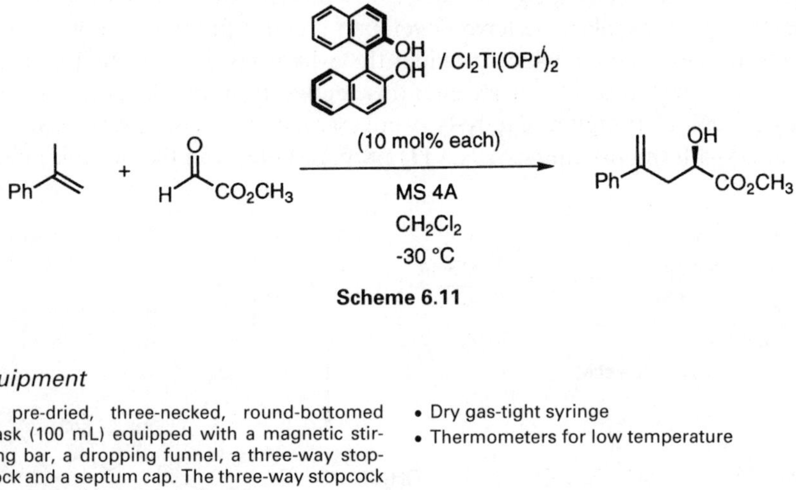

Scheme 6.11

Equipment
- A pre-dried, three-necked, round-bottomed flask (100 mL) equipped with a magnetic stirring bar, a dropping funnel, a three-way stopcock and a septum cap. The three-way stopcock is connected to a vacuum/argon source
- Dry gas-tight syringe
- Thermometers for low temperature

Materials
- Molecular sieves 4A,[a] powder, <5 mm, activated, 2.0 g — irritant, hygroscopic
- (R)-(+)- or (S)-(−)-1,1'-bi-2-naphthol (FW 286.3), 100 mg, 0.35 mmol — irritant
- 0.2 M solution of di-isopropoxytitanium(IV) dibromide in toluene, 1.75 mL, 0.35 mmol — moisture sensitive, irritant
- α-Methylstyrene (FW 118.2), 14.0 mL, 108 mmol — corrosive, lachrymator
- Freshly distilled methyl glyoxylate[b] (FW 88.1), 6.16 g, 70 mmol — lachrymator, irritant
- Dry dichloromethane, 45 mL — toxic, irritant

102

6: Chiral titanium complexes for enantioselective catalysis

1. Add the powder molecular sieves 4A (2.0 g) and (R)-(+)-1,1'-bi-2-naphthol (100 mg, 0.35 mmol) to the three-necked flask, purge with argon, and suspend with CH_2Cl_2 (20 mL). Stir for 15 min at room temperature.

2. Add a 0.2 M toluene solution of di-isopropoxytitanium dibromide (0.33 mL, 0.10 mmol) (prepared as described in Protocol 1) by syringe at room temperature.

3. After stirring for 1 h at room temperature, cool to −30°C. Add the α-methylstyrene (14.0 mL, 108 mmol) in CH_2Cl_2 (5 mL) by syringe. Add the methyl glyoxylate (6.16 g, 70 mmol) dissolved in CH_2Cl_2 (20 mL) dropwise over 30 min.

4. Stir for 6 h at −30°C. The reaction temperature must be kept in the range of −35°C to −30°C. The progress of the reaction is monitored by TLC.

5. Pour the CH_2Cl_2 solution into a beaker containing 10 mL of saturated $NaHCO_3$ and 50 mL of ether. Filter off molecular sieves 4A through a pad of Celite and rinse the filter cake three times with ethyl acetate. Separate the phases and extract the aqueous phase three times with 80 mL of ethyl acetate. Wash the combined organic phases twice with 50 mL of brine, dry over magnesium sulfate, and filter the solution.

6. Concentrate the crude product under vacuum and distil the residue at 0.2 mmHg to collect 12.1 g (84%) of methyl 2-hydroxy-4-phenyl-4-pentenoate boiling at 105–106°C. The product displays the appropriate spectral characteristics and high resolution mass spectral data.

7. The enantiomeric purity is determined by the HPLC analysis using a CHIRALPAK AS as a chiral stationary phase column with 25% i-PrOH/hexane as a mobile phase (94% ee).

[a] Purchased from Aldrich Chemical Company, Inc. and used without activation.
[b] Immediately prior use, the methyl glyoxylate is depolymerized by vacuum distillation from phosphorus pentoxide (10% weight) b.p. 62°C/60 mmHg, bath temperature 120–140°C.

This limitation has been overcome by the use of vinylic sulfides and selenides instead of mono- and 1,2-disubstituted olefins. With these substrates, the ene products are formed with virtually complete enantioselectivity and high diastereoselectivity.[75] The synthetic utility of the vinylic sulfide and selenide approach is exemplified by the synthesis of enantiopure (R)-(−)-ipsdienol, an insect aggregation pheromone (Scheme 6.12).[76–78]

The synthetic potential of the asymmetric catalytic carbonyl-ene reaction depends greatly on the functionality which is possible in the carbonyl enophile. However, the types of enophile that can be used in the asymmetric catalytic ene reaction have previously been limited to aldehydes such as glyoxylate[68,69] and chloral.[70] Thus, it is highly desirable to develop other types of carbonyl enophiles to provide enantio-enriched molecules with a wider range

Table 6.2 Asymmetric catalytic glyoxylate-ene reactions with various olefins[a]

Run	olefin	$X_2Ti(OPr^i)_2$ (X)	catalyst (mol%)	time (h)	products	%yield	%ee[b]
A	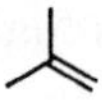	Cl	10	8	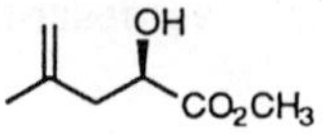	72	95 *R*
		Cl	10	8		68	95 *S*[c]
		Cl	1.0	8		78	93 *R*
		Br	10	3		87	94 *R*
B	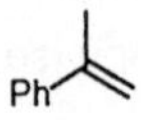	Cl	1.0	8	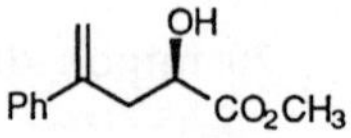	97	97 (*R*)
		Br	1.0	3		98	95 (*R*)
C	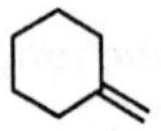	Cl	10	8	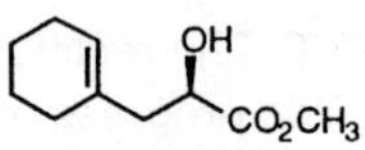	82	97 (*R*)
		Br	5	3		89	98 (*R*)
D	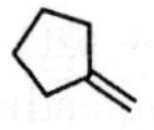	Cl	10	8	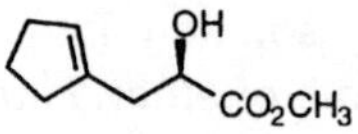	87	88 (*R*)
		Br	5	3		92	89 (*R*)

[a] All reactions were carried out using 1.0 mmol of methyl glyoxylate, 2.0 mmol of olefin, and indicated amount of BINOL-Ti complex at −30 °C.
[b] Determined by LIS-NMR analysis after conversion to the corresponding α-methoxy esters. The configuration in parentheses could be assigned by the similarity in shift pattern seen in the LIS-NMR spectra using (+)-Eu(dppm)$_3$ as a chiral shift reagent.
[c] (*S*)-BINOL was used instead of the (*R*)-counterpart.

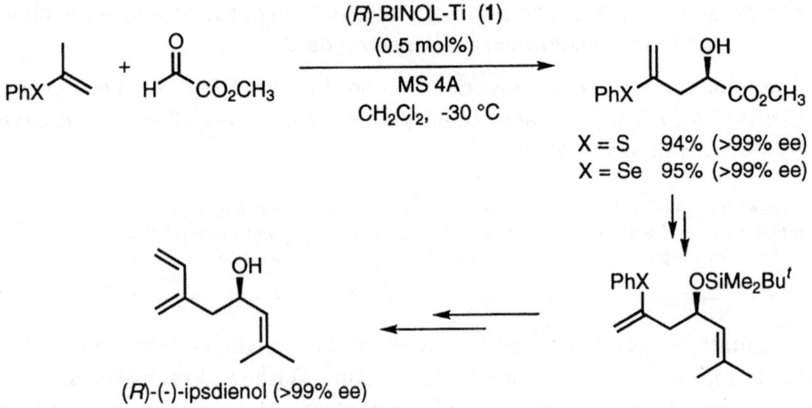

Scheme 6.12

of functionalities. We have developed as asymmetric catalytic fluoral-ene reaction,[79–81] which provides an efficient approach for the asymmetric synthesis of some fluorine-containing compounds of biological and synthetic importance.[82] The reaction of fluoral with 1,1-disubstituted and trisubstituted olefins proceeds quite smoothly under the catalysis by the BINOL–Ti complex (**1**) to provide the corresponding homoallylic alcohol with extremely high enantioselectivity (>95% ee) and *syn*-diastereoselectivity (>90%) (Scheme 6.13).

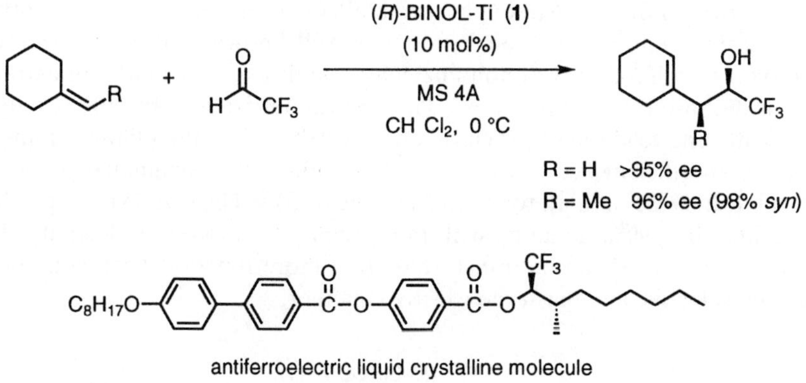

antiferroelectric liquid crystalline molecule
Scheme 6.13

The sense of asymmetric induction in the fluoral-ene reaction is exactly the same as observed for the glyoxylate-ene reaction; (R)-BINOL–Ti (**1**) provides the (R)-α-CF_3 alcohol. The *syn*-diastereomers of α-trifluoromethyl-β-methyl-substituted compounds thus synthesized with *two stereogenic centres* show anti-ferroelectric properties preferentially to the *anti*-diastereomers.[83–86]

The BINOL–Ti catalysis can also be used for the carbonyl-ene reaction with formaldehyde or vinyl and alkynyl analogues of glyoxylates in an asymmetric catalytic desymmetrization (see below) approach to the asymmetric synthesis of isocarbacycline analogues (Scheme 6.14).[87,88]

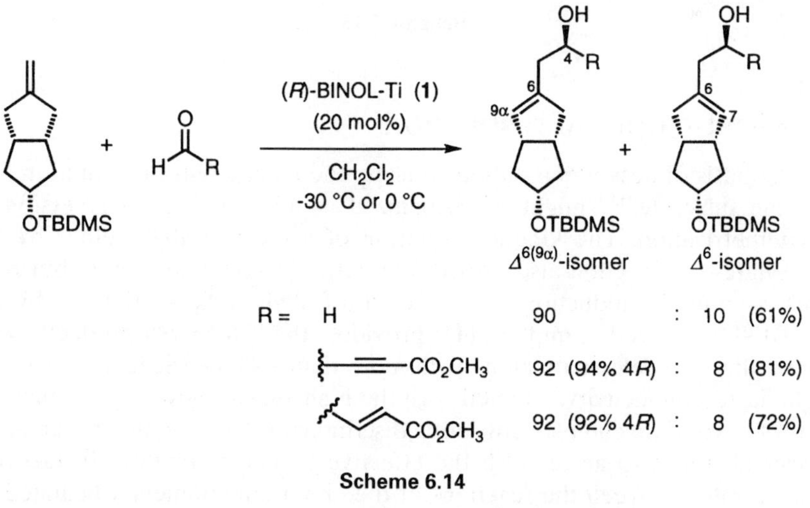

Scheme 6.14

4. Asymmetric catalytic desymmetrization

Desymmetrization of an achiral, symmetrical molecule through a catalytic process is a potentially powerful but relatively unexplored concept for asymmetric synthesis. Although the ability of enzymes to differentiate between enantiotopic functional groups is well known,[89] little has been explored on a similar ability of non-enzymatic catalysts, particularly for carbon–carbon bond-forming processes. The desymmetrization by the catalytic glyoxylate–ene reaction of prochiral ene substrates with the planar symmetry provides an efficient access to remote[90] and internal[91] asymmetric induction which is otherwise difficult to attain (Scheme 6.15).[92] The (2R,5S)-*syn*-product is obtained in >99% ee along with more than 99% diastereoselectivity. The diene thus obtained can be transformed to a more functionalized compound in a regioselective and diastereoselective manner.

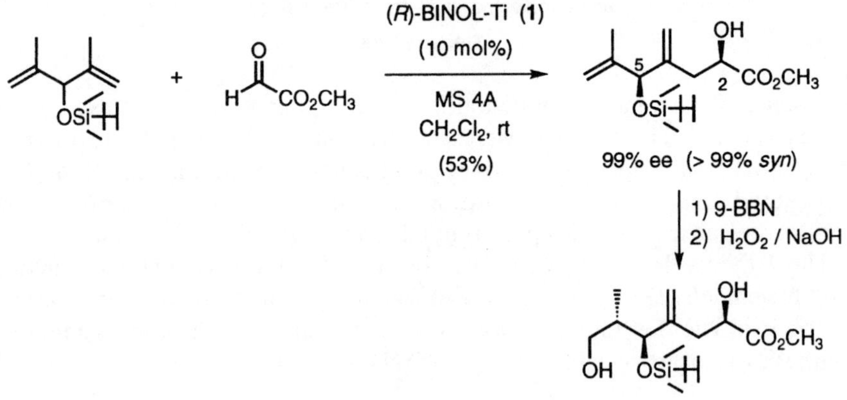

Scheme 6.15

5. Kinetic optical resolution

On the basis of desymmetrization concept, the kinetic optical resolution of a racemic substrate[93,94] might be recognized as an intermolecular version of desymmetrization. The kinetic resolution of a racemic allylic ether by the glyoxylate–ene reaction also provides an efficient access to remote but relative[91] asymmetric induction. The reaction of allylic ethers catalysed by the (R)-BINOL-derived complex (**1**) provides the 2R,5S-*syn*-products with >99% diastereoselectivity along with more than 95% ee (Scheme 6.16). The high diastereoselectivity, coupled with the high ee, strongly suggests that the catalyst/glyoxylate complex efficiently discriminates between the two enantiomeric substrates to accomplish the effective kinetic resolution. In fact, the relative rates between the reactions of the either enantiomers, calculated by the equation $(1n[(1-c)(1-ee_{recov})] \times \{1n[(1-c)(1 + ee_{recov})]\}^{-1}$, $c = (ee_{recov}) \times (ee_{recov} + ee_{prod})^{-1}$, $0<c$, ee<1 where c is the fraction of consumption), were approx. 700 for R = *i*-Pr and 65 for R = Me. As expected, the double asymmetric induction[95,96] in the reaction of (R)-ene component using the

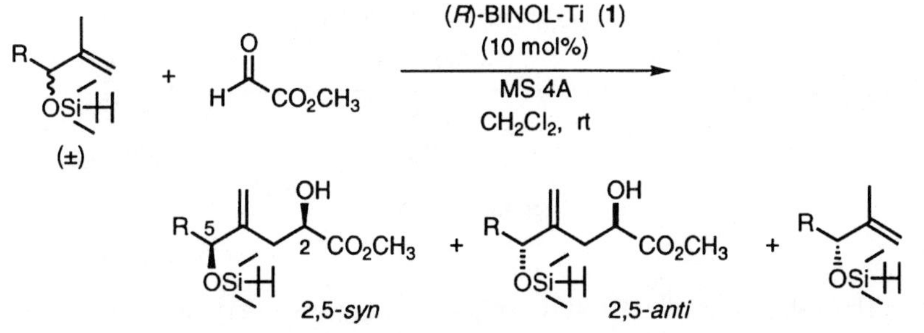

R	ene-product	recovered ene	relative rate (k_R/k_S)
i-Pr	99.6% ee (>99% *syn*)	37.8% ee	720
Me	96.2% ee (>99% *syn*)	22.0% ee	64

Scheme 6.16

catalyst (*S*)-**1** ('matched' catalytic system) leads to the complete (>99%) 2,5-*syn*-diastereoselectivity in high chemical yield, whereas the reaction of (*R*)-ene using (*R*)-**1** ('mis-matched' catalytic system) produces a diastereomeric mixture in quite low yield (Scheme 6.17).

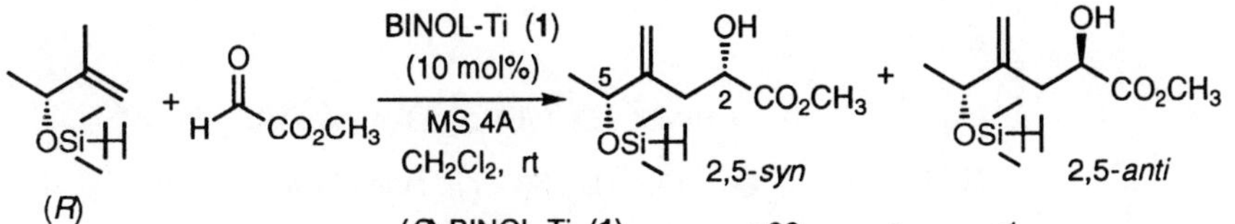

	2,5-*syn*		2,5-*anti*
(*S*)-BINOL-Ti (**1**)	>99	:	<1
(*R*)-BINOL-Ti (**1**)	50	:	50

Scheme 6.17

6. Positive non-linear effect of non-racemic catalysts

Deviation from the linear relationship, namely 'non-liner effect' (NLE) is sometimes observed between the enantiomeric purity of chiral catalysts and the optical yields of the products. Among, the convex deviation, which Kagan[97,98] and Mikami[99,100] independently refer to as positive non-linear effect (abbreviated as (+)-NLE (asymmetric amplification[101])) has attracted current attention to achieve higher level of asymmetric induction than the enantio-purity of the non-racemic (partially resolved) catalysts.[97,98,101–103] In turn, (−)-NLE stands for the opposite phenomenon of concave deviation, namely negative non-linear effect.

We have observed a remarkable level of (+)-NLE in the catalytic ene reaction. For instance, in the glyoxylate–ene reaction, the use of a catalyst prepared from BINOL of 33.0% ee provides the ene product with 91.4% ee in 92% chemical yield (Scheme 6.18).[99,100] The ee thus obtained is not only much

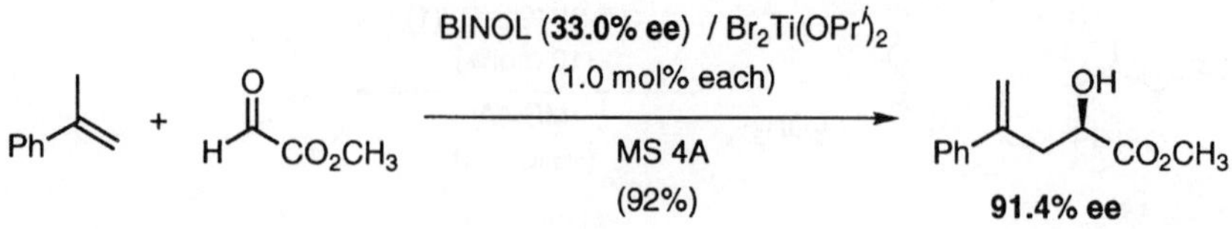

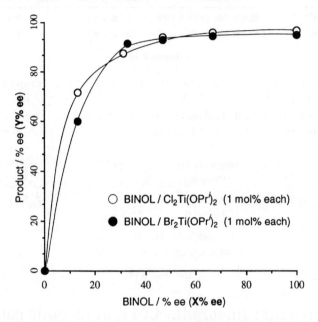

BINOL (**33.0% ee**) / Br$_2$Ti(OPr^i)$_2$
(1.0 mol% each)

MS 4A
(92%)

91.4% ee

Scheme 6.18

higher than the ee of the BINOL employed, but also very close to the value (94.6% ee) obtained using enantiomerically pure BINOL (Fig. 6.1).

○ BINOL / Cl$_2$Ti(OPr^i)$_2$ (1 mol% each)
● BINOL / Br$_2$Ti(OPr^i)$_2$ (1 mol% each)

Fig. 6.1. (+)-NLE in assymetric glyoxylate-ene reaction catalysed by BINOL-Ti complex (2).

7. Enantiomer-selective activation of racemic catalysts

Whereas non-racemic catalysts can generate non-racemic products with or without NLE, racemic catalysts inherently produce only racemic products. A strategy whereby a racemic catalyst is enantiomer-selectively de-activated by a chiral molecule has been shown to yield non-racemic products.[104,105] However, the level of asymmetric induction does not exceed the level attained by the enantiopure catalyst (Fig. 6.2a). Recently, 'chiral poisoning'[106] has been named for such a *de-activating* strategy. In contrast, we have reported an alternative but conceptually opposite strategy to asymmetric catalysis by

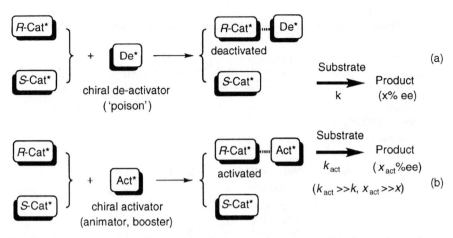

Fig. 6.2. Asymmetric de-activation (a) and asymmetric activation (b) of racemic catalysts.

racemic catalysts. A *chiral activator* selectively activates one enantiomer of a racemic chiral catalyst. Higher enantioselectivity might be attained than that achieved by an enantio-pure catalyst (% ee_{act} >> % $ee_{enantio-pure}$), in addition to a higher level of catalytic efficiency (k_{act} >> $k_{enantio-pure}$) (Fig. 6.2b). However, this still remains to be examined.

Catalysis with racemic BINOL-Ti(OPr^i)$_2$ (**2**) achieves extremely high enantioselectivity by adding another diol for the enantiomer-selective activation (Table 6.3).[107–110] Significantly, a remarkably high enantioselectivity (90% ee, *R*) was achieved using just a half-molar amount (5 mol%) of (*R*)-BINOL activator added to a *racemic* (±)-BINOL–Ti(OPr^i)$_2$ complex (**2**) (10 mol%).

The activation of the enantiopure (*R*)-BINOL–Ti(OPr^i)$_2$ catalyst (**2**) was investigated by further addition of (*R*)-BINOL (Table 6.4). The reaction proceeded quite smoothly to provide the carbonyl-ene product in higher chemical yield (82%) and enantioselectivity (97% ee) than without additional BINOL (95% ee, 20%). Comparing the results of enantiomer-selective activation of the racemic catalyst (90% ee, *R*) with those of the enantiopure catalyst (with (97% ee, *R*) or without activator (95% ee, *S*)), the reaction catalysed by the (*R*)-BINOL–Ti(OPr^i)$_2$/(*R*)-BINOL complex (**2'**) is calculated to be 26.3 times as fast as that catalysed by the (*S*)-BINOL–Ti(OPr^i)$_2$ (**2**) in the racemic case (Fig. 6.3). Indeed, kinetic studies according to a rapid-quench GC analysis show that the reaction catalysed by the (*R*)-BINOL–Ti(OPr^i)$_2$/(*R*)-BINOL complex (**2'**) is 25.6 times as fast as that catalysed by the (*R*)-BINOL–Ti(OPr^i)$_2$ (**2**). These results imply that the racemic (±)-BINOL–Ti(OPr^i)$_2$ (**2**) and half-molar amount of (*R*)-BINOL assemble preferentially into the (*R*)-BINOL–Ti(OPr^i)$_2$/(*R*)-BINOL complex (**2'**) and unchanged (*S*)-BINOL–Ti(OPr^i)$_2$ (**2**). In contrast, the enantiomeric

Table 6.3 Enantiomer selective activation of racemic BINOL–Ti(OPr^i)$_2$ (2)

Run	Chiral activator	%yield	% ee
1	none	1.6	0
2	(biphenol OH OH)	20	0
3	(dichloro dimethyl biphenol OH OH, Cl, Cl)	38	81
4	(binaphthol OH OH)	52	90
5[a]		35	80

[a] 2.5 mol% of (R)-BINOL was used as a chiral activator.

Table 6.4 Asymmetric activation of enantiopure (R)-BINOL–Ti(OPr^i)$_2$ (2)

Run	BINOL	Time (min)	%yield	% ee
1	none	60	20	95
2		1	1.6	95
3	(R)-BINOL	60	82	97
4		1	41	97
5		0.5	24	97
6	(S)-BINOL	60	48	86
7		0.5	2.6	86
8	(±)-BINOL	60	69	96

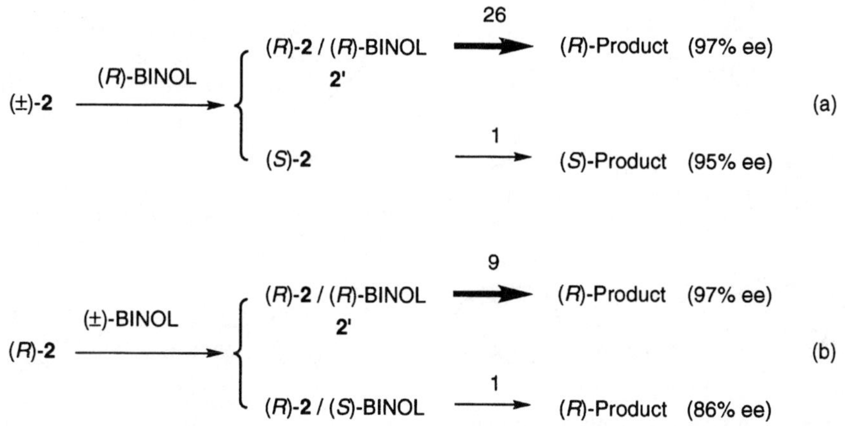

26

(R)-2 / (R)-BINOL ⟶ (R)-Product (97% ee)

(R)-BINOL 2'

(±)-2 ⟶ { (a)

1

(S)-2 ⟶ (S)-Product (95% ee)

9

(R)-2 / (R)-BINOL ⟶ (R)-Product (97% ee)

(±)-BINOL 2'

(R)-2 ⟶ { (b)

1

(R)-2 / (S)-BINOL ⟶ (R)-Product (86% ee)

Fig. 6.3. Kinetic features of asymmetric activation of BINOL-Ti $(OPr^6)_2$ (3).

form of the additional chiral ligand ((S)-BINOL) activates the (R)-BINOL–Ti(OPr^i)$_2$ (2) to a smaller degree, thus providing the carbonyl-ene product in lower optical (86% ee, R) and chemical (48%) yields than (R)-BINOL does.

The great advantage of asymmetric activation of the racemic BINOL–Ti(OPr^i)$_2$ complex (2) is highlighted in a catalytic version (Table 6.3, Run 5). High enantioselectivity (80.0% ee) is obtained by adding less than the stoichiometric amount (0.25 molar amount) of additional (R)-BINOL. A similar phenomenon on enantiomer-selective activation has been observed in aldol and (hetero) Diels–Alder reactions catalyzed by a racemic BINOL–Ti(OPr^i)$_2$ catalyst (2) (see below).

Another possibility was explored using racemic BINOL as an activator. Racemic BINOL was added to the (R)-BINOL–Ti(OPr^i)$_2$ (2), giving higher yield and enantioselectivity (96% ee, 69%) than those obtained by the original catalyst (R)-BINOL–Ti(OPr^i)$_2$ (2) without additional BINOL (95% ee, 20%). Comparing the results (96% ee, R) by the racemic activator with those of enantiopure catalyst, (R)-BINOL–Ti(OPr^i)$_2$/(R)-BINOL (2') (97% ee, R) or (R)-BINOL–Ti(OPr^i)$_2$/(S)-BINOL (86% ee, R), the reaction catalysed by the (R)-BINOL-Ti(OPr^i)$_2$ catalyst/(R)-BINOL complex (2') is calculated to be 8.8 times as fast as that catalysed by the (R)-BINOL–Ti(OPr^i)$_2$/(S)-BINOL (Fig. 6.3). A rapid-quench GC analysis revealed the reaction catalysed by the (R)-BINOL–Ti(OPr^i)$_2$/(R)-BINOL complex (2') to be 9.2 times as fast as that catalysed by the (R)-BINOL–Ti(OPr^i)$_2$/(S)-BINOL.

8. Ene cyclization

Conceptually, intramolecular ene reactions[111–114] (ene cyclizations) can be classified into six different modes (Fig. 6.4).[54,115] In the ene cyclizations, the

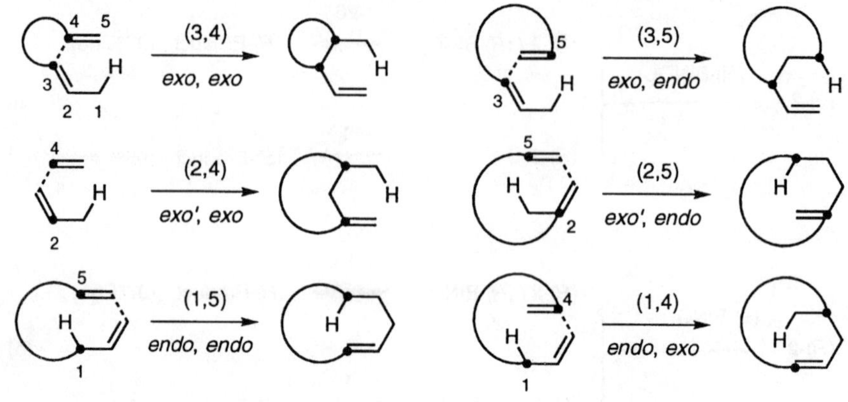

Fig. 6.4. Classification of ene cyclizations.

carbon numbers where the tether connects the [1,5]-hydrogen shift system, are expressed in (m,n) type. A numerical prefix stands for the forming ring size.

Asymmetric catalysis of ene reactions was initially explored in the intramolecular cases, since the intramolecular versions are much more facile than their intermolecular counterparts. The first example of an enantioselective 6-(3,4) carbonyl-ene cyclization was reported using a BINOL-derived zinc reagent.[116,117] However, this was successful only when using an excess of the zinc reagent (at least 3 equivalents). Recently, an enantioselective 6-(3,4) olefin-ene cyclization has been developed using a stoichiometric amount of a TADDOL-derived chiral titanium complex (Scheme 6.19).[118] In this ene reaction, a hetero Diels–Alder product was also obtained, the ratio depending critically on the solvent system employed. In both cases, geminal disubstitution is required in order to obtain high ee's values. However, neither reaction constitutes an example of a truly catalytic asymmetric ene cyclization.

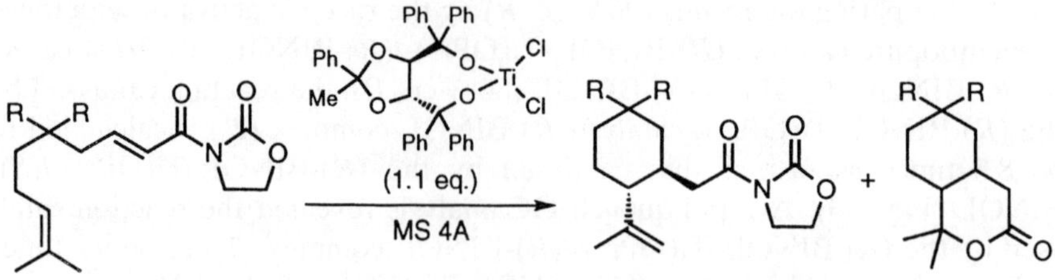

toluene	R = H	20 days	17%		37%
	R = Me	4 days	39% (82% ee)		36% (92% ee)
mesitylene			32% (86% ee)		37% (>98% ee)
CFCl₂CF₂Cl			63% (>98% ee)		25% (-)

Scheme 6.19

112

We reported the first examples of asymmetric catalysis of intramolecular carbonyl-ene reactions of types (3,4) and (2,4), using the BINOL-derived titanium complex (**1**).[115,119] The catalytic 7-(2,4) carbonyl-ene cyclization gives the oxepane with high ee, and gem-dimethyl groups are not required (Scheme 6.20). In a similar catalytic 6-(3,4) ene cyclization, the *trans*-tetrahydropyran is preferentially obtained with high ee (Scheme 6.21). The sense of asymmetric induction is exactly the same as observed for the glyoxylate-ene reaction: the (*R*)-BINOL–Ti catalyst provides the (*R*)-cyclic alcohol. Therefore, the chiral BINOL–Ti catalyst works efficiently for both the chiral recognition of the enantioface of the aldehyde and the discrimination of the diastereotopic protons of the ene component in a truly catalytic fashion.

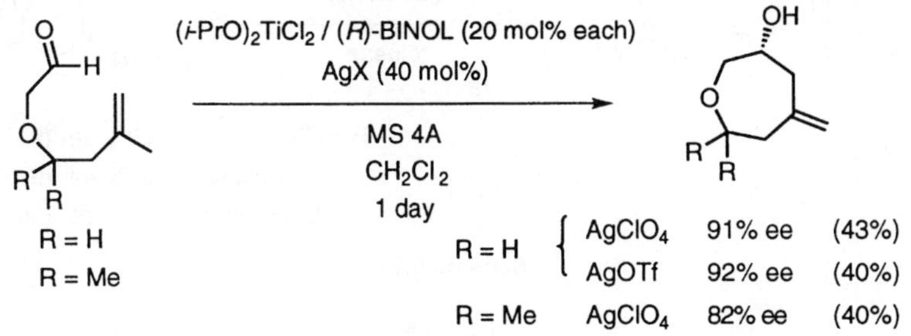

Scheme 6.20

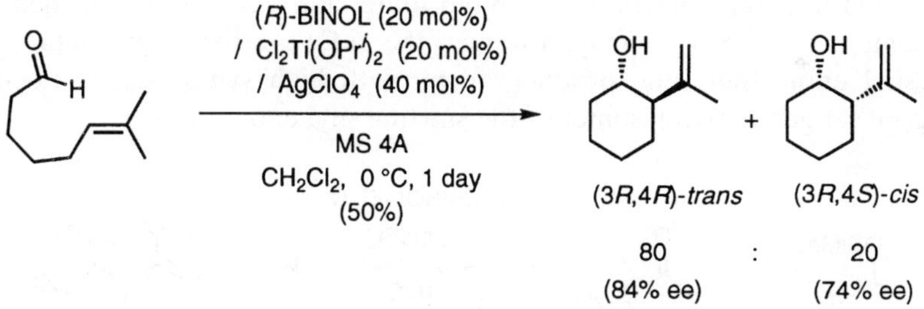

Scheme 6.21

9. Aldol reaction

The aldol reaction constitutes one of the most fundamental bond construction processes in organic synthesis.[120,121] Therefore, much attention has been focused on the development of asymmetric catalysts for aldol reactions using silyl enol ethers of ketones or esters as storable enolate components (the Mukaiyama aldol condensation).

Reetz reported the catalysis by BINOL–TiCl$_2$ of aldol reactions with aliphatic aldehydes.[122] In his case, BINOL-TiCl$_2$ was prepared by treatment of the lithium salt of BINOL with TiCl$_4$ in ether. After removal of the ether, the residue was treated with dry benezene and the solid was separated under nitrogen. Removal of the solvent provided the red–brown complex, which was used as the catalyst for the aldol reaction to give 8% ee. Later, Mukaiyama reported the use of BINOL-Ti oxide prepared from $(i\text{-PrO})_2\text{Ti=O}$ and BINOL, giving moderate to high levels of enantioselectivity (Scheme 6.22).[123,124]

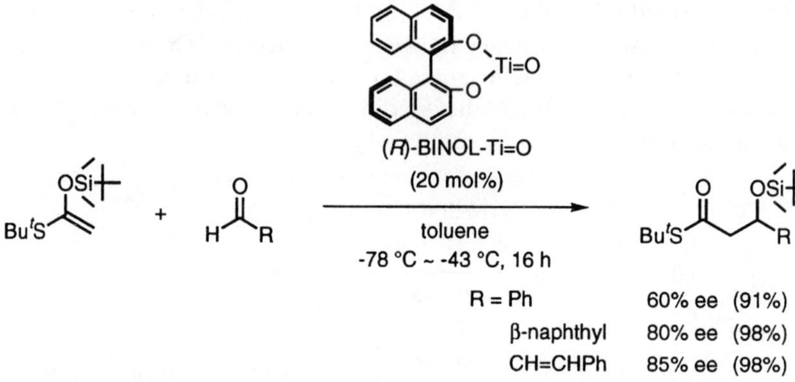

(*R*)-BINOL-Ti=O
(20 mol%)

toluene
-78 °C ~ -43 °C, 16 h

R = Ph	60% ee	(91%)
β-naphthyl	80% ee	(98%)
CH=CHPh	85% ee	(98%)

Scheme 6.22

We have found that the BINOL-derived titanium complex serves as an efficient catalyst for the Mukaiyama-type aldol reaction of ketone silyl enol ethers with good control of both absolute and relative stereochemistry (Scheme 6.23).[125] Surprisingly, however, the aldol products were obtained in the silyl enol ether (ene product) form, with high *syn*-diastereoselectivity from either geometrical isomer of the starting silyl enol ethers.

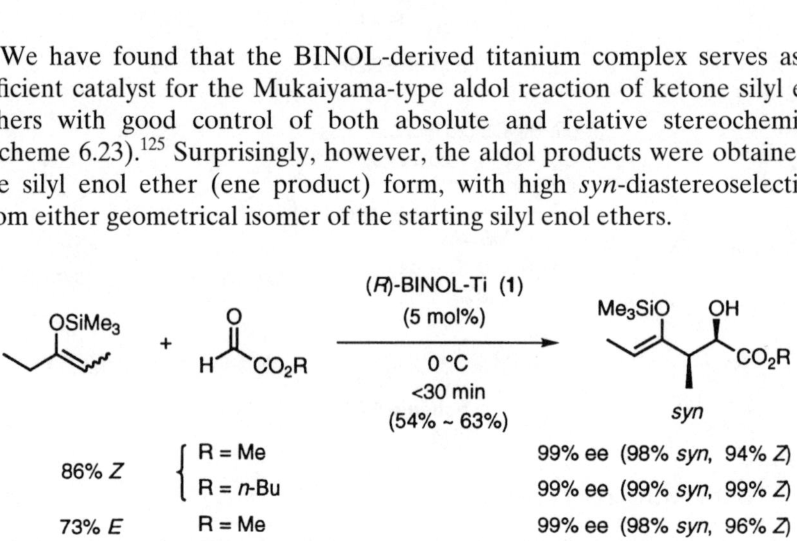

(*R*)-BINOL-Ti (1)
(5 mol%)

0 °C
<30 min
(54% ~ 63%)

syn

86% Z	R = Me	99% ee (98% *syn*, 94% *Z*)
	R = *n*-Bu	99% ee (99% *syn*, 99% *Z*)
73% E	R = Me	99% ee (98% *syn*, 96% *Z*)

Scheme 6.23

It appears likely that the reaction proceeds through an ene reaction pathway. Such an ene reaction pathway has not been previously recognized as a possible mechanism in the Mukaiyama aldol condensation. Usually an acyclic antiperiplanar transition state model has been used to explain the formation of the *syn*-diastereomer from either (*E*)- or (*Z*)-silyl enol ethers.[126,127] How-

114

ever, the cyclic ene mechanism now provides another rationale for the *syn*-diastereoselection irrespective of the enol silyl ether geometry (Scheme 6.24).

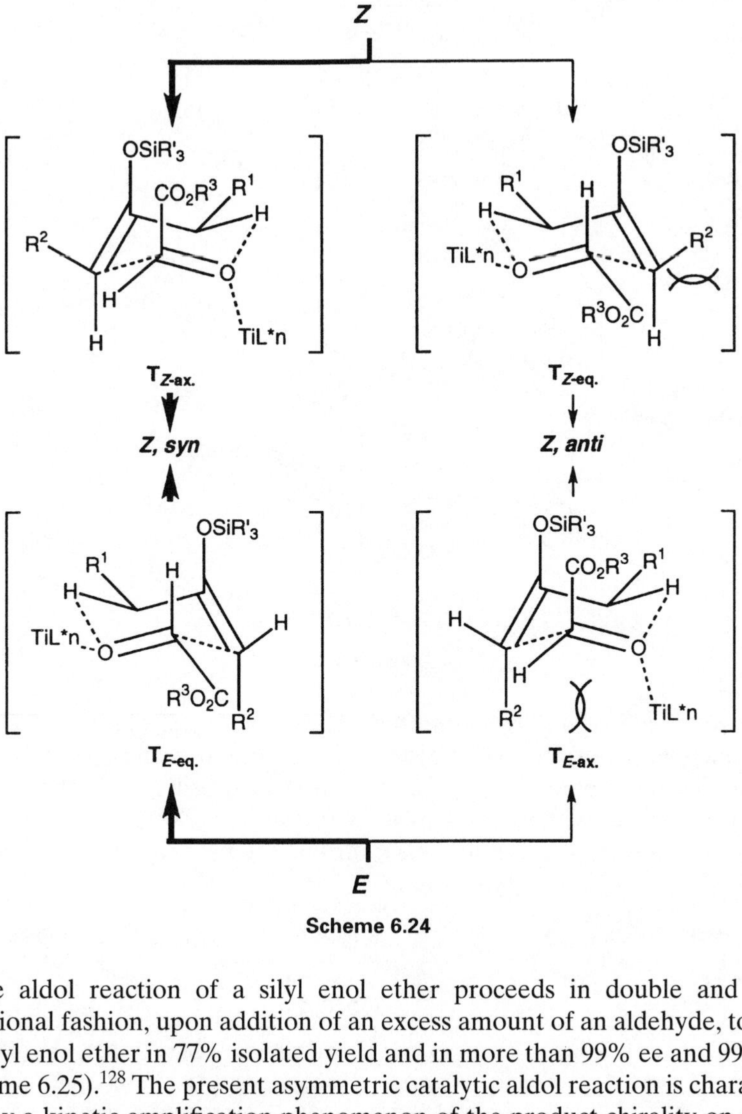

Scheme 6.24

The aldol reaction of a silyl enol ether proceeds in double and two-directional fashion, upon addition of an excess amount of an aldehyde, to give the silyl enol ether in 77% isolated yield and in more than 99% ee and 99% de (Scheme 6.25).[128] The present asymmetric catalytic aldol reaction is characterized by a kinetic amplification phenomenon of the product chirality on going from the one-directional aldol intermediate to the two-directional product. Further transformation of the *pseudo* C_2 symmetric product whilst still being protected as the silyl enol ether leads to a potent analogue of an HIV protease inhibitor.

The silatropic ene pathway, that is direct silyl transfer from an enol silyl ether to an aldehyde, may be involved as a possible mechanism in the

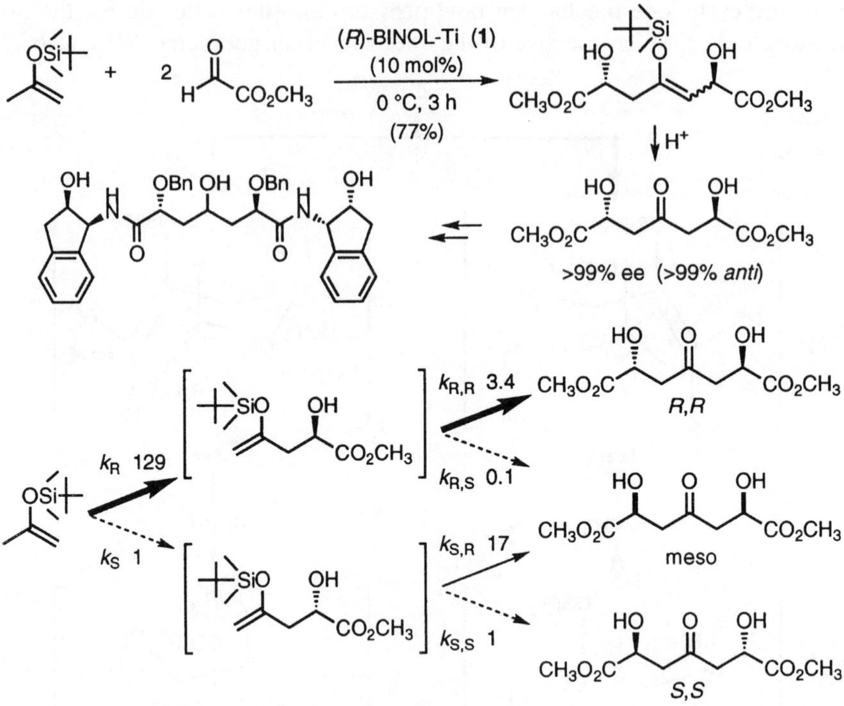

Scheme 6.25

Mukaiyama aldol-type reaction. Indeed, *ab initio* calculations show the silatropic ene pathway involving the cyclic (boat and chair) transition states for the BH$_3$-promoted aldol reaction of the trihydrosilyl enol ether derived from acetaldehyde with formaldehyde to be favoured.[129] Recently, we have reported the possible intervention of a silatropic ene pathway in the asymmetric

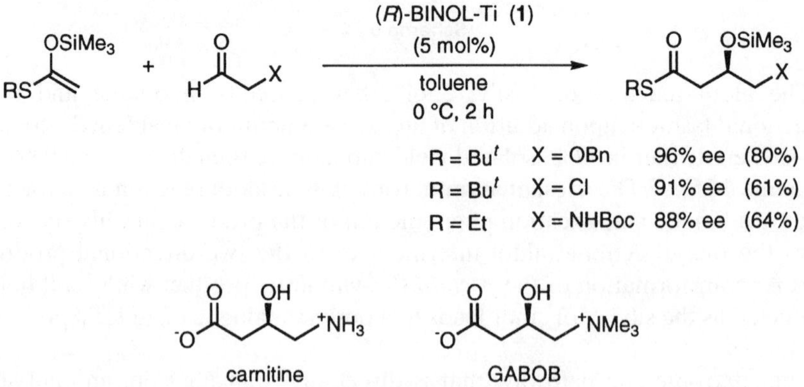

Scheme 6.26

116

catalytic aldol-type reaction of silyl enol ethers of thioesters.[130] Chloro and amino compounds thus obtained are useful intermediates for the synthesis of carnitine and GABOB (Scheme 6.26).[131,132]

There is a dichotomy in the sense of *syn-* vs. *anti*-diastereofacial preference, dictated by the bulkiness of the migrating group.[129] The sterically demanding silyl group shows *syn*-diastereofacial preference but the less demanding proton leads to *anti*-preference (Scheme 6.27). The *anti*-diastereoselectivity in carbonyl-ene reactions can be explained by the Felkin–Anh-like cyclic transition state model (T_1) (Fig. 6.5). In the aldol reaction, by contrast, the inside-crowded transition state (T_1') is less favourable than T_2', because of the steric repulsion between the trimethylsilyl group and the inside methyl group of aldehyde (T_1'). Therefore, the *syn*-diastereofacial selectivity is visualized by the anti-Felkin–Anh-like cyclic transition state model (T_2').

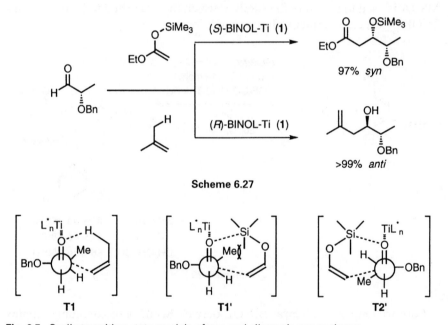

Scheme 6.27

Fig. 6.5. Cyclic transition state models of ene and silatropic ene pathway.

An aldol reaction with chiral β-benzyloxy aldehyde provides a method for the stereodivergent synthesis of both *syn-* and *anti*-diastereomers[133–136] with high levels of diastereoselectivity dictated primarily by the chirality of a BINOL-Ti catalyst (1) rather than a β-benzyloxy aldehyde (Scheme 6.28).[137–139] The aldol products can be used as useful key intermediates for β-lactone synthesis.[140]

Keck[141] and Carreira[142,143] have independently reported catalytic asymmetric Mukaiyama aldol reactions. Keck *et al.* also reported an aldol reaction of α-benzyloxy aldehyde with Danishefsky's diene. The aldol product was trans-

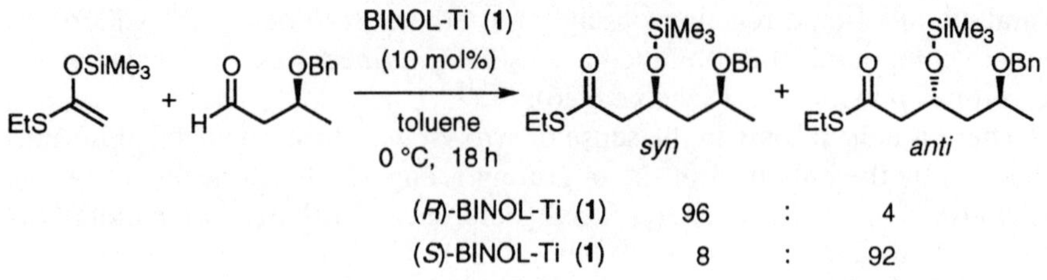

Scheme 6.28

formed to the hetero Diels–Alder type product through acid-catalysed cyclization. In their method, the catalyst is prepared using 1:1 and 2:1 stoichiometry of BINOL and $Ti(OPr^i)_4$ (Scheme 6.29).[144] In their cases, oven dried MS 4A is used to generate the catalyst, which they reported to be of BINOL-$Ti(OPr^i)_2$ structure, under refluxing conditions.

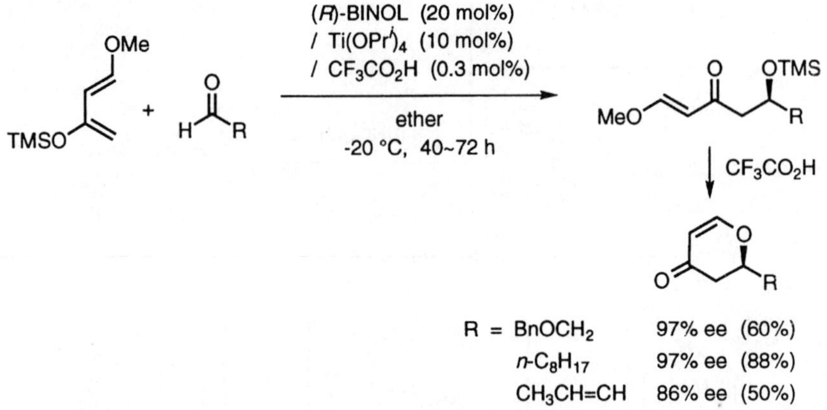

R =	BnOCH$_2$	97% ee	(60%)
	n-C$_8$H$_{17}$	97% ee	(88%)
	CH$_3$CH=CH	86% ee	(50%)

Scheme 6.29

Carreira employed a chiral BINOL-derived Schiff base–titanium complex as a catalyst for aldol reactions with acetate-derived ketene silyl acetals (Scheme 6.30).[142,143] The catalyst was prepared in toluene in the presence of salicylic acid, which was reported to be crucial to attain a high enantioselectivity.

10. (Hetero) Diels–Alder reaction

The (hetero) Diels–Alder (D–A) reaction also constitutes one of the most efficient carbon–carbon bond forming processes in the construction of six-membered rings by virtue of its high level of regioselectivity and the high

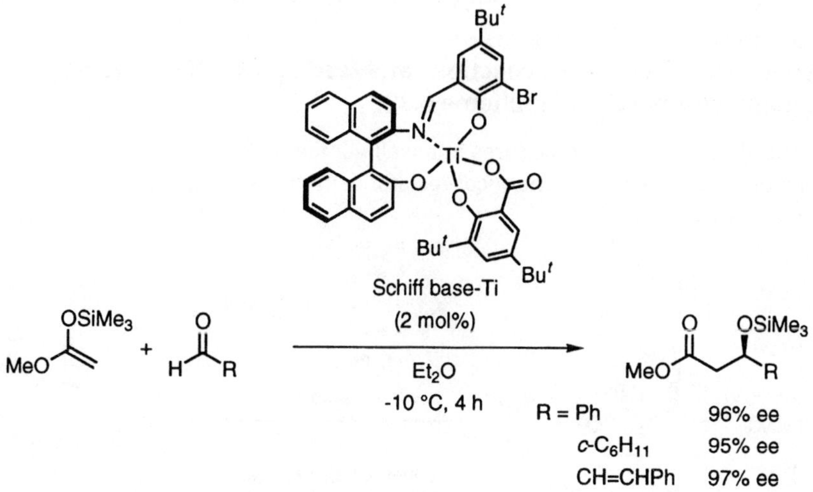

Scheme 6.30

potential for control of the absolute stereochemistry of the up to four newly created chiral centres.[145,146]

Narasaka has reported that a TADDOL–Ti dichloride—in the presence of MS 4A, acts as a good catalyst for the asymmetric catalytic D–A reactions with oxazolidinone derivatives of acrylates, giving extremely high enantio-selectivities (Scheme 6.31) (Protocol 4).[147–153]

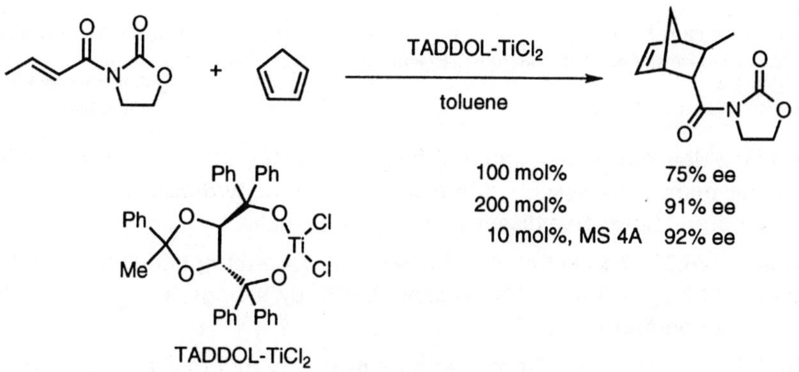

100 mol%	75% ee
200 mol%	91% ee
10 mol%, MS 4A	92% ee

TADDOL-TiCl₂

Scheme 6.31

Protocol 4.
Asymmetric Diels–Alder reaction catalysed by TADDOL-derived titanium complex[147,148] (Scheme 6.32)

Caution! Carry out all procedures in a well-ventilated hood, and wear disposable vinyl or latex gloves and chemical-resistant safety goggles.

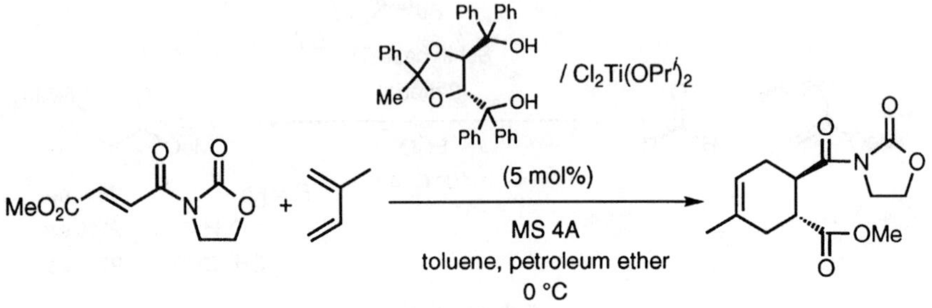

Scheme 6.32

Equipment

- A pre-dried, two-necked, round-bottomed flask (30 mL) equipped with a magnetic stirring bar, a three-way stopcock, and a septum cap. The three-way stopcock is connected to a vacuum/argon source
- Dry gas-tight syringe

- A pre-dried, two-necked, round-bottomed flask (500 mL) equipped with a three-way stopcock, a magnetic stirring bar, and a septum cap. The three-way stopcock is connected to a vacuum/argon source
- Double-ended cannula

Materials

- Molecular sieves 4A, powder, <5 μm, activated, 3.74 g — **irritant, hygroscopic**
- (2R,3R)-2,3-O-(1-phenylethylidene)-1,1,4,4-tetraphenylbutane-1,2,3,4-tetraol (FW 528.6), 1.32 g, 2.5 mmol
- Di-isopropoxytitanium(IV) dichloride (FW 237.0), 540 mg, 2.3 mmol — **moisture sensitive, irritant**
- 3-((E)-3-(Methoxycarbonyl)propenoyl)-1.3-oxazolidin-2-one (FW 199.0), 9.10 g, 46 mmol — **irritant**
- Isoprene (FW 68.1), 50 mL — **cancer suspect agent, flammable liquid**
- Dry toluene (distilled from P_2O_5 and from CaH_2 and stored over MS 4A) — **flammable liquid, toxic**
- Dry petroleum ether, 150 mL — **flammable liquid, toxic**

1. Add the di-isopropoxytitanium dichloride (540 mg, 2.3 mmol) (prepared in a similar manner as described in protocol 1) to the two-necked flask (30 mL) under argon flow, then dissolve in toluene (5 mL).

2. Add (2R,3R)-2,3-O-(1-phenylethylidene)-1,1,4,4-tetraphenylbutane-1,2,3,4-tetraol (1.32 g, 2.5 mmol) in toluene (5 mL) by syringe, then stir for 1 h at room temperature.

3. In the separate flask (500 mL), add toluene (175 mL) to the powdered molecular sieves 4A (3.74 g) to give a suspension.

4. Add, *via* the double-ended cannula, the titanium catalyst solution to the suspension of molecular sieves 4A at room temperature.
5. Cool to 0°C, then add 3-((E)-3-(methoxycarbonyl)propenoyl)-1.3-oxazolidin-2-one (9.10 g, 46 mmol), petroleum ether (150 mL), and isoprene (50 mL) in this order.
6. After stirring overnight at 0°C, add pH 7 phosphate buffer. Filter off molecular sieves 4A through a pad of celite and rinse the filter cake with ethyl acetate. Separate the phases and extract the aqueous phase three times with ethyl acetate. Dry over Na_2SO_4, and filter the solution.
7. Concentrate the crude product under vacuum and purify the crude product by silica-gel column chromatography with hexane-ethyl acetate (4:1) as eluent to obtain 3-(((4′R,5′R)-1′-methyl-5′-(methoxycarbonyl)cyclohexen-4′-yl)-carbonyl)-1,3-oxazolidin-2-one (11.6 g, 94% yield, 93% ee). The product displays the appropriate spectral characteristics and elemental analysis. The enantiomeric purity is determined by NMR analysis with a chiral NMR shift reagent, Eu(hfc)$_3$ (MeO signal separated).
8. Recrystallize from hexane–ethyl acetate to give the optically pure product in 64% yield.

We have also reported that the hetero D–A reactions of glyoxylates with 1-methoxy-1,3-butadienes proceed smoothly under catalysis by BINOL–Ti complex to give the *cis*-product with high ee (Scheme 6.33).[154] The hetero D–A products thus obtained can be transformed into monosaccarides.[155,156] Furthermore, the hetero D–A product can readily be converted into the lactone portion of HMG-Co A inhibitors such as mevinolin or compactin[157] in few steps.

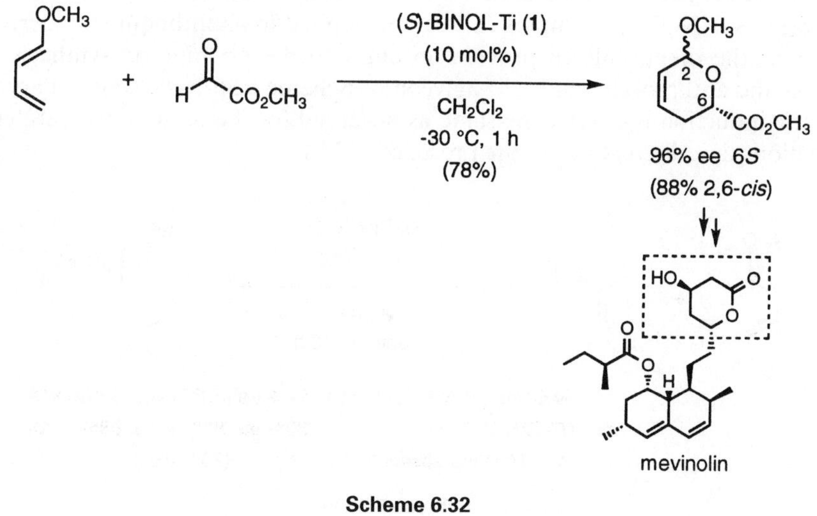

Scheme 6.32

121

The chiral titanium complex derived from 6-Br-BINOL affords higher enantioselectivity and catalytic activity than the parent BINOL–Ti catalyst in the hetero D–A reactions of 1-methoxydienes with glyoxylate, but not with bromoacrolein (Scheme 6.34).[158]

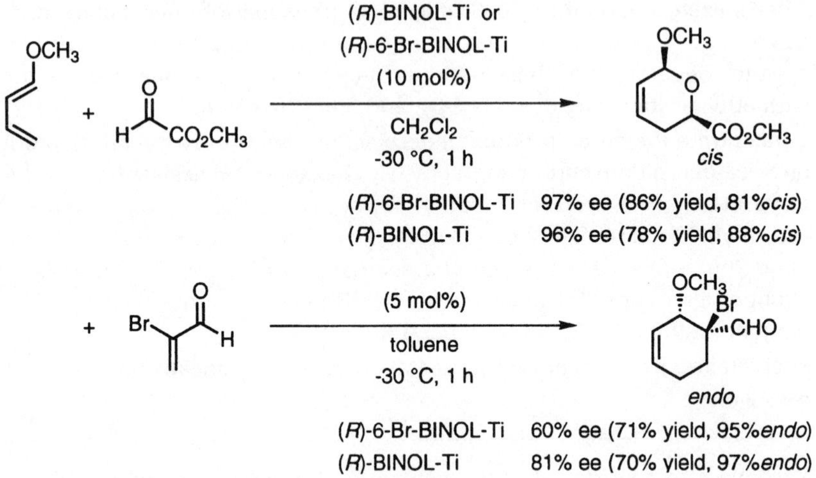

(R)-BINOL-Ti or
(R)-6-Br-BINOL-Ti
(10 mol%)
CH2Cl2
-30 °C, 1 h
OCH3
CO2CH3
cis
(R)-6-Br-BINOL-Ti 97% ee (86% yield, 81%cis)
(R)-BINOL-Ti 96% ee (78% yield, 88%cis)
(5 mol%)
toluene
-30 °C, 1 h
Br
CHO
endo
(R)-6-Br-BINOL-Ti 60% ee (71% yield, 95%endo)
(R)-BINOL-Ti 81% ee (70% yield, 97%endo)

Scheme 6.34

The D–A reaction of methacrolein with 1,3-dienol derivatives can also be catalysed by the BINOL-derived titanium complex. However, the catalyst must be freed from MS to give the *endo*-adduct with high enantioselectivity (Scheme 6.35) (Protocol 5).[159,160] Because MS works as an achiral catalyst for the D–A reaction. The asymmetric D–A reaction catalysed by the MS-free (MS-(−)) BINOL–Ti complex (**1′**) can be applied to naphthoquinone derivatives as the dienophile to provide an entry to the asymmetric synthesis of tetra- and anthracyclinone[161–163] aglycones (Scheme 6.36). The sense of asymmetric induction is exactly the same as observed for the asymmetric catalytic reactions described above in the presence of MS.

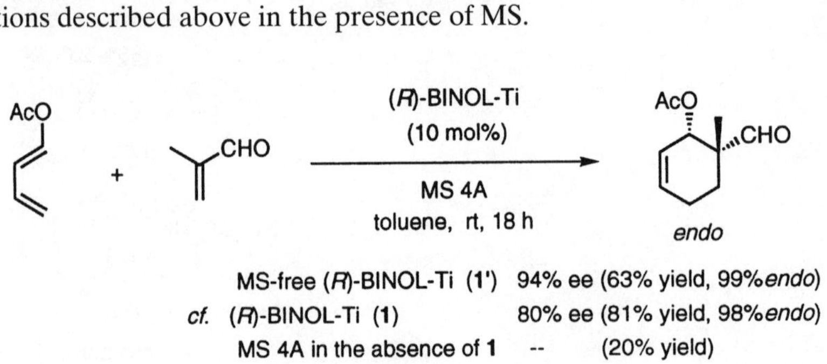

AcO
CHO
(R)-BINOL-Ti
(10 mol%)
MS 4A
toluene, rt, 18 h
endo
MS-free (R)-BINOL-Ti (1') 94% ee (63% yield, 99%endo)
cf. (R)-BINOL-Ti (1) 80% ee (81% yield, 98%endo)
MS 4A in the absence of 1 -- (20% yield)

Scheme 6.35

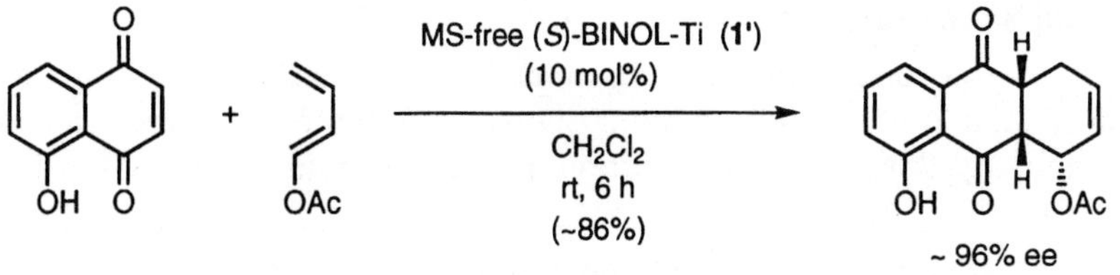

Scheme 6.36

Protocol 5.
Asymmetric Diels–Alder reaction catalyzed by MS-free BINOL-derived titanium complex[159,160] (Scheme 6.37)

Caution! Carry out all procedures in a well-ventilated hood, and wear disposable vinyl or latex gloves and chemical-resistant safety goggles.

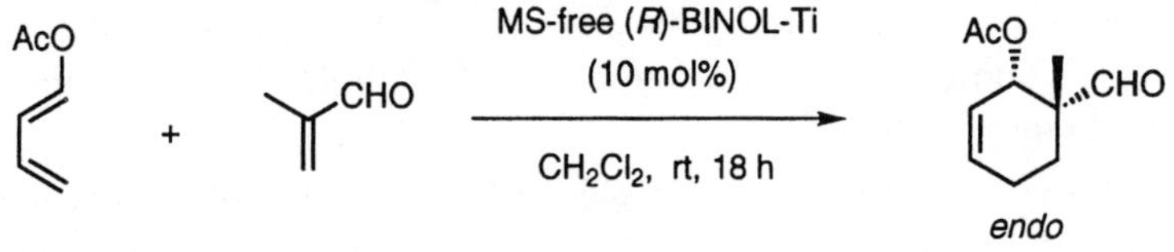

Scheme 6.37

Equipment

- A pre-dried, two-necked, round-bottomed flask (100 mL) equipped with a magnetic stirring bar, a three-way stopcock, and a septum cap. The three-way stopcock is connected to an argon source
- A pre-dried, two-necked, round-bottomed flask (100 mL) equipped with a magnetic stirring bar, a three-way stopcock, and a septum cap. The three-way stopcock is connected to a vacuum/argon source
- A pre-dried distillation apparatus

- Two pre-dried centrifuge tubes (40 mL) equipped with a septum cap
- A pre-dried, two-necked, round-bottomed flask (20 mL) equipped with a magnetic stirring bar, a three-way stopcock, and a septum cap. The three-way stopcock is connected to an argon source
- Dry gas-tight syringe
- Double-ended cannula
- Thermometers for low temperature

Materials

- Molecular sieves 4A,^{a} powder, <5 μm, activated, 10 g — **irritant, hygroscopic**
- (*R*)-(+)- or (*S*)-(−)-1,1′-bi-2-naphthol (FW 286.3), 0.573 g, 2.0 mmol — **irritant**
- Di-isopropoxytitanium(IV) dibromide (FW 325.9), 0.652 g, 2.0 mmol — **moisture sensitive, irritant**
- 1-Acetoxy-1,3-butadiene (mixture of *trans/cis*, 6:4) (FW 112.1), 198 mL, 1.67 mmol (*trans* isomer 1.0 mmol) — **flammable liquid, toxic**
- Freshly distilled methacrolein (FW 70.1), 82.8 mL, 1.0 mmol — **flammable liquid, corrosive**
- Dry dichloromethane — **toxic, irritant**
- Dry toluene — **flammable liquid, toxic**
- Triethylamine (FW 101.2), 70 mL, 0.5 mmol — **flammable liquid, corrosive**

1. Add the powdered molecular sieves 4A (10 g) and (*R*)-(+)-1,1′-bi-2-naphthol (0.573 g, 2.00 mmol) to the pre-dried two-necked flask, purge with argon, and add the dry CH$_2$Cl$_2$ (60 mL). Stir for 15 min at room temperature.

Protocol 5. *Continued*

2. Add di-isopropoxytitanium dibromide (0.652 g, 2.0 mmol) (prepared as described in protocol 1) to the suspension in one portion under a flow of argon. At this point the reaction mixture will turn into a red–brown suspension.

3. After stirring for 1 h at room temperature, transfer the suspension with the double-ended cannula to two pre-dried centrifuge tubes capped with a septum by pressurizing with argon.

4. Sediment the molecular sieves 4A by centrifugation at 4000 r.p.m. for 20 min. Transfer the resulting supernatant solvent with the double-ended cannula into the pre-dried two-necked round-bottomed flask. Exchange the septum cap with a distillation apparatus under a flow of argon.

5. Slowly evaporate the solvent *in vacuo* at 0°C. Control the speed of the distillation by regulation of the vacuum.

6. Dry the resulting dark-red solid *in vacuo* and obtain 0.58–0.62 g of the BINOL–Ti complex.[b] The MS-free BINOL–Ti complex may be stored under argon in refrigerator.

7. Add the MS-free BINOL–Ti complex (43 mg, 0.1 mmol[b] to the pre-dried two-necked round-bottomed flask. Dissolve the BINOL–Ti complex with toluene (3 mL).

8. Add freshly distilled methacrolein (82.8 mL, 1 mmol) and a solution of 1-acetoxy-1,3-butadiene (198 mL, 1.67 mmol) in toluene (1 mL) at room temperature.

9. After stirring for 18 h at room temperature, dilute the resulting solution with ether (5 mL) and quench with a solution of triethylamine (70 mL, 0.5 mmol) in hexane (10 mL). At this point, a yellow precipitate will form.

10. Filter off the precipitate through a pad of Celite and florisil, and rinse the filter cake three times with ether.

11. Concentrate the crude product *in vacuo* and purify the residue by silica gel chromatography to collect 115 mg (63%) of 2-acetoxy-1-methylcyclohex-3-enecarbaldehyde. The product displays the appropriate spectral characteristics and high resolution mass spectral data.

12. The enantiomeric purity is determined by the capillary GC analysis using a CP-cyclodextrin-β-2,3,6-M-19 as a chiral stationary phase column (94% ee).

---

[a] Purchased from Aldrich Chemical Company, Inc. and used without activation.
[b] Exact molecular weight of the MS-free BINOL–Ti complex has not been determined yet. However, on the basis of elemental analysis of the BINOL–Ti complex, there exists almost any bromine (<0.1%) and the content of titanium is ranging from 10.8% to 11.2%. The catalyst mol% is therefore calculated by its titanium content (averaged value 11.0%) wherein 434 mg of the BINOL–Ti complex equal 1 mmol of titanium.

Table 6.5 NLE in asymmetric D–A reaction of 1-acetoxy-1,3-butadiene and methacrolein catalyzed by MS-free BINOL-Ti (1')

Run	MS-free BINOL-Ti (1') (% ee)	%yield	%*endo*	%ee
1[a]	52	41	98	76
2[b]	50	50	99	74
3[c]	50	62	99	40
4[d]	50	67	99	60
5[e]	50	62	95	29
6[f]	–	20	–	–
7[g]	50	52	99	53

[a] Prepared from partially resolved BINOL (52% ee) and Cl$_2$Ti(OPr^i)$_2$.
[b] MS-free (R)-1' and MS-free (±)-1' (1:1).
[c] MS-free (R)-1' and MS-free (S)-1' (3:1).
[d] Prepared from MS-free (R)-1' and MS-free (S)-1' (3:1) in the presence of MS which was filtered prior to the reaction.
[e] MS-free (R)-1' and MS-free (±)-1' (1:1) in the presence of MS 4A.
[f] No MS-free catalyst (1') in the presence of MS 4A.
[g] MS-free (R)-1' and MS-free (S)-1' (3:1) in CH$_2$Cl$_2$.

The mode of preparation of the MS-free BINOL-Ti catalyst (1') determines the presence or the absence of a non-linear effect (NLE) (Table 6.5, Fig. 6.6). When the MS-free catalyst (1') was prepared from partially resolved BINOL, a (+)-NLE was observed (Run 1). The combined use of an enantiopure (R)-1' and (±)-1' catalysts in a ratio of 1:1 results in a similar (+)-NLE (Run 2). By contrast, mixing enantiopure (R)- and (S)-1' catalysts in a ratio of 3:1 leads to a linearity (no NLE) (Run 3). However, a MS-free catalyst obtained by mixing (R)- and (S)-1' catalysts in the same ratio of 3:1 *in the presence of MS*, which was filtered off prior to the reaction, showed a (+)-NLE (Run 4). These experimental facts can be explained if the complex consists of oligomers which do not interconvert *in the absence of MS* in toluene but do interconvert in dichloromethane (see Run 7 for the 9(+)-NLE in CH$_2$Cl$_2$). When the reaction was carried out *in the presence of MS*, however, a (−)-NLE was observed (Run 5), because MS acts as an achiral catalyst for the D–A reaction (Run 6). Moreover, in dichloromethane, the combined use of (R)- and (S)-1' catalysts (3:1), even without prior treatment with MS, exhibited a (+)-NLE (Run 7).

Asymmetric activation of the BINOL-Ti(OPr^i)$_2$ (2) by (6-Br)-BINOLs is essential to provide higher levels of enantioselectivity than those attained by the enantiopure BINOL-Ti catalyst (5% ee) in the D–A reaction of glyoxylates with the Danishefsky diene (Scheme 6.38) (Protocol 6)[110]

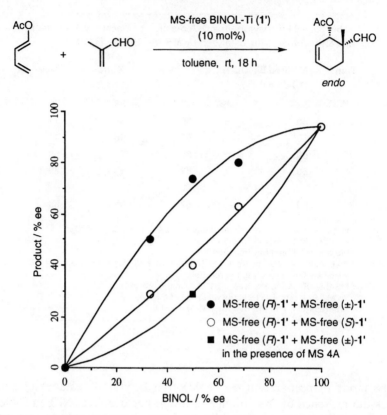

Fig. 6.6. (+)–, (–)-NLE, and linear relationships depending on the catalyst preparation.

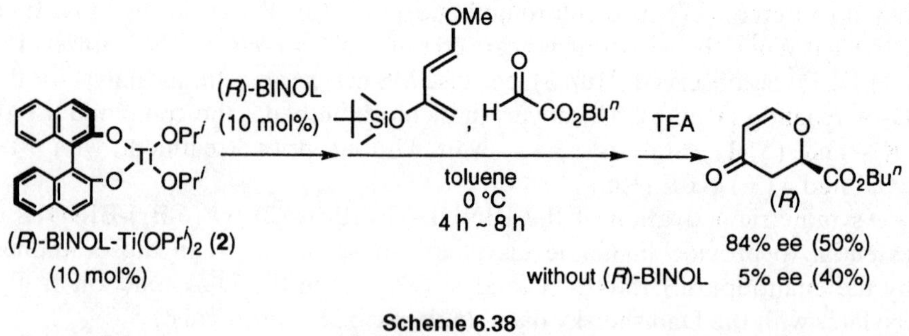

Scheme 6.38

126

Protocol 6.
Asymmetric hetero Diels–Alder reaction catalysed by (R)-BINOL-Ti(OPr^i)$_2$/(R)-BINOL complex: Asymmetric activation of (R)-BINOL-Ti(OPr^i)$_2$ by additional (R)-BINOL[110] (Scheme 6.39)

Caution! Carry out all procedures in a well-ventilated hood, and wear disposable vinyl or latex gloves and chemical-resistant safety goggles.

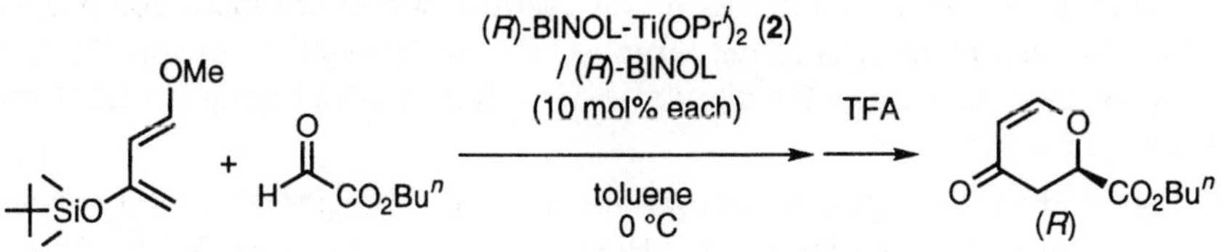

Scheme 6.39

Equipment
- A pre-dried, two-necked, round-bottomed flask (100 mL) equipped with a magnetic stirring bar, a three-way stopcock, and a septum cap. The three-way stopcock is connected to an argon source
- Dry gas-tight syringe
- Thermometers for low temperature

Materials
- (R)-(+)-1,1'-bi-2-naphthol (FW 286.3), 28.6 mg, 0.1 mmol × 2   irritant
- Titanium(IV) isopropoxide (FW 284.3), 0.297 mL, 0.1 mmol   moisture sensitive, irritant
- (E)-1-Methoxy-3-((tert-butyldimethylsilyl)oxy)-1,3-butadiene
  (FW 214.4), 257 mg, 1.2 mmol   flammable liquid, toxic
- Freshly distilled n-butyl glyoxylate^a (FW 130.1), 130.1 mg, 1.0 mmol   lachrymator, irritant
- Dry toluene   flammable liquid, toxic
- Trifluoroacetic acid (FW 114.0), 0.1 mL, 1.3 mmol   corrosive, toxic

1. Add (R)-(+)-1,1'-bi-2-naphthol (28.6 mg, 0.1 mmol) to the pre-dried two-necked flask, purge with argon, and add the dry toluene (2 mL) and the titanium(IV) isopropoxide (0.297 mL, 0.1 mmol) in this order. Stir for at room temperature 20 min. At this point the mixture will turn into a yellow–orange solution.

2. Add additional amount of (R)-(+)-1,1'-bi-2-naphthol (28.6 mg), 0.1 mmol) to the solution in one portion under a flow of argon. The mixture will turn immediately into a red–brown solution.

3. After stirring at room temperature for 20 min, cool the catalyst solution to 0°C. Add the (E)-1-methoxy-3-((tert-butyldimethylsilyl)oxy)-1,3-butadiene (257 mg, 1.2 mmol) and the freshly distilled n-butyl glyoxylate (130.1 mg, 1.0 mmol) in this order to the catalyst solution.

4. After stirring at 0°C for 2 h, add the trifluoroacetic acid (0.1 mL, 1.3 mmol) to the reaction mixture at that temperature.

Protocol 6. *Continued*

5. Stir for an additional 5 min at 0°C and then add saturated NaHCO₃ (10 mL) at that temperature. Filter the resulting yellowish suspension through a pad of Celite and rinse the filter cake with ether.

6. Separate the phases of the filtrate and extract the aqueous phase three times with ether (15 mL). Wash the combined organic phases twice with brine, dry over magnesium sulfate, and filter the solution.

7. Concentrate the crude product under vacuum and purify the crude product by silica-gel column chromatography with hexane–ethyl acetate (10:1) as eluent to obtain butyl 3,4-dihydro-4-oxo-2*H*-pyran-2-carboxylate (99.1 mg, 50% yield).

8. The product displays the appropriate spectral characteristics. The enantiomeric purity is determined by the HPLC analysis using a CHIRALPAK AS as a chiral stationary phase column with 10% *i*-PrOH/hexane as a mobile phase (84 %ee).

[a] Immediately prior use, the *n*-butyl glyoxylate is depolymerized by vacuum distillation from phosphorus pentoxide (10% weight) bp. 80°C/35 mmHg, bath temperature 120–140 °C.

As shown above, asymmetric catalysis of the D–A reactions have been attained using chiral titanium complexes bearing chiral diol ligands.[164–166] Whereas Yamamoto has reported a chiral helical titanium complex derived from Ti(OPr^i)₄ and a BINOL-derived tetraol ligand (Scheme 6.40).[167] The

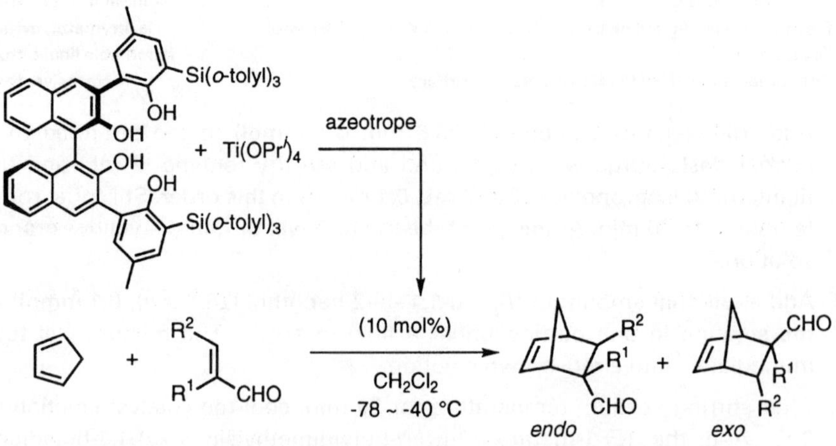

R¹ = H, R² = H   96% ee (85% *endo*)
R¹ = Me, R² = H   94% ee (99% *exo*)
R¹ = H, R² = Me   95% ee (70% *endo*)

Scheme 6.39

D–A products are obtained with uniformly high enantioselectivity, irrespective of the substituent pattern of α,β-unsaturated aldehydes.

Corey has also reported a new type of chiral titanium complex, which is derived from an amino alcohol ligand[168] (Scheme 6.41). The chiral titanium complex serves as an efficient asymmetric catalyst for the reaction of 2-bromoacrolein to yield the D–A product in high enantioselectivity.

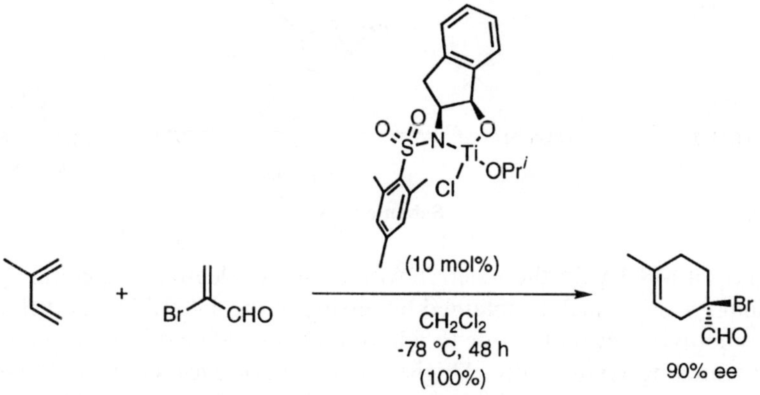

Scheme 6.41

11. Cyanohydrine formation

In the addition reaction of cyanotrimethylsilane[169] to aliphatic aldehydes, another synthetic application of a BINOL-Ti catalyst was reported by Reetz.[122] In his case, however, BINOL-TiCl$_2$ was reported to be prepared by treatment of the lithium salt of BINOL with TiCl$_4$ in ether (see above). The BINOL-TiCl$_2$ thus obtained was used as a catalyst for the cyanosilylation reaction to give the cyanohydrins in <82% ee (Scheme 6.42).

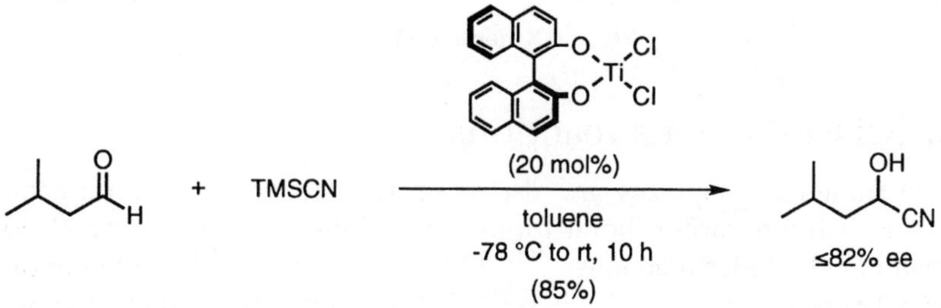

Scheme 6.42

Narasaka has also reported that TADDOL-Ti dichloride acts as a good catalyst for the asymmetric addition of trimethylsilylcyanide to aromatic and aliphatic aldehydes (Scheme 6.43).[170] The reactions proceeded only in the

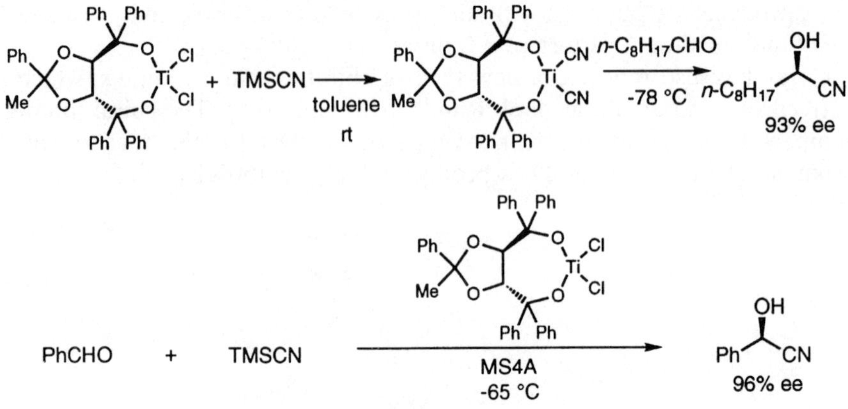

Scheme 6.43

presence of MS 4A. In the reaction with aromatic aldehydes, a chiral cyano-titanium species, which is obtained by mixing of the TADDOL-Ti dichloride and trimethylsilylcyanide prior to addition of the aldehydes, acts as a better chiral cyanating agent to afford higher enantiomeric excesses. Chiral titanium complexes obtained after addition of a salicylaldehyde-type Schiff bases have been reported to catalyse the asymmetric addition of hydrogen cyanide[171] or trimethylsilylcyanide[172-174] to aromatic and aliphatic aldehydes with high enantioselectivity (Scheme 6.44).

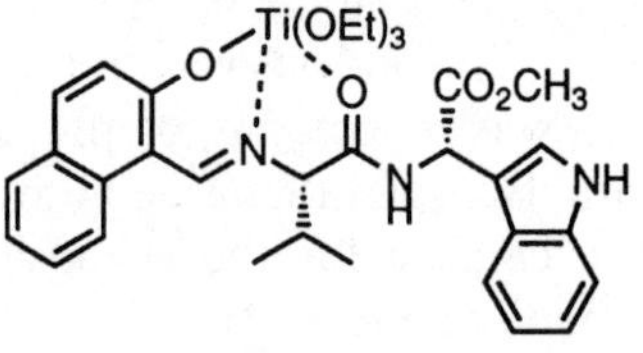

Amino acid-derived Schiff base 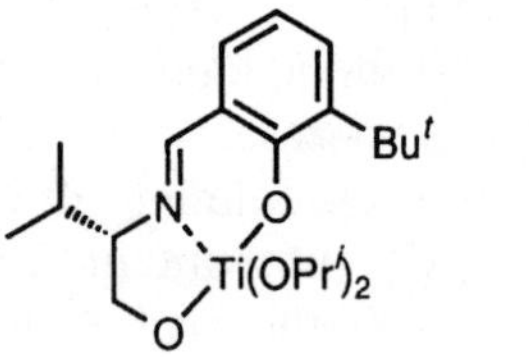   Amino alcohol-derived Schiff base

Scheme 6.44

12. Miscellaneous reactions

Chiral titanium complexes are also used as effective asymmetric catalysts for other carbon–carbon bond forming reactions, such as inverse electron-demand Diels–Alder reactions,[175-177] [2+2] additions,[178,179] [2+3] additions,[180] Michael additions,[181,182] and others[183-187] From a practical point of view, the development of more active and efficient catalysts is important, for the molecular design of asymmetric catalysts is the key to the basis of the structure–catalytic activity relationship. Although structure determination of active titanium species has been quite limited so far,[188-194] any progress along this line is highly promising and worth the effort.

Acknowledgements

We are grateful to Drs Darren Smyth and Thorsten Volk for their eagle-eyed proofreading of the manuscript.

References

1. Noyori, R. *Science* **1990**, *248*, 1194–1199.
2. Noyori, R. *Asymmetric Catalysis in Organic Synthesis*. Wiley, New York, **1994**.
3. Kagan, H. B. *Asymmetric Synthesis*. Georg Thieme Verlag, Stuttgart, **1997**.
4. Reetz, M. T. *Organotitanium Reagents in Organic Synthesis*. Springer-Verlag, Berlin, **1986**.
5. Seebach, D. *Angew. Chem., Int. Ed. Engl.* **1990**, *29*, 1320–1367.
6. Brown, I. D. *Acta Cryst.* **1988**, *B44*, 545–553.
7. Duthaler, R. O.; Hafner, A.; Riediker, M. *Organic Synthesis via Organometallics*. Friedr. Vieweg, Braunschweig, **1991**, p 285–309.
8. Duthaler, R. O.; Hafner, A.; Riediker, M. *Pure Appl. Chem.* **1990**, *62*, 631–642.
9. Mikami, K.; Terada, M.; Nakai, T. *Kikan Kagaku Sosetsu No. 17: Organic Chemistry of the Early Transition Metals*. Gakkai Shuppan Center, Tokyo, **1993**, p 87–98.
10. Mikami, K.; Nakai, T. *Kagaku Zoukann No. 124*. Kagaku Doujinn, Kyoto, **1995**, p 177–192.
11. Duthaler, R. O.; Hafner, A. *Chem. Rev.* **1992**, *92*, 807–832.
12. Bradley, D. C.; Mehrotra, R. C.; Gaur, D. P. *Metal Alkoxide*. Academic Press, New York, **1978**.
13. Feld, R.; Cowe, P. L. *The Organic Chemistry of Titanium*. Butterworths, London, **1965**.
14. Katsuki, T.; Martin, V. S. *Org. React.* **1996**, *48*, 1–299.
15. Johnson, R. A.; Sharpless, K. B. In *Comprehensive Organic Synthesis* (ed. B. M. Trost, I. Fleming), Pergamon Press, Oxford, **1991**, Vol. 7, pp 389–436.
16. Finn, M. G.; Sharpless, K. B. In *Asymmetric Synthesis* (ed. J. D. Morrison) Academic Press, New York, **1985**, Vol. 5, pp 247–308.
17. Rossiter, B. E. In *Asymmetric Synthesis* (ed. J. D. Morrison). Academic Press, New York, **1985**, Vol. 5, pp 193–246.
18. Halterman, R. L. *Chem. Rev.* **1992**, *92*, 965–994.
19. Carter, M. B.; Schiott, B.; Gutierrez, A.; Buchwald, S. L. *J. Am. Chem. Soc.* **1994**, *116*, 11667–11670.
20. Willoughby, C. A.; Buchwald, S. L. *J. Am. Chem. Soc.* **1994**, *116*, 11703–11714.
21. Almqvist, F.; Torsyensson, L.; Gudmundsson, A.; Frejd, T. *Angew. Chem., Int. Ed. Engl.* **1997**, *36*, 367–377.
22. Lindsley, C. W.; DiMare, M. *Tetrahedron Lett.* **1994**, *35*, 5141–5144.
23. Giffels, G.; Dreisbach, C.; Kragl, U.; Wegerding, M.; Waldmann, H.; Wandrey, C. *Angew. Chem., Int. Ed. Engl.* **1995**, *34*, 2005–2006.
24. Mikami, K.; Terada, M.; Narisawa, S.; Nakai, T. *Org. Synth.* **1992**, *71*, 14–21.
25. Dijkgraff, C.; Rousseau, J. P. G. *Spectrochim. Acta* **1968**, *2*, 1213–1217.
26. Reetz, M. T.; Kyung, S.-H.; Westermann, J. *Organometallics* **1984**, *3*, 1716–1717.
27. Mikami, K.; Motoyama, Y. In *Encyclopedia of Reagents for Organic Synthesis* (ed. L. A. Paquette), Wiley, Chichester, **1995**, Vol. 1, pp 397–403.

28. Mikami, K. In *Encyclopedia of Reagents for Organic Synthesis* (ed. L. A. Paquette), Wiley, Chichester, **1995**, Vol. 1, pp 407–408.
29. Wang, J.-T.; Fan, X.; Feng, X.; Qian, Y.-M. *Synthesis* **1989**, 291–292.
30. Olivero, A. G.; Weidmann, B.; Seebach, D. *Helv. Chim. Acta* **1982**, *64*, 2485–2488.
31. Braun, M. *Angew. Chem., Int. Ed. Engl.* **1996**, *35*, 519–522.
32. Dahinden, R.; Beck, A. K.; Seebach, D. In *Encyclopedia of Reagents for Organic Synthesis* (ed. L. A. Paquette), Wiley, Chichester, **1995**, Vol. 3, pp 2167–2170.
33. Narasaka, K.; Iwasawa, N.; Inoue, M.; Yamada, T.; Nakashima, M.; Sugimori, J. *J. Am. Chem. Soc.* **1989**, *111*, 5340–5344.
34. Beck, A. K.; Bastani, B.; Plattner, D. A.; Petter, W.; Seebach, D. *Chimia* **1991**, *45*, 238–244.
35. Hafner, A.; Duthaler, R. O.; Marti, R.; Rihs, G.; Rothe-Streit, P.; Schwarzenbach, F. *J. Am. Chem. Soc.* **1992**, *114*, 2321–2336.
36. Seebach, D.; Behrendt, L.; Felix, D. *Angew. Chem., Int. Ed. Engl.* **1991**, *30*, 1008–1009.
37. Bussche-Hünnefeid, J. L.; Seebach, D. *Tetrahedron* **1992**, *48*, 5719–5730.
38. Takahashi, H.; Kawakita, T.; Ohno, M.; Yoshioka, M.; Kobayashi, S. *Tetrahedron* **1992**, *48*, 5691–5700.
39. Yoshioka, M.; Kawakita, T.; Ohno, M. *Tetrahedron Lett.* **1989**, *30*, 1657–1660.
40. Takahashi, H.; Kawakita, T.; Yoshioka, M.; Kobayashi, S.; Ohno, M. *Tetrahedron Lett.* **1989**, *30*, 7095–7098.
41. Eisenberg, C.; Knochel, P. *J. Org. Chem.* **1994**, *59*, 3760–3761.
42. Rozema, M. J.; Sidduri, A.; Knochel, P. *J. Org. Chem.* **1992**, *57*, 1956–1958.
43. Knochel, P. In *Comprehensive Organic Synthesis* (eds. B. M. Trost, I. Fleming). Pergamon, London, **1991**, Vol. 1, pp 211–229.
44. Mikami, K. In *Encyclopedia of Reagents for Organic Synthesis* (ed. L. A. Paquette), Wiley, Chichester, **1995**, Vol. 1, pp 403–406.
45. Aoki, S.; Mikami, K.; Terada, M.; Nakai, T. *Tetrahedron* **1993**, *49*, 1783–1792.
46. Roush, W. R.; Blizzad, T. A. *Tetrahedron Lett.* **1982**, *23*, 2331–2334.
47. Still, W. C.; Ohmizu, H. *J. Org. Chem.* **1981**, *46*, 5242–5244.
48. Mikami, K.; Matsukawa, S. *Tetrahedron Lett.* **1994**, *35*, 3133–3166.
49. Costa, A. L.; Piazza, M. G.; Tagliavini, E.; Trombini, C.; Umani-Ronchi, A. *J. Am. Chem. Soc.* **1993**, *115*, 7001–7002.
50. Keck, G. E.; Tarbet, K. H.; Geraci, L. S. *J. Am. Chem. Soc.* **1993**, *115*, 8467–8468.
51. Keck, G. E.; Krishnamurthy, D.; Grier, M. C. *J. Org. Chem.* **1993**, *58*, 6543–6544.
52. Weigand, S.; Brückner, R. *Chem. Eur. J.* **1996**, *2*, 1077–1084.
53. Gauthier, Jr, D. R.; Carreira, E. M. *Angew. Chem., Int. Ed. Engl.* **1996**, *35*, 2363–2365.
54. Mikami, K.; Shimizu, M. *Chem. Rev.* **1992**, *92*, 1021–1050.
55. Snider, B. B. In *Comprehensive Organic Synthesis* (ed. B. M. Trost, I. Fleming). Pergamon, London, **1991**, Vol. 2, pp 527–561, Vol. 5, pp 1–27.
56. Mikami, K.; Terada, M.; Shimizu, M.; Nakai, T. *J. Synth. Org. Chem. Jpn* **1990**, *48*, 292–303.
57. Hoffmann, H. M. R. *Angew. Chem., Int. Ed. Engl.* **1969**, *8*, 556–577.
58. Mikami, K. *Advances in Asymmetric Synthesis*. JAI Press, Greenwich, Connecticut, **1995**, Vol. 1, pp 1–44.
59. Mikami, K.; Loh, T.-P.; Nakai, T. *Tetrahedron Lett.* **1988**, *29*, 6305–6308.
60. Mikami, K.; Loh, T.-P.; Nakai, T. *J. Chem. Soc., Chem. Comm.* **1988**, 1430–1431.

61. Omura, S. *J. Synth. Org. Chem. Jpn.* **1986**, *44*, 127.
62. Hanessian, S. *Total Synthesis of Natural Products: The 'Chiron' Approach.* Pergamon, Oxford, **1983**.
63. Mori, K. *The Total Synthesis of Natural Products.* Wiley, New York, **1981**, Vol. 4.
64. Seebach, D.; Hungerbughler, E. *Modern Synthetic Methods* (ed. R. Scheffold). Otto Salle Verlag, Frankfurt am Main, **1980**, Vol. 2, pp 91–172.
65. Mikami, K. *Pure Appl. Chem.* **1996**, *68*, 639–644.
66. Mikami, K.; Terada, M.; Nakai, T. In *Advances in Catalytic Processes*, (ed. M. P. Doyle). JAI Press, London, **1995**, Vol. 1, p 123–149.
67. Mikami, K.; Terada, M.; Narisawa, S.; Nakai, T. *Synlett* **1992**, 255–265.
68. Mikami, K.; Terada, M.; Nakai, T. *J. Am. Chem. Soc.* **1989**, *111*, 1940–1941.
69. Mikami, K.; Terada, M.; Nakai, T. *J. Am. Chem. Soc.* **1990**, *112*, 3949–3954.
70. Maruoka, K.; Hoshino, Y.; Shirasaka, T.; Yamamoto, H. *Tetrahedron Lett.* **1988**, *29*, 3967–3970.
71. Mikami, K.; Terada, M.; Nakai, T. *Chem. Express* **1989**, *4*, 589–592.
72. Mikami, K.; Motoyama, Y.; Terada, M. *Inorg. Chim. Acta* **1994**, *222*, 71–75.
73. Terada, M.; Motoyama, Y.; Mikami, K. *Tetrahedron Lett.* **1994**, *35*, 6693–6696.
74. Terada, M.; Mikami, K. *J. Chem. Soc. Chem. Commun.* **1995**, 2391–2392.
75. Terada, M.; Matsukawa, S.; Mikami, K. *J. Chem. Soc., Chem. Commun.* **1993**, 327–328.
76. Mori, K.; Takigawa, H. *Tetrahedron* **1991**, *47*, 2163–2168.
77. Brown, H. C.; Randad, R. S. *Tetrahedron* **1990**, *46*, 4463–4472.
78. Ohloff, G.; Giersch, W. *Helv. Chim. Acta* **1977**, *60*, 1496–1500.
79. Mikami, K.; Yajima, T.; Terada, M.; Uchimaru, T. *Tetrahedron Lett.* **1993**, *34*, 7591–7594.
80. Mikami, K.; Yajima, T.; Terada, M.; Kato, E.; Maruta, M. *Tetrahedron Asymm.* **1994**, *5*, 1087–1090.
81. Mikami, K.; Yajima, T.; Takasaki, T.; Matsukawa, S.; Terada, M.; Uchimaru, T.; Maruta, M. *Tetrahedron* **1996**, *52*, 85–98.
82. Welch, J. T.; Eswarakrishnan, S. *Flurine in Bioorganic Chemistry.* Wiley, New York, **1990**.
83. Mikami, K.; Siree, N.; Yajima, T.; Terada, M.; Suzuki, Y. Annual Meeting of the Chemical Society of Japan, Kyoto, March 28–31, 1995; Abstract No. 3H218.
84. Mikami, K.; Yajima, T.; Siree, N.; Terada, M.; Suzuki, Y.; Kobayashi, I. *Synlett* **1996**, 837–838.
85. Mikami, K.; Yajima, T.; Terada, M.; Kawauchi, S.; Suzuki, Y.; Kobayashi, I. *Chem. Lett.* **1996**, 861–862.
86. Mikami, K.; Yajima, T.; Terada, M.; Suzuki, Y.; Kobayashi, I. *Chem. Commun.* **1997**, 57–58.
87. Mikami, K.; Yoshida, A. *Synlett* **1995**, 29–31.
88. Mikami, K.; Yoshida, A.; Matsumoto, Y. *Tetrahedron Lett.* **1996**, *37*, 8515–8518.
89. Ward, R. S. *Chem. Soc. Rev.* **1990**, *19*, 1–19.
90. Mikami, K.; Shimizu, M. *J. Synth. Org. Chem. Jpn.* **1993**, 21–31.
91. Bartlett, P. A. *Tetrahedron* **1980**, *36*, 2–72.
92. Mikami, K.; Narisawa, S.; Shimizu, M.; Terada, M. *J. Am. Chem. Soc.* **1992**, *114*, 6566–6568; 9242–9242.
93. Kagan, H. B.; Fiaud, J. C. *Topics in Stereochemistry* Interscience, New York, **1988**, Vol. 18.

Koichi Mikami and Masahiro Terada

94. Brown, J. M. *Chem. Ind. (London)* **1988**, 612–617.
95. Masamune, S.; Choy, W.; Peterson, J.; Sita, L. R. *Angew. Chem., Int. Ed. Engl.* **1985**, *24*, 1–30.
96. Heathcock, C. H. In *Asymmetric Synthesis* (ed. J. D. Morrison). Academic Press: New York, **1985**, Vol. 3, pp 111–212.
97. Guillaneux, D.; Zhao, S.-H.; Samuel, O.; Rainford, D.; Kagan, H. B. *J. Am. Chem. Soc.* **1994**, *116*, 9430–9439.
98. Puchot, C.; Samuel, O.; Dunach, E.; Zhao, S.; Agami, C.; Kagan, H. B. *J. Am. Chem. Soc.* **1986**, *108*, 2353–2357.
99. Mikami, K.; Terada, M. *Tetrahedron* **1992**, *48*, 5671–5680.
100. Terada, M.; Mikami, K.; Nakai, T. *J. Chem. Soc., Chem. Commun.* **1990**, 1623–1624.
101. Oguni, N.; Matsuda, Y.; Kaneko, T. *J. Am. Chem. Soc.* **1988**, *110*, 7877–7877.
102. Noyori, R.; Kitamura, M. *Angew. Chem. Int. Ed. Engl.* **1991**, *30*, 49–69.
103. Kitamura, M.; Suga, S.; Niwa, M.; Noyori, R. *J. Am. Chem. Soc.* **1995**, *117*, 4832–4842.
104. Alcock, N. W.; Brown, J. M.; Maddox, P. J. *J. Chem Soc., Chem. Commun.* **1986**, 1532–1533.
105. Maruoka, K.; Yamamoto, H. *J. Am. Chem. Soc.* **1988**, *111*, 789–790.
106. Faller, J. W.; Parr, J. *J. Am. Chem. Soc.* **1993**, *115*, 804–805.
107. Mikami, K.; Matsukawa, S. *Nature* **1997**, *385*, 613–615.
108. Matsukawa, S.; Mikami, K. *Tetrahedron Asymmetry* **1995**, *6*, 2571–2574.
109. Matsukawa, S.; Mikami, K. *Enantiomer* **1996**, *1*, 69–73.
110. Matsukawa, S.; Mikami, K. *Tetrahedron Asymmetry* **1997**, *8*, 815–816.
111. Taber, D. F. *Intramolecular Diels–Alder and Alder Ene Reactions.* Springer Verlag, Berlin, **1984**.
112. Fujita, Y.; Suzuki, S.; Kanehira, K. *J. Synth. Org. Chem. Jpn.* **1983**, *41*, 1152–1167.
113. Oppolzer, W.; Snieckus, V. *Angew. Chem. Int. Ed. Engl.* **1978**, *17*, 476–486.
114. Conia, J. M.; Le Perchec, P. *Synthesis* **1975**, 1–19.
115. Mikami, K.; Sawa, E.; Terada, M. *Tetrahedron Asymmetry* **1991**, *2*, 1403–1412.
116. Sakane, S.; Maruoka, K.; Yamamoto, H. *Tetrahedron* **1986**, *42*, 2203–2209.
117. Sakane, S.; Maruoka, K.; Yamamoto, H. *Tetrahedron Lett.* **1985**, *26*, 5535–5538.
118. Narasaka, K.; Hayashi, Y.; Shimada, S. *Chem. Lett.* **1988**, 1609–1612.
119. Mikami, K.; Terada, M.; Sawa, E.; Nakai, T. *Tetrahedron Lett.* **1991**, *32*, 6571–6574.
120. Evans, D. A.; Nelson, J. V.; Taber, T. R. *Topics in Stereochemistry.* Vol. 13, Interscience, New York, **1982**.
121. Mukaiyama, T. *Org. React.* **1982**, *28*, 203–331.
122. Reetz, M. T.; Kyung, S.-H.; Bolm, C.; Zierke, T. *Chem. Ind. (London)* **1986**, 824–824.
123. Mukaiyama, T.; Inubushi, A.; Suda, S.; Hara, R.; Kobayashi, S. *Chem. Lett.* **1990**, 1015–1018.
124. Kitamoto, D.; Imma, H.; Nakai, T. *Tetrahedron Lett.* **1995**, *36*, 1861–1864.
125. Mikami, K.; Matsukawa, S. *J. Am. Chem. Soc.* **1993**, *115*, 7039–7040.
126. Murata, S.; Suzuki, M.; Noyori, R. *J. Am. Chem. Soc.* **1980**, *102*, 3248–3249.
127. Yamamoto, Y.; Maruyama, K. *Tetrahedron Lett.* **1980**, *21*, 4607–4610.
128. Mikami, K.; Matsukawa, S.; Nagashima, M.; Funabashi, H.; Morishima, H. *Tetrahedron Lett.* **1997**, *38*, 579–582.

134

6: *Chiral titanium complexes for enantioselective catalysis*

129. Mikami, K.; Matsukawa, S.; Sawa, E.; Harada, A.; Koga, N. *Tetrahedron Lett.* **1997**, *38*, 1951–1954.
130. Mikami, K.; Matsukawa, S. *J. Am. Chem. Soc.* **1994**, *116*, 4077–4078.
131. Kolb, H. C.; Bennari, Y. L.; Sharpless, K. B. *Tetrahedron Asymmetry* **1993**, *4*, 133–141.
132. Larcheveque, M.; Henrot, S. *Tetrahedron* **1990**, *46*, 4277–4282.
133. Reetz, M. T.; Jung, A. *J. Am. Chem. Soc.* **1983**, *105*, 4833–4835.
134. Evans, D. A.; Duffy, J. L.; Dart, M. J. *Tetrahedron Lett.* **1994**, *35*, 8537–8540.
135. Evans, D. A.; Dart, M. J.; Duffy, J. L.; Yang, M. G.; Livingston, A. B. *J. Am. Chem. Soc.* **1995**, *117*, 6619–6629.
136. Evans, D. A.; Dart, M. J.; Duffy, J. L.; Yang, M. G. *J. Am. Chem. Soc.* **1996**, *118*. 4322–4343.
137. Matsukawa, S.; Mikami, K. Annual Meeting of the Chemical Society of Japan, Kyoto, March 28–31, 1995; Abstract No. 3H106.
138. Mikami, K.; Yajima, T.; Takasaki, T.; Matsukawa, S.; Terada, M.; Uchimaru, T.; Maruta, M. *Tetrahedron* **1996**, *52*, 85–98.
139. Mikami, K.; Takasaki, T.; Matsukawa, S.; Maruta, M. *Synlett* **1995**, 1057–1058.
140. Pommier, A.; Pons, J.-M. *Synthesis* **1995**, 729–744.
141. Keck, G. E.; Krishnamurthy, D. *J. Am. Chem. Soc.* **1995**, *117*, 2363–2364.
142. Carreira, E. M.; Singer, R. A.; Lee, W. *J. Am. Chem. Soc.* **1994**, *116*, 8837–8838.
143. Carreira, E. M.; Lee, W.; Singer, R. A. *J. Am. Chem. Soc.* **1995**, *117*, 3649–3650.
144. Keck, G. E.; Li, X.-Y.; Krishnamurthy, D. *J. Org. Chem.* **1995**, *60*, 5998–5999.
145. Review: Kagan, H. B.; Riant, O. *Chem. Rev.* **1992**, *92*, 1007–1019.
146. Narasaka, K. *Synthesis* **1991**, 1–11.
147. Narasaka, K.; Iwasawa, N.; Inoue, M.; Yamada, T.; Nakashima, M.; Sugimori, J. *J. Am. Chem. Soc.* **1989**, *111*, 5340–5345.
148. Narasaka, K.; Tanaka, H.; Kanai, F. *Bull. Chem. Soc. Jpn.* **1991**, *64*, 387–391.
149. Corey, E. J.; Matsumura, Y. *Tetrahedron Lett.* **1991**, *32*, 6289–6292.
150. Seebach, D.; Beck, A. K.; Imwinkelried, R.; Roggo, S.; Wonnacott, A. *Helv. Chim. Acta* **1987**, *70*, 954–975.
151. Engler, T. A.; Letavic, M. A.; Takusagawa, F. *Tetrahedron Lett.* **1992**, *33*, 6731–6734.
152. Wada, E.; Pei, W.; Kanemasa, S. *Chem. Lett.* **1994**, 2345–2348.
153. Quinkert, G.; Grosso, M. D.; Döring, A.; Döring, W.; Schenkel, R. I.; Bauch, M.; Dambacher, G. T.; Bats, J. W.; Zimmermann, G.; Dürner, G. *Helv. Chim. Acta* **1995**, *78*, 1345–1391.
154. Terada, M.; Mikami, K.; Nakai, T. *Tetrahedron Lett.* **1991**, *32*, 935–938.
155. Konowal, A.; Jurczak, J.; Zamojski, A. *Tetrahedron* **1976**, *32*, 2957–2959.
156. Danishefsky, S. J.; DeNinno, M. P. *Angew. Chem., Int. Ed. Engl.* **1987**, *26*, 15–23.
157. Rosen, T.; Heathcock, C. H. *Tetrahedron* **1986**, *42*, 4909–4951.
158. Motoyama, Y.; Terada, M.; Mikami, K. *Synlett* **1995**, 967–968.
159. Mikami, K.; Motoyama, Y.; Terada, M. *J. Am. Chem. Soc.* **1994**, *116*, 2812–2820.
160. Mikami, K.; Terada, M.; Motoyama, Y.; Nakai, T. *Tetrahedron Asymmetry* **1991**, *2*, 643–646.
161. Krohn, K. *Tetrahedron* **1990**, *46*, 291–318.
162. Krohn, K. *Angew. Chem., Int. Ed. Engl.* **1986**, *25*, 790–807.
163. Broadhurst, M. J.; Hassall, C. H.; Thomas, G. *J. Chem. Ind. (London)* **1985**, *18*, 106–112.

164. Devine, P. N.; Oh, T. *J. Org. Chem.* **1992**, *57*, 396–399.
165. Ketter, A.; Glahsl, G.; Herrmann, R. *J. Chem. Research (M)* **1990**, 2118–2156.
166. Chapuis, C. Jurczak, J. *Helv. Chim. Acta* **1987**, *70*, 436–440.
167. Maruoka, K.; Murase, N.; Yamamoto, H. *J. Org. Chem.* **1993**, *58*, 2938–2939.
168. Corey, E. J.; Roper, T. D.; Ishihara, K.; Sarakinos, G. *Tetrahedron Lett.* **1993**, *34*, 8399–8402.
169. Rasmussen, J. K.; Heilmann, S. M.; Krepski, L. R. *The Chemistry of Cyanotrimethylsilane, Advances in Silicon Chemistry*, JAI Press, Greenwich, **1991**, Vol. 1.
170. Minamikawa, H.; Hayakawa, S.; Yamada, T.; Iwasawa, N.; Narasaka, K. *Bull. Chem. Soc. Jpn*, **1988**, *61*, 4379–4383.
171. Nitta, H.; Yu, D.; Kudo, M.; Mori, A.; Inoue, S. *J. Am. Chem. Soc.* **1992**, *114*, 7969–7975.
172. Hayashi, M.; Miyamoto, Y.; Inoue, T.; Oguni, N. *J. Org. Chem.* **1993**, *58*, 1515–1522.
173. Hayashi, M.; Inoue, T.; Miyamoto, Y.; Oguni, N. *Tetrahedron* **1994**, *50*, 4385–4398.
174. Hayashi, M.; Matsuda, T.; Oguni, N. *J. Chem. Soc., Perkin Trans. 1* **1992**, 3135–3140.
175. Posner, G. H.; Eydoux, F.; Lee, J.-K.; Bull, D. S. *Tetrahedron Lett.* **1994**, *35*, 7541–7544.
176. Posner, G. H.; Dai, H.; Bull, D. S.; Lee, J.-K.; Eydoux, F.; Ishihara, Y.; Welsh, W.; Pryor, N.; Petr, Jr. S. *J. Org. Chem.* **1996**, *61*, 671–676.
177. Wada, E.; Yasuoka, H.; Kanemasa, S. *Chem. Lett.* **1994**, 1637–1640.
178. Engler, T. A.; Letavic, M. A.; Reddy, J. P. *J. Am. Chem. Soc.* **1991**, *113*, 5068–5070.
179. Narasaka, K.; Hayashi, Y.; Shimadzu, H.; Niihata, S. *J. Am. Chem. Soc.* **1992**, *114*, 8869–8885.
180. Gothelf, K. V.; Thomsen, I.; Jørgensen, K. A. *J. Am. Chem. Soc.* **1996**, *118*, 59–64.
181. Kobayashi, S.; Suda, S.; Yamada, M.; Mukaiyama, T. *Chem. Lett.* **1994**, 97–100.
182. Bernardi, A.; Karamfilova, K.; Boschin, G.; Scolastico, C. *Tetrahedron Lett.* **1995**, *36*, 1363–1366.
183. Inoue, T.; Kitagawa, O.; Oda, Y.; Taguchi, T. *J. Org. Chem.* **1996**, *61*, 8256–8263.
184. Inoue, T.; Kitagawa, O.; Ochiai, O.; Shiro, M.; Taguchi, T. *Tetrahedron Lett.* **1995**, *36*, 9333–9336.
185. Hayashi, M.; Inoue, T.; Oguni, N. *J. Chem. Soc, Chem. Commun.* **1994**, 341–342.
186. Corey, E. J.; Rao, S. A.; Noe, M. C. *J. Am. Chem. Soc.* **1994**, *116*, 9345–9346.
187. Yamazaki, S.; Tanaka, M.; Yamabe, S. *J. Org. Chem.* **1996**, *61*, 4046–4050.
188. Williams, I. D.; Pedersen, S. F.; Sharpless, K. B.; Lippard, S. L. *J. Am. Chem. Soc.* **1984**, *106*, 6430–6431.
189. Bachand, B. Wuest, J. D. *Organometallics* **1991**, *10*, 2015–2025.
190. Boyle, T. J.; Eilerts, N. W.; Heppert, J. A.; Takusagawa, F. *Organometallics* **1994**, *13*, 2218–2229.
191. Corey, E. J.; Letavic, M. A.; Noe, M. C.; Sarshar, S. *Tetrahedron Lett*, **1994**, *35*, 7553–7556.
192. Nugent, W. A.; Harlow, R. L. *J. Am. Chem. Soc.* **1994**, *116*, 6142–6148.
193. Gothelf, K. V.; Hazell, R. G.; Jørgensen, K. A. *J. Am. Chem. Soc.* **1995**, *117*, 4435–4436.
194. Terada, M.; Matsumoto, Y.; Nakamura, Y.; Mikami, K. *Chem. Commun.* **1997**, 281–282.

7

Tin Lewis acid

SHŪ KOBAYASHI

1. Introduction

The element tin has played an increasingly important role in organic chemistry as well as organometallic chemistry, serving as a source of new reagents for selective transformations.[1-3] The main activity in these fields has been focused for a long time on tin(IV) compounds, and tin(II) compounds have been used primarily as reductants of aromatic nitro compounds to aromatic amines.[4] During the last decade, asymmetric synthesis has been developed increasingly, and in this field tin(II) reagents have served main roles rather than tin(IV) reagents.

Electronegativity of tin(II) and tin(IV) is shown in Table 7.1.[5] Tin(II) is more electronpositive and hence cationic than tin(IV), and is expected to co-ordinate with nucleophilic ligands. On the other hand, covalent radius and ionic radius of tin(II) are longer than those of tin(IV) (Tables 7.2 and 7.3).[6-13] This is due to electronic repulsion caused by unpaired electrons of tin(II), which weakens the δ-bond because tin(II) uses p-orbital for bonding. These

---

Table 7.1. Comparison of electronegativity

Oxidation state	Electronegativity (Pauling)	Electronegativity (Sanderson)
Sn(II)	1.80	1.58
Sn(IV)	1.96	2.02

Table 7.2. Sn-X bond distance (Å) in gas-phase

SnX$_2$	Sn-X	SnX$_4$	Sn-X	References
SnR$_2$	2.28	SnR$_4$	2.17	4
SnCl$_2$	2.42~2.43	SnCl$_4$	2.28~2.31	5–8
SnBr$_2$	2.55	SnBr$_4$	2.44	5,6,9
SnI$_2$	2.73~2.78	SnI$_4$	2.64	5,6,9

Table 7.3. Ionic radii (Å)

Oxidation state	Ionic radii
Sn(II)	1.02
Sn(IV)	0.71

characteristics are utilized in the following asymmetric reactions: to form a rigid complex with chiral diamines and to create an efficient asymmetric field, which has enough space to include several reactants. This chapter surveys use of tin(II) and tin(IV) Lewis acids, especially chiral Lewis acid, in organic synthesis.

2. Asymmetric aldol reaction

2.1 Preparation of chiral tin(II) Lewis acids

Chiral tin(II) Lewis acids are prepared *in situ* by mixing tin(II) triflate and chiral diamines, which are readily prepared from (*S*)-prolin in an appropriate solvent (Scheme 7.1). Tin(II) has three vacant orbitals, and after co-ordination of two nitrogen atoms one vacant orbital still remains.[14] Therefore, chiral diamine-co-ordinated tin(II) has a rigid bicyclo[3.3.0.]octane-like structure consisting of two fused five-membered rings, and can activate an aldehyde using the vacant orbital without changing the rigid structure.[15]

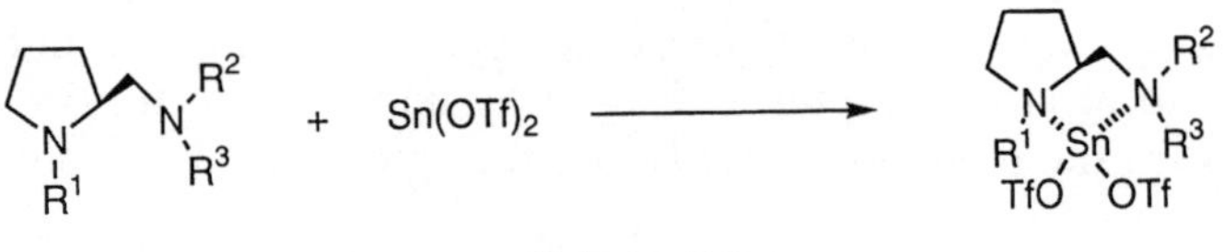

Scheme 7.1

2.2 Asymmetric aldol reactions of silyl enol ethers derived from acetic and propionic acid derivatives

In the presence of a stoichiometric amount of tin(II) triflate, chiral diamine **1**, and tributyltin fluoride (Bu$_3$SnF), 1-*S*-ethyl-1-trimethylsiloxyethene or 1-*S*-*t*-butyl-1-trimethylsiloxyethene reacts with aldehydes to afford the corresponding adducts in high yields with high enantioselectivities.[16] No chiral induction is observed without using Bu$_3$SnF. Although the precise function of Bu$_3$SnF is not yet clarified, it is believed that the fluoride connects the chiral tin(II) Lewis acid with the nucleophile, the silyl enol ether.[17,18]

In the reactions with the propionate derivatives, which provide synthetic-ally useful α-methyl-β-hydroxy ester derivatives, combination of tin(II) triflate, chiral diamine **5**, and Bu$_2$Sn(OAc)$_2$ gives better results.[17,19] The asymmetric

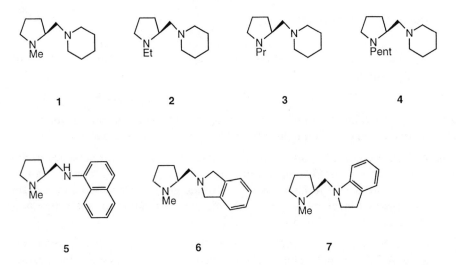

aldol reactions proceed with higher enantioselectivity, and in addition, the reactions proceed faster using $Bu_2Sn(OAc)_2$ than using Bu_3SnF as an additive.[20] A wide variety of aldehydes including aliphatic, aromatic, and α,β-unsaturated aldehydes are applicable to this reaction, and in all cases the aldol adducts are obtained in high yields with perfect *syn*-selectivities, and the enantiomeric excesses of these *syn*-adducts are more than 98%.

Protocol 1.
Asymmetric aldol reaction of (Z)-1-S-ethyl-1-trimethylsiloxypropene (8Z) with an aldehyde

Equipment
- Two-necked, round-bottomed flask (30 mL) fitted with a magnetic stirring bar. A three-way stopcock is fitted to the top of the flask and connected to a vacuum/argon source
- Vacuum/inert gas source (argon source may be an argon balloon)

Materials
- Tin(II) triflate (FW 416.82), 166.3 mg, 0.40 mmol
- Distilled chiral diamine **5** (FW 240.35), 115.4 mg, 0.48 mmol
- Distilled $Bu_2Sn(OAc)_2$ (FW 351.03), 154.5 mg, 0.44 mmol
- Distilled benzaldehyde (FW 106.12), 38.2 mg, 0.36 mmol
- Distilled (Z)-1-S-ethyl-1-trimethylsiloxypropene (**8Z**) (FW 190.38), 76.2 mg, 0.40 mmol
- Dry, distilled dichloromethane

1. Flame dry the reaction vessel with a stirring bar under dry argon.
2. Add tin(II) triflate to the flask in dry box under argon atmosphere and dry at 100°C for 1 h under reduced pressure (0.5 mmHg).

Protocol 1. *Continued*

3. Add 1 ml of dry dichloromethane to the flask with stirring. Add chiral diamine **5** in 0.5 mL of dichloromethane and then $Bu_2Sn(OAc)_2$ in 0.5 mL of dichloromethane.

4. After stirring the mixture for 5 min at room temperature, cool to −78°C.

5. Add (*Z*)-1-*S*-ethyl-1-trimethylsiloxypropene (**8Z**) in 0.5 mL of dichloromethane and benzaldehyde in 0.5 mL of dichloromethane successively.

6. After stirring for 20 h, add saturated sodium hydrogen carbonate. Warm the mixture to the room temperature, filter the suspension through a pad of Celite, and wash the filter cake three times with dichloromethane. Extract the aqueous layer three times with dichloromethane and wash the combined extracts with brine.

7. Concentrate the crude product under reduced pressure and apply to preparative TLC (silica gel). Elute with a mixed solvent of hexane–ethyl acetate (4:1) to obtain the product. Only *syn* isomer is produced and the optical purity is determined by HPLC analysis using a chiral column.

2.3 Asymmetric synthesis of 2-substituted malate derivatives

2-Substituted malic acids and their esters are widely distributed in nature as a variety of natural sources produced by plants or micro-organisms. Of more interest is their common inclusion as carboxylic acid components in biologically active alkaloids, and intense efforts have been made to prepare their carboxylic acid residues as optically active forms. Concerning asymmetric synthesis of 2-substitued malates, asymmetric aldol reactions of acetic acid derivatives with α-ketoesters are one of the most prospective methods.[21–23]

In the presence of tin(II) triflate, (*S*)-1-pentyl-2-[(piperidin-1-yl)methyl]-pyrrolidine (**4**), and Bu_3SnF, 1-*S*-ethyl-1-trimethylsiloxyethene reacts with methyl pyruvate to give the desired adduct in 92% ee. Methyl isopropyl-glyoxylate and methyl phenylglyoxylate also react with 1-*S*-ethyl-1-trimethyl-siloxyethene to give the corresponding 2-substituted malates in good yields with excellent enantioselectivities.[24]

Recently, a series of pyrrolizidine alkaloids has attracted much attention due to their potent hepatotoxic, carcinogenic, and mutagenic properties.[25–30] These alkaloids, especially 11- or 12-membered pyrrolizidine dilactones such as integerrimine, senecionine, fulvine, crispatine, etc., possess unique common structures of the α-methyl- β-hydroxy- β-alkyl units, and rather complicated multistage transformations have often been required for the stereoselective construction of these successive asymmetric centres including quaternary carbons.

140

Table 7.4. Asymmetric aldol reactions of silyl enolates with aldehydes

$$R^1CHO \ + \ \overset{OSiMe_3}{\underset{R^3}{\diagup}}SR^2 \ \xrightarrow[CH_2Cl_2,\ -78\ °C]{Sn(OTf)_2 + \text{Chiral diamine} + Bu_3SnF} \ R^1\overset{OH\ \ O}{\diagup\diagup}SR^2$$

Entry	R^1	R^2	R^3	Chiral diamine	Yield(%)	*syn/anti*	ee (%)
1	Ph	Et	H	1	78	—	82
2	Ph	Et	H	5	52	—	92
3	Ph	t-Bu	H	1	73	—	86
4	Ph(CH$_2$)$_2$	Et	H	1	70	—	78
5	Ph(CH$_2$)$_2$	Et	H	5	50	—	81
6	Ph(CH$_2$)$_2$	t-Bu	H	1	71	—	85
7	i-Pr	Et	H	1	77	—	95
8	t-Bu	Et	H	1	90	—	>98
9	Ph	Et	Me	5	85	100/0	>98
10	p-Cl Ph	Et	Me	5	96	100/0	>98
11	p-M$_3$ Ph	Et	Me	5	92	100/0	>98
12	p-MeO$_3$ Ph	Et	Me	5	95	100/0	>98
13	CH$_3$(CH$_2$)$_6$	Et	Me	5	90	100/0	>98
14	c-C$_6$H$_{11}$	Et	Me	5	90	100/0	>98
15	i-Pr	Et	Me	5	70	100/0	>98
16	i-Bu	Et	Me	5	86	100/0	>98
17	(E)-CH$_3$CH=CH	Et	Me	5	92	100/0	>98
18	(E)-PhCH=CH	Et	Me	5	91	100/0	>98
19	(E)-n-PrCH=CH	Et	Me	5	91	100/0	>98
20		Et	Me	5	93	100/0	>98
21		Et	Me	5	92	100/0	>98

Protocol 2.
Asymmetric aldol reaction of 1-*S*-ethyl-1-trimethylsiloxyethene with an α-ketoester

Equipment

- Two-necked, round bottomed flask (30 mL) fitted with a magnetic stirring bar. A three-way stopcock is fitted to the top of the flask and connected to a vacuum/argon source
- Vacuum/inert gas source (argon source may be an argon balloon)

Materials

- Tin(II) triflate (FW 416.82), 166.7 mg, 0.40 mmol
- Distilled chiral diamine **4** (FW 238.42), 114.4 mg, 0.48 mmol
- Bu$_3$SnF (FW 309.05), 136.0 mg, 0.44 mmol

Shū Kobayashi

Protocol 2. *Continued*

- Distilled methylpyruvate (FW 102.09), 36.8 mg, 0.36 mmol
- Distilled 1-S-ethyl-1-trimethylsiloxyethene (FW 176.35), 70.5 mg, 0.40 mmol
- Dry, distilled dichloromethane

1. Flame dry the reaction vessel with a stirring bar under dry argon.
2. Add tin(II) triflate to the flask in dry box under argon atmosphere and dry at 100 °C for 1 h under reduced pressure (0.5 mmHg).
3. Add 1 ml of dry dichloromethane to the flask with stirring. Add chiral diamine 4 in 1 mL of dichloromethane and then Bu₃SnF.
4. After stirring the mixture for 5 min at room temperature, cool to −78 °C.
5. Add 1-S-ethyl-1-trimethylsiloxyethene in 0.5 mL of dichloromethane and methylpuruvate in 0.5 mL of dichloromethane successively.
6. Adter stirring for 20 h, add saturated sodium hydrogen carbonate. Warm the mixture to the room temperature and filter the suspension through a pad of Celite, and wash the filter cake three times with dichloromethane. Extract the aqueous layer three times with dichloromethane and wash the combined extracts with brine.
7. Concentrate the crude product under reduced pressure and apply to preparative TLC (silica gel). Elute with a mixed solvent of hexane–ethyl acetate (4:1) to obtain the product. The optical purity is determined by HPLC analysis using a chiral column.

When (Z)-1-S-ethyl-1-trimethylsiloxypropene (**8Z**) is treated with methyl pyruvate in the presence of tin(II) triflate, (S)-1-pentyl-2-[(piperidin-1-yl)methyl]pyrrolidine (**4**), and Bu₃SnF, the reaction proceeds smoothly to give the *syn* isomer in high yield with high diastereo- and enantioselectivities (Scheme 7.2).[31] Similarly, (Z)-1-S-ethyl-1-trimethylsiloxypropene (**8Z**) reacts

Table 7.5. Asymmetric aldol reactions of α-Ketoesters

Entry	R	Chiral diamine	Yield (%)	ee (%)
1	Me	1	73	83
2	Me	4	78	92
3	i-Pr	1	76	>98
4	i-Pr	4	81	>98
5	Ph	1	74	>98
6	Ph	4	74	>98

142

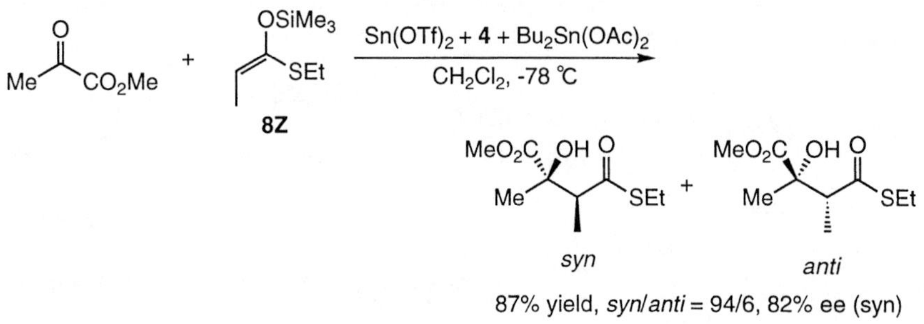

Scheme 7.2

with methyl phenylglyoxylate smoothly to give the corresponding *syn* adduct in a high enantiomeric excess (Scheme 7.3). The successive asymmetric centres including quarternary carbons are constructed efficiently with high selectivities by using this methodology.[32] On the other hand, (*E*)-1-*S*-ethyl-1-trimethylsiloxypropene reacts with methyl phenylglyoxylate or methyl pyruvate very slowly under the same reaction conditions.

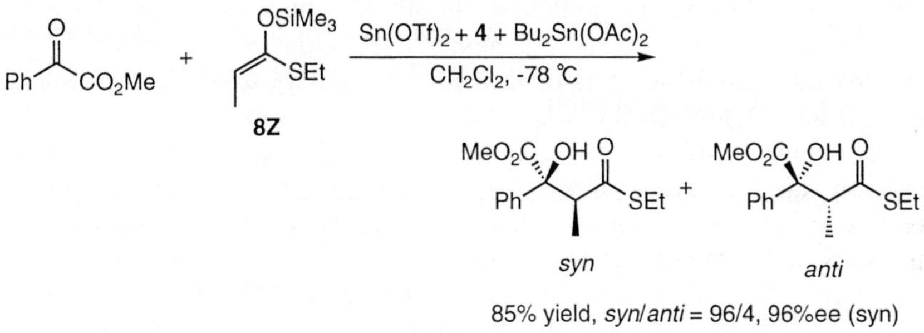

Scheme 7.3

2.4 Asymmetric synthesis of *syn*- and *anti*-1,2diol derivatives

Optically active 1,2-diol units are often observed in nature such as carbohydrates, macrolides, polyethers, etc. Recently several excellent asymmetric oxidation reactions of olefins using osmium tetroxide with a chiral ligand has been developed to achieve high enantiomeric excesses.[33-36]

Asymmetric synthesis of 1,2-diol derivatives based on asymmetric aldol reactions of α-alkoxy silyl enol ethers with aldehydes has been developed. The reaction of (*Z*)-2-benzyloxy-1-*S*-ethyl-1-trimethylsiloxyethene (**9Z**) with benzaldehyde is carried out in dichloromethane at −78 °C by using a chiral promoter consisting of tin(II) triflate, (*S*)-1-methyl-2-[(*N*-1-naphthylamino) methyl]-pyrrolidine (**5**), and Bu₃SnF, to afford the corresponding aldol adduct

Table 7.6. Asymmetric synthesis of *anti*-1,2-diol derivatives

RCHO + (9Z, OSiMe₃/BnO SEt) $\xrightarrow[\text{CH}_2\text{Cl}_2, -78\,°C]{\text{Sn(OTf)}_2 + 2 + \text{Bu}_2\text{Sn(OAc)}_2}$ (syn) + (anti)

Entry	R	Yield (%)	syn/anti	ee (%) (anti)
1	Ph	83	1/99	96
2	CH₃CH₂	72	2/98	97
3	c-C₆H₁₁	59	9/91	96
4	(E)-PhCH=CH	88	2/98	98
5	(E)-CH₃CH=CH	85	2/98	97
6	(E,E)-CH₃CH=CHCH=CH	83	2/98	95

in a 69% yield with *anti* preference. The enantiomeric excesses of *syn* and *anti* aldols are 30% and 97%, respectively. Examination of several chiral diamines improves the diastereoselectivity, and when (S)-1-ethyl-2-[(piperidin-1-yl)-methyl]pyrrolidine (**2**) is employed, the aldol adduct is obtained in a 54% yield with excellent diastereo- and enantioselectivities. Furthermore, the yield is improved without any loss of the stereoselectivity by the combination of tin(II) triflate, **2**, and Bu₂Sn(OAc)₂.

The results employing several kinds of aldehydes such as aromatic, aliphatic, α,β-unsaturated aldehydes and a dienal of the present asymmetric aldol reaction are shown in Table 7.6.[37] In all cases, *anti*-α,β-dihydroxy thioesters are obtained in high yields with excellent diastereo- and enantioselectivities. The aldol adducts thus obtained, optically active *anti*-α,β-dihydroxy ester derivatives, are generally difficult to prepare by the conventional asymmetric oxidative procedure because starting materials, *cis*-α,β-unsaturated ester equivalents, are sometimes difficult to obtain. Moreover, a consideration of the mechanistic model of the asymmetric dioxyosmylation was recently reported and it was shown that preparation of *anti*-1,2-diols in high enantiomeric excesses is difficult.[38] According to the present aldol methodology, two hydroxyl groups can be introduced in 1,2-*trans* position stereoselectively during new carbon–carbon bond formation.

In contrast to the fact the aldol reactions of the silyl enol ether derived from *S*-ethyl propanethioate (**8Z**) with aldehydes using the above chiral promoter proceed with *syn* preference in excellent diastereo- and enantioselectivities (see Table 7.4), excellent *anti* selectivities have been achieved in reactions of (Z)-2-benzyloxy-1-*S*-ethyl-1-trimethylsiloxyethene (**9Z**) with aldehydes. The consideration of the transition states of these aldol reactions has led to the assumption that (i) co-ordination of the oxygen atom of the α-benzyloxy group of silyl enol ether **9Z** to the tin(II) atom of tin(II) triflate is essential in

Table 7.7. Asymmetric synthesis of *syn*-1,2-diol derivatives

$$\text{RCHO} + \underset{\text{TBSO}}{\overset{\text{OSiMe}_3}{\diagdown}}\text{SEt} \xrightarrow[\text{CH}_2\text{Cl}_2, -78\,°\text{C}]{\text{Sn(OTf)}_2 + \mathbf{3} + \text{Bu}_2\text{Sn(OAc)}_2} \underset{\text{OTBS}}{\overset{\text{HO}\quad\text{O}}{R\diagup\diagdown}}\text{SEt} + \underset{\text{OTBS}}{\overset{\text{HO}\quad\text{O}}{R\diagup\diagdown}}\text{SEt}$$

Entry	R	Yield (%)	syn/anti	ee (%) (anti)
1	Ph	86	88/12	90
2	CH_3CH_2	46	92/8	82
3	(furyl)	93	94/6	93
4	(E)-PhCH=CH	76	90/10	92
5	(E)-CH₃CH=CH	75	97/3	94
6	(E,E)-CH₃CH=CHCH=CH	83	93/7	94

the *anti* selective transition state, leading to the different course in the diastereofacial selectivity compared with that of the *syn* selective reaction of **8Z**, (ii) *syn* α,β-dihydroxy thioesters would be formed when this co-ordination is restrained.

According to this hypothesis, the *t*-butyldimethylsilyl group has been chosen as a sterically hindered functional group, which would forbid the co-ordination of the oxygen atom to tin(II) atom, and (Z)-2-*t*-butyldimethyl-siloxy-1-*S*-ethyl-1-trimethylsiloxyethene (**10Z**) has been prepared. As expected, the *syn* aldol adduct has been obtained under the same reaction conditions. In the presence of tin(II) triflate, chiral diamine (*S*)-1-propyl-2-[(piperidin-1-yl)methyl]pyrrolidine (**3**), and Bu₂Sn(OAc)₂, the reaction of **9Z** with benzaldehyde proceeds smoothly to give the corresponding aldol adduct in high yield with high *syn* selectivity.

Several examples of the *syn* selective aldol reactions are shown in Table 7.7. In all cases, the reactions proceed smoothly to afford the aldol adducts in good yields with very high *syn* selectivities, and the enantiomeric excesses of these *syn* isomers are more than 90% in most cases.[31,39]

Now it becomes possible to control the enantiofacial selectivity of the silyl enol ethers derived from α-alkoxy thioesters **9Z** and **10Z** by choosing the appropriate protective groups of the alkoxy parts of the silyl enol ethers, and both diastereomers of optically active α,β-dihydroxy thioesters can be synthesized.

It is also possible to synthesize both enantiomers including 1,2-diol units with perfect stereochemical control by using similar types of chiral sources, based on Chiral Lewis Acid-Controlled Synthesis (CLAC Synthesis).[40-43] The key is to use two new types of chiral diamines, **6** and **7**. In the presence of tin(II) triflate, chiral diamine **6**, and Bu₂Sn(OAc)₂, (Z)-2-(*t*-butyldimethyl-siloxy)-1-ethylthio-1-trimethylsiloxyethene (**10Z**) reacts with aldehydes to afford the desired aldol adducts with a *2S, 3R* configuration. On the other hand, when a similar chiral diamine, **7**, is used, the reaction also proceeds smoothly, but the absolute configuration of the adducts is the reverse (*2R, 3S*). In both cases, the selectivities are very high; almost perfect *syn*-selectivities and more than 98% enantiomeric excesses are obtained. Diamines **6** and **7** are readily prepared from *S*-proline and the absolute configurations of the 2-position were *S* in both cases. The difference is the position of the benzene ring connected to the pyrrolidine moiety. It is exciting that the slight difference in the structure of the chiral sources completely reverses the enantiofacial selectivities.

Protocol 3.
Synthesis of both enantiomers of 1,2-diol units by choosing similar chiral ligands in asymmetric aldol reactions of the silyl enol ethers with aldehydes (chiral Lewis acid-controlled synthesis)

Equipment
- Two-necked, round-bottomed flask (30 mL) fitted with a magnetic stirring bar. A three-way stopcock is fitted to the top of the flask and connected to a vacuum/argon source
- Vacuum/inert gas source (argon source may be an argon balloon)

Materials
- Tin(II) triflate (FW 416.82), 166.7 mg, 0.40 mmol
- Distilled chiral diamine **6** (FW 216.33), 103.8 mg, 0.48 mmol
- Distilled chiral diamine **7** (FW 216.33), 103.8 mg, 0.48 mmol
- Distilled $Bu_2Sn(OAc)_2$ (FW 351.03), 154.5 mg, 0.44 mmol
- Distilled benzaldehyde (FW 106.12), 38.2 mg, 0.36 mmol
- Distilled (*Z*)-2-(*t*-butyldimethylsiloxy)-1-ethylthio-1-trimethylsiloxyethene (**10Z**) (FW 278.53), 111.4 mg, 0.40 mmol
- Dry, distilled dichloromethane

1. Flame dry the reaction vessel with a stirring bar under dry argon.

2. Add tin(II) triflate to the flask in dry box under argon atmosphere and dry at 100°C for 1 h under reduced pressure (0.5 mmHg).

3. Add 1 mL of dry dichloromethane to the flask with stirring. Add chiral diamine **6** or **7** in 0.5 mL of dichloromethane and then $Bu_2Sn(OAc)_2$ in 0.5 mL of dichloromethane.

4. After stirring the mixture for 5 min at room temperature, cool to −78°C.

5. Add (*Z*)-2-(*t*-butyldimethylsiloxy)-1-ethylthio-1-trimethylsiloxyethene (**10Z**) in 0.5 ML of dichloromethane and benzaldehyde in 0.5 mL of dichloromethane successively.

6. After stirring for 20 h, add saturated sodium hydrogencarbonate. Warm the mixture to the room temperature and filter the suspension through a pad of

Celite, and wash the filter cake three times with dichloromethane. Extract the aqueous layer three times with dichloromethane and wash the combined extracts with brine.

7. Concentrate the crude product under reduced pressure and apply to preparative TLC (silica gel). Elute with a mixed solvent of hexane–ethyl acetate (4:1) to obtain the product. Only *syn* isomer is produced and the optical purity is determined by HPLC analysis using a chiral column.

2.5 Catalytic asymmetric aldol reactions

As shown in the previous sections, highly diastereo- and enantioselective aldol reactions of silyl enol ethers with aldehydes using the novel promoter

Table 7.8. Synthesis of both enantiomers

R	Chiral diamine	Yield (%)	2S,3R/2R,3S	syn/anti	ee (%)[a]
Ph	6	86	98/2	99.0/1.0	(98)
C$_2$H$_5$	6	61	>99/1	99.0/1.0	(98)
(propenyl)	6	86	>99/1	99.0/1.0	(98)
Ph(cinnamyl)	6	84	>99/1	99.5/0.5	(99)
(furyl)	6	94	>99/1	98.5/1.5[b]	(97)
(pentadienyl)	6	86	98/2	98.0/2.0	(96)
Ph	7	82	99/1	1.0/99.0	(98)
C$_2$H$_5$	7	63	>99/1	1.0/99.0	(98)
(propenyl)	7	81	>99/1	1.0/99.0	(98)
Ph(cinnamyl)	7	80	99/1	1.0/99.0	(98)
(furyl)	7	86	>99/1	<0.5/>99.5[b]	(>99)
(pentadienyl)	7	78	>99/1	<0.5/>99.5	(>99)

[a] Enantiomeric excesses of *syn*-adducts.
[b] 2S,3S/2R,3R.

system, combined use of stoichiometric amounts of tin(II) triflate, a chiral diamine, and a tin(IV) compound (Bu_3SnF or $Bu_2Sn(OAc)_2$), have been developed. According to these reactions, optically active aldol adducts are easily prepared from both achiral aldehydes and silyl enol ethers, but stoichiometric use of the chiral source still remains a problem in terms of practical use. Based on the investigations to characterize the above promoter system as well as to clarify the mechanism of these reactions toward a truly catalytic aldol process,[44] the following catalytic cycle is postulated (Scheme 7.4).

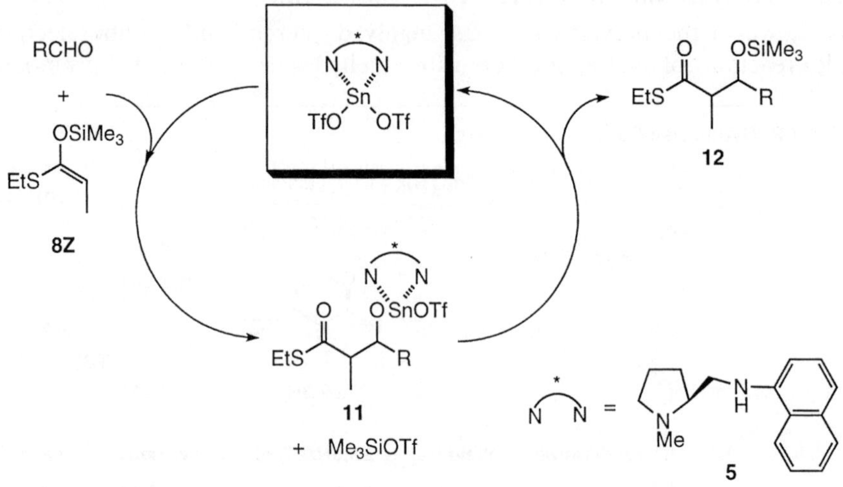

Scheme 7.4. The catalytic cycle of the asymmetric aldol reaction

A chiral diamine co-ordinated tin(II) triflate (chiral tin(II) Lewis acid) interacts with an aldehyde, and tin(II) alkoxide **11** and trimethylsilyl triflate (TMSOTf) are initially produced by the attack of silyl enol ether **8Z** to the activated aldehyde. When the metal exchange between tin(II) and silicon of the above product **11** takes place smoothly, the corresponding aldol adduct can be obtained as its trimethylsilyl ether **12** together with the regeneration of the catalyst. If the above-mentioned metal exchange step is slow, undesirable TMSOTf-promoted reaction[45,46] (to afford the achiral aldol adduct) proceeds to result in lowering the selectivity.

On the basis of these considerations, a slow addition of the substrates to the solution of the catalyst has been performed to keep the concentration of TMSOTf as low as possible during the reaction. A dichloromethane solution of silyl enol ether **8Z** and an aldehyde is added for over 9 h to a dichloromethane solution of the catalyst (20 mol%) (Table 7.9).[47] As expected, aldol adducts are obtained in good yields with excellent enantiomeric excesses and high diastereoselectivities in some cases, but the selectivities are not so high

Table 7.9. Catalytic asymmetric aldol reactions (1) (solvent : CH_2Cl_2)

Entry	R	Yield (%)	syn/anti	ee (%) (anti)
1	Ph	86	93/7	91
2	p-Cl Ph	80	93/7	93
3	p-CH$_3$Ph	82	78/22	80
4	CH$_3$(CH$_2$)$_6$	75	100/0	>98
5	c-C$_5$H$_{11}$	31	>99/1	74
6	(E)-CH$_3$CH=CH	51	84/16	77
7	(E)-CH$_3$(CH$_2$)$_2$CH=CH	62	81/19	74

when p-tolualdehyde, α,β-unsaturated aldehydes, and cyclohexanecarbox-aldehyde are used.

The lower selectivities are considered to be ascribed to the incompleteness of the above catalytic cycle, especially the metal exchange reaction of the initially formed aldol adduct **11** with TMSOTf (metal exchange between tin(II) and silicon). In order to accelerate this metal exchange step, various polar solvents with low melting points (below −78°C) have been carefully examined by taking the reaction of **8Z** with benzaldehyde as a model, and finally propionitrile (C_2H_5CN) has been found to be an excellent solvent.[48] The examination of the addition time (addition of the reactants to the solution of the catalyst) reveals that the rate of the metal exchange in propionitrile is faster than that in dichloromethane. Although 9 h (addition time) is necessary to attain the best result in dichloromethane, comparable selectivities are achieved when the substrates are added to the catalyst for 3 h in propionitrile. It is noted that a tin(II) triflate is more soluble in propionitrile than in dichloromethane indicating that the co-ordination of the nitrile group to tin(II) is expected to be rather strong, but that the ligand exchange of the nitrile for the diamine takes place smoothly to form the desired chiral Lewis acid when the chiral diamine is added to this propionitrile solution of tin(II) triflate. Although tin(II) triflate is also soluble in tetrahydrofuran (THF) or 1,2-dimethoxyethane (DME), the above aldol reaction does not proceed at −78°C after the addition of chiral diamine **5** in these solvents.

Several aldehydes including aromatic, aliphatic, and α,β-unsaturated aldehydes, are applicable to this reaction, and the desired products are obtained in good yields with high selectivities (>90% ee, Table 7.10). In particular, the lower yields or selectivities observed in the reaction of p-tolualdehyde, (E)-crotonaldehyde, (E)-2-hexenal, and cyclohexanecarboxaldehyde, are remarkably improved by using propionitrile as a solvent. High selectivities are also achieved even when 10 mol% of the catalyst was used.[49]

The key step in the asymmetric aldol reactions using the chiral tin(II) Lewis acid is believed to be the metal exchange process from tin(II) to silicon on the initially produced aldol adduct **11**. If this step is slow, the TIMSOTf-promoted aldol reaction of aldehydes with silyl enol ethers proceed to result in lower

Shū Kobayashi

Table 7.10. Catalytic asymmetric aldol reactions (2) (solvent : C_2H_5CN)

Entry	R	Addition time (h)	Yield (%)	syn/anti	ee (%) (syn)
1	Ph	3	77	92/8	90
2	p-Cl Ph	4.5	83	87/13	90
3	p-CH₃Ph	3	75	89/11	91
4	CH₃(CH₂)₆	4.5	80	100/0	>98
5	c-C₅H₁₁	3	71	100.0	>98
6	(E)-CH₃CH=CH	3	76	96/4	93
7	(E)-CH₃(CH₂)₂CH=CH	3	73	97/3	93

selectivities. Slow addition of the substrates to the catalyst in propionitrile has been successfully performed to suppress the undesirable reaction. It is also expected that higher stereoselectivities would be obtained when the Lewis acidity of TMSOTf is reduced by using an additive. This idea has been obtained from the recent report on the catalytic asymmetric aldol reaction using TMSOTf, a chiral diamine, and tin(II) oxide.[50] In this reaction, the lone-pair electrons of tin(II) oxide interact with TMSOTf to weaken the Lewis acidity of TMSOTf, and as a result high selectivities have been obtained. It would be possible to reduce the Lewis acidity of various triflates by using this interaction. On the basis of this idea, it has been found that in the presence of a novel chiral catalyst system consisting of tin(II) triflate, a chiral diamine, and tin(II) oxide, highly enantioselective aldol reactions of the silyl enol ether of S-ethyl ethanethioate or S-ethyl propanethioate with aldehydes proceed smoothly to afford the aldol adducts in high yields.

Protocol 4.
Catalytic asymmetric aldol reactions of the silyl enol ether derived from (Z)-1-S-ethyl-1-trimethylsiloxypropene with an aldehyde

Equipment

- Two-necked, round-bottomed flask (30 mL) fitted with a magnetic stirring bar. A three-way stopcock is fitted to the top of the flask and connected to a vacuum/argon source
- Vacuum/inert gas source (argon source may be an argon balloon)

Materials

- Tin(II) triflate (FW 416.82), 33.4 mg, 0.08 mmol
- Tin(II) oxide (FW 134.71), 21.6 mg, 0.16 mmol
- Distilled chiral diamine 5 (FW 240.35), 23.1 mg, 0.096 mmol
- Distilled benzaldehyde (FW 106.12), 42.5 mg, 0.40 mmol
- Distilled (Z)-1-S-ethyl-1-trimethylsiloxypropene (FW 190.38), 76.2 mg, 0.40 mmol
- Dry, distilled propionitrile

1. Flame dry the reaction vessel with a stirring bar under dry argon.

2. Add tin(II) triflate and tin(II) oxide to the flask in dry box under argon atmosphere and dry at 100 °C for 1 h under reduced pressure (0.5 mmHg).

3. Add 1 mL of dry propionitrile to the flask with stirring. Add chiral diamine **4** in 0.5 mL of propionitrile.

4. Ater stirring the mixture for 5 min at room temperature, cool to −78°C.

5. Slowly add a mixture of (Z)-1-S-ethyl-1-trimethylsiloxypropene in 0.5 mL of propionitrile and benzaldehyde in 0.5 mL of propionitrile over 4 h (syringe pump).

6. After stirring for 1 h (after addition), add saturated sodium hydrogen carbonate. Warm the mixture to the room temperature and filter the suspension through a pad of Celite, and wash the filter cake three times with dichloromethane. Extract the aqueous layer three times with dichloromethane and wash the combined extracts with brine. Dry the combined organic layer over (Na_2SO_4).

7. Concentrate the crude product under reduced pressure and apply to preparative TLC (silica gel). Elute with a mixed solvent of hexane–ethyl acetate (6:1) to obtain the product. Only *syn* isomer is produced and the optical purity is determined by HPLC analysis using a chiral column. After a usual work-up, the aldol-type adduct was isolated as the corresponding trimethylsilyl ether. The enantiomeric excess of this aldol adduct was determined after converting it into the corresponding alcohol (tetrahydrofuran–1M HCl (20:1) at 0°C) to be 94% by HPLC analysis using a chiral stationary phase column.

3. Asymmetric cyanation reaction

Asymmetric cyanation reaction of aldehydes is an important process in organic synthesis and recently some excellent works have been made in this

Table 7.11. Catalytic asymmetric aldol reactions using a novel catalyst system

Entry	R^1	R^2	With SnO			Without SnO		
			Yield (%)	*syn/anti*	ee (%)	Yield (%)	*syn/anti*	ee (%)
1	Ph	Me	78	95/5	93	77	93/7	90
2	$CH_3(CH_2)_6$	Me	81	100/0	>98	80	100/0	>98
3	$CH_3CH=CH$	Me	85	99/1	95	76	96/4	93
4	$CH_3(CH_2)_2CH=CH$	Me	81	100/0	94	73	97/3	93
5	$CH_3(CH_2)_3$	H	71	—	92	79	—	91
6	*i*-Pr	H	50	—	92	48	—	90
7	*c*-C_6H_{11}	H	78	—	94	81	—	92
8	$CH_3(CH_2)_2CH=CH$	H	83	—	84	65	—	72

area,[51-55] however, some significant problems still remain such as stoichiometric use of chiral sources, lower enantioselectivities in the reaction with aliphatic aldehydes, etc.

For the synthesis of cyanohydrin trimethylsilyl ethers,[56,57] one of the most convenient preparative methods is addition reactions of trimethylsilyl cyanide (TMS-CN) with aldehydes under the influence of Lewis acids such as zinc iodide (ZnI_2), aluminium chloride ($AlCl_3$), or under almost neutral conditions using anionic catalysis.[58,59] It has been found that in the presence of a catalytic amount of Lewis base such as amine, phosphine, arsine, or antimony, TMS-CN smoothly reacts with aldehydes to afford the corresponding cyanohydrin trimethylsilyl ethers in excellent yields.

Asymmetric version of this reaction is investigated. In the presence of a catalytic amount of (+)-cinchonine (10 mol%), cyclohexanecarboxyaldehyde reacts with TMS-CN smoothly at −78°C to give the corresponding cyanohydrin trimethylsilyl ether in a 94% yield with 25% ee. (+)-Cinchonine trimethylsilyl ether also works with the same level of ee (25%). The importance of the silyl ether is revealed as almost no chiral induction is observed when (+)-cinchonine acetate or benzoate is used as a catalyst. The proposed mechanism is that the tertiary amine part of (+)-cinchonine interacts with the silicon atoms of TMS-CN to form the pentavalent silicate and, at the same time, the Lewis acidic silicon atom of the silyl ether activates an aldehyde. This 'double activation' would provide a desirable transition state in the chiral induction using chiral catalysts. Based on this hypothesis, high levels of enantiomeric excesses are obtained when tin(II) ether is introduced instead of the silyl ether and a novel tin(II) Lewis acid, tin(II) monoalchoxymonotriflate containing (+)-cinchonine as a chiral source, has been designed. This chiral tin(II) Lewis acid is easily prepared from 1,1′-dimethylstannocene,[60] triflic acid, and (+)-cinchonine. In the presence of this Lewis acid, the reactions of TMS-CN with aldehydes proceed smoothly at −78°C in dichloromethane to give the corresponding cyanohydrin trimethylsilyl ether in high yields with good to excellent ees (Scheme 7.5). In the present reaction, the products are

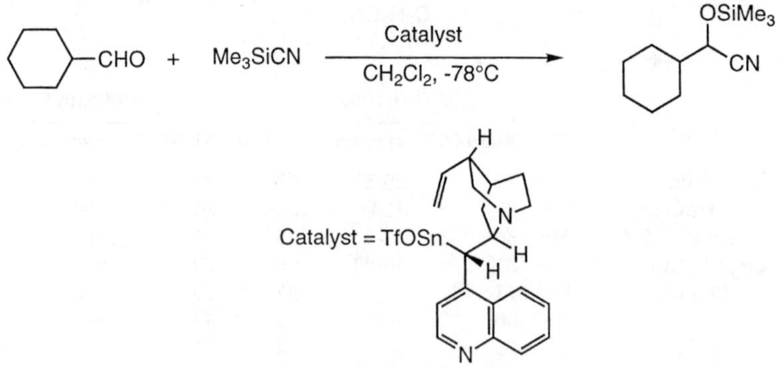

Scheme 7.5

152

isolated as trimethylsilyl ether form and the reaction smoothly proceeds in the presence of 30 mol% of the tin(II) Lewis acid. The catalyst, tin(II) mono-alkoxymonotriflate, is supposed to be regenerated from the initially produced tin(II) alkoxide and trimethylsilyl triflate.

4. Asymmetric protonation

Enantioselective protonation of prochiral enol derivatives provides a convenient route for the preparation of optically active carbonyl compounds.[61–89] Several examples of protonation of metal enolates by chiral proton sources[64–79] and hydrolysis of enol esters catalysed by enzymes[80,81] or antibodies[82] under basic or neutral conditions, have been reported. The acid-promoted hydrolysis of enol ethers is an interesting alternative which has been little investigated for enantioselectivity.[83–86] A new Lewis acid-assisted chiral Brønsted acid (LBA)[87–89] for enantioselective protonation of prochiral silyl enol esters and ketene bis(trialkylsilyl) acetals is reported.[90]

The LBA (12) is generated *in situ* from (R)-(+)-1,1'-binaphthol (BINOL) and tin tetrachloride (SnCl₄) in toluene or dichloromethane at −78 °C. The protons in (R)-BINOL are activated by the co-ordination of tin tetrachloride. In the presence of a stoichiometric amount of 12 in dichloromethane, the C-protonation of the trimethylsilyl enol ether derived from 2-phenylcyclo-hexanone proceeds even at −78 °C to from (S)-2-phenylcyclohexanone with good enantioselectivity (79% ee). The enantioselectivity of the reaction is dramatically increased by using sterically bulky O-substituents. The protonation of the *tert*-butyldimethylsilyl enol ether (Ar = Ph, SiR₃ = Si'BuMe₂) occurs with excellent enantioselectivity (93% ee), and the best result (96% ee) is achieved by using toluene as solvent. (R)-BINOL is converted into a mixture of 2,2'-disiloxy-1,1'-binaphthyl and 2-siloxy-2'-hydroxy-1,1'-binaphthyl in dichloromethane solution, whereas it is only converted into the latter product in toluene. It is noted that the enantioselective protonation proceeded with a catalytic amount (10 mol %) of tin tetrachloride.

The enantioselective protonation of a variety of silyl enol ethers derived from 2-arylcyclohexanones with 12 under optimum conditions is summarized in Table 7.12. The reactions are generally complete after 1 h at −78 °C. Excellent enantioselectivity was achieved in the reactions of the silyl enol ethers of 2-arylcyclohexanones, and both enantiomers can be obtained from racemic 2-arycyclohexanones depending on the choice of optically active BINOL : (S)- and (R)-arylcyclohexanones are obtained using (R)- and (S)-12, respectively. Enantioselectivity is low in the case of the protonation of the silyl enol ether derived from 2-methylcycloxanone (42% ee).

Representative results of application of 12 to enantioselective protonation of ketene bis(trialkylsilyl) acetals derived from 2-arylcarboxylic acids are summarized in Table 7.13. The crude carboxylic acids, which are formed with excellent enantiomeric excesses, are isolated as the corresponding methyl

Table 7.12. The enantioselective protonation of silyl enol ethers from 2-arylcyclohexanones with **12**

AR	ee (%)
Ph	96
p-MePh	93
p-MeOPh	85
2-naphthyl	91

Table 7.13. Enantioselective protonation of ketene bis(trialkyl silyl) acetals derived from 2-arylcarboxylic acids

R¹	R²	SiR₃	ee (%)	config.
Ph	Me	SiMe₃	92	S
Ph	Me	SiEt₃	95	S
Ph	Et	SiMe₃	60	S
p-(*i*Bu)C₆H₄	Et	SiMe₃	94	S
(MeO-naphthyl)	Me	SiMe₃	92 (98)[a]	S
Ph	OMe	SiEt₃	87	S

[a] After recrystallization from dichloromethane-hexane.

esters in quantitative yields. The enantioselectivity is independent of the steric features of trialkylsilyl substituents. Simple recrystallization of the (*S*)-methyl ester of naproxen can be used to upgrade the optical purity. It is noteworthy that the protonation of the ketene silyl acetal, diastereomer ratio (*E* and *Z*) = 63:37, derived from methyl 6-methoxy-α-methyl-2-naphthalene acetate and trimethylsilyl chloride by (*R*)-**12** gives the (*S*)-methyl ester of naproxen with decreased enantioselectivity (79% ee).

References

1. Poller, R. C. Organic compounds of group IV metals. In *Comprehensive organic chemistry*, Vol. 3 (ed. D. N. Jones). Pergamon Press, Oxford, **1979**, p 1073.
2. Davies, A. G.; Smith, P. J. Tin. In *Comprehensive organometallic chemistry*, Vol. 2 (ed. G. Wilkinson). Pergamon Press, Oxford, **1982**, p 519.
3. Pereyre, M.; Quintard, J.-P.; Rahm, A. *Tin in organic synthesis*, Butterworths, London, **1987**.
4. Mukaiyama, T. *Pure Appl. Chem.* **1986**, *58*, 505.
5. Huheey, J. E. *Inorganic chemistry*, 3rd edn. Harper & Row, New York, **1983**.
6. Veith, M.; Recktenwald, O. *Top. Curr. Chem.* **1982**, *104*, 1.
7. Goldberg, D. E. *J. Chem. Soc., Chem. Commun.* **1976**, 261.
8. Lister, M. W.; Sutton, L. E. *Trans. Far. Soc.* **1941**, *37*, 406.
9. Akshin, P. A.; Spiridonov, V. P.; Khodschenko, A. N. *Zh. Fiz. Khim.* **1958**, *32*, 1679.
10. Fuji, H.; Kimura, M. *Bull. Chem. Soc. Jpn* **1970**, *43*, 1933.
11. Livingstone, R. L.; Rao, C. N. R. *J. Chem. Phys.* **1959**, *30*, 399.
12. Lister, M. W.; Sutton, L. E. *Trans. Far. Soc.* **1941**, *37*, 393.
13. Dean, J. A. *Lange's handbook of chemistry*, 3rd edn, McGraw-Hill, New York, **1985**, pp 3–121.
14. Donaldson, J. D. *Progress in Inorganic Chemistry*, (ed. F. A. Cotton, Vol. 8. John Wiley, New York, **1967**, p 287.
15. Shields, K. G.; Seccombe, R. C.; Kennard, C. H. L. *J. Chem. Soc., Dalton* **1973**, 741.
16. Kobayashi, S.; Mukaiyama, T. *Chem. Lett.* **1989**, 297.
17. Kobayashi, S.; Uchiro, H.; Fujishita, Y.; Shiina, I.; Mukaiyama, T. *J. Am. Chem. Soc.* **1991**, *113*, 4247.
18. Mukaiyama, T.; Kobayashi, S. *J. Organomet. Chem.*, **1990**, *382*, 39.
19. Mukaiyama, T.; Uchiro, H.; Kobayashi, S. *Chem. Lett.* **1989**, 1757.
20. Mukaiyama, T.; Uchiro, H.; Kobayashi, S. *Chem. Lett.* **1989**, 1001.
21. Ojima, I.; Yoshida, K.; and Inaba, S. *Chem. Lett.* **1977**, 429.
22. Mioskowski, C.; Solladie, G. *Tetrahedron* **1980**, *36*, 227.
23. Stevens, R. W.; Mukaiyama, T. *Chem. Lett.* **1983**, 1799.
24. Kobayashi, S.; Fujishita, Y.; Mukaiyama, T. *Chem. Lett.* **1989**, 2069.
25. Robins, D. J. *Fortschr. Chem. Org. Naturst.* **1982**, *41*, 115.
26. Mulzer, J.; Kirstein, H. M.; Buschmann, J.; Lehmann, C.; Luger, P. *J. Am. Chem. Soc.* **1991**, *13*, 910.
27. Hirota, M.; Sakurai, Y.; Horita, K.; Yonemitsu, O. *Tetrahedron Lett.* **1990**, *31*, 6367.
28. White, J. D.; Jayasinghe, L. R. *Tetrahedron Lett.* **1988**, *29*, 2139.
29. Niwa, H.; Miyachi, Y.; Uosaki, T.; Yamada, K. *Tetrahedron Lett.* **1986**, *27*, 4601.
30. Narasaka, K.; Sakakura, T.; Uchimaru, T.; Guedin-Vuong, D. *J. Am. Chem. Soc.* **1984**, *106*, 2954.
31. Kobayashi, S.; Horibe, M.; Saito, Y. *Tetrahedron* **1994**, *50*, 9629.
32. Kobayashi, S.; Hachiya, I. *J. Org. Chem.* **1992**, *57*, 1324.
33. Tomioka, K.; Nakajima, M.; Koga, K. *J. Am. Chem. Soc.* **1987**, *109*, 6213.
34. Jacobsen, E. N.; Marco, I.; Mungall, W. S.; Schroder, G.; Sharpless, K. B. *J. Am. Chem. Soc.* **1988**, *110*, 1968.

35. Wai, J. S. M.; Marco, I.; Svendsen, S.; Finn, M. G.; Jacobsen, E. N.; Sharpless, K. B. *J. Am. Chem. Soc.* **1989**, *111*, 1123.
36. Johnson, R. A.; Sharpless, K. B. Catalytic asymmetric dihydroxylation. In *Catalytic asymmetric synthesis* (ed. I. Ojima). VCH, Weinheim, Germany, **1993**, pp 227–272.
37. Mukaiyama, T.; Uchiro, H.; Shiina, I.; Kobayashi, S. *Chem. Lett.* **1990**, 1019.
38. Corey, E. J.; Jardine, P. D.; Virgil, S.; Yuen, P.-W.; Connell, R. D. *J. Am. Chem. Soc.* **1989**, *111*, 9243.
39. Mukaiyama, T.; Shiina, I.; Kobayashi, S. *Chem. Lett.* **1991**, 1902.
40. Kobayashi, S.; Horibe, M. *J. Am. Chem. Soc.* **1994**, *116*, 9805.
41. Kobayashi, S.; Horibe, M. *Chem. Eur. J.* **1977**, *3*, 1472.
42. Kobayashi, S.; Hayashi, T. *J. Org. Chem.* **1995**, *60*, 1098.
43. Kobayashi, S.; Horibe, M.; Matsumura, M. *Synlett* **1995**, 675.
44. Kobayashi, S.; Harada, T.; Han, J. S. *Chem. Express* **1991**, *6*, 563.
45. Murata, S.; Suzuki, M.; Noyori, R. *Tetrahedron* **1988**, *44*, 4259.
46. Mukai, C.; Hashizume, S.; Nagami, K.; Hanaoka, M. *Chem. Pharm. Bull.* **1990**, *38*, 1509.
47. Mukaiyama, T.; Kobayashi, S.; Uchiro, H.; Shiina, I. *Chem. Lett.* **1990**, 129.
48. Kobayashi, S.; Fujishita, Y.; Mukaiyama, T. *Chem. Lett.* **1990**, 1455.
49. Kobayashi, S.; Uchiro, H.; Shiina, I.; Mukaiyama, T. *Tetrahedron* **1993**, *49*, 1761.
50. Mukaiyama, T.; Uchiro, H.; Kobayashi, S. *Chem. Lett.* **1990**, 1147.
51. Minamikawa, H.; Hayakawa, S.; Yamada, T.; Iwasawa, N.; Narasaka, K. *Bull. Chem. Soc. Jpn.* **1988**, *61*, 4379.
52. Tanaka, K.; Mori, A.; Inoue, S. *J. Org. Chem.* **1990**, *55*, 181.
53. Almsick, A.; Buddrus, J.; H-Schmidt, P.; Laumen, K.; Schneider, M. P. *J. Chem. Soc. Commun.* **1989**, 1391.
54. Ziegler, T.; Horsch, B.; Ellenberger, F. *Synthesis* **1990**, 575.
55. Hayashi, M.; Matsuda, T.; Oguni, N. *J. Chem. Soc., Chem. Commun.* **1990**, 1364.
56. Weber, W. P. *Silicon reagents for organic synthesis*. Springer-Verlag, Berlin, **1983**.
57. Colvin, E. W. *Silicon in organic synthesis*, Butterworths, London, **1981**.
58. Evans, D. A.; Truesdale, L. K. *Tetrahedron Lett.*, **1973**, 4929.
59. Evans, D. A.; Truesdale, L. K.; Carroll, G. L. *J. Chem. Soc., Chem. Commun.* **1973**, 55.
60. Mukaiyama, T.; Ichikawa, J.; Asami, M. *Chem. Lett.* **1983**, 293.
61. Duhamel, L.; Duhamel, P.; Launay, J.; Plaquevent, J. C. *Bull. Soc. Chim, Fr.* **1984**, II–421.
62. Fehr, C. *Chimia* **1991**, *45*, 253.
63. Waldmann, H. *Nachr. Chem. Tech. Lab.* **1991**, *39*, 413.
64. Duhamel, L.; Fouquay, S.; Plaquevent, J. C. *Tetrahedron Lett.* **1986**, *27*, 4975.
65. Hogeveen, H.; Eleveld, M. B. *Tetrahedron Lett.* **1986**, *27*, 631.
66. Gerlach, U.; Hunig, S. *Angew. Chem., Int. Ed. Engl.* **1987**, *26*, 1283.
67. Potin, D.; Williams, K.; Rebek, J.; Jr. *Angew. Chem., Int. Ed. Engl.* **1990**, *29*, 1420.
68. Kumar, A.; Salunkhe, R. V.; Rane, R. A.; Dike, S. Y. *J. Chem. Soc., Chem. Commun.* **1991**, 485.
69. Henin, F.; Muzart, J.; Pete, J-p.; Piva, O. *New J. Chem.* **1991**, *15*, 611.
70. Matsumoto, K.; Ohat, H. *Tetrahedron Lett.* **1991**, *32*, 4729.
71. Henin, F.; Muzart, J. *Tetrahedron: Asymmetry* **1992**, *3*, 1161.
72. Takeuchi, S.; Miyoshi, N.; Hirata, K.; Hayashida, H. Ohgo, Y. *Bull. Chem. Soc. Jpn.* **1992**, *65*, 2001.

73. Yasukata, T.; Koga, K. *Tetrahedron: Asymmetry* **1993**, *4*, 35.
74. Fuji, K.; Tanaka, K.; Miyamoto, H. *Tetrahedron: Asymmetry* **1993**, *4*, 247.
75. Haubenreich, T.; Hunig, S.; Schultz, H-J. *Angew. Chem., Int. Ed. Engl.* **1993**, *32*, 398.
76. Fehr, C.; Stempf, I.; Galindo, J. *Angew. Chem., Int. Ed. Engl.* **1993**, *32*, 1044.
77. Vedejs, E.; Lee, N.; Sakata, S. T. *J. Am. Chem. Soc.* **1994**, *116*, 2175.
78. Yanagisawa, A.; Kuribayashi, T.; Kikuchi, T.; Yamamoto, H. *Angew. Chem. Int. Ed. Engl.* **1994**, *33*, 107.
79. Aboulhoda, S. J.; Henin, F.; Muzart, J.; Thorey, C.; Behnen, W.; Martens, J.; Mehler, T. *Tetrahedron: Asymmetry* **1994**, *5*, 1321.
80. Matsumoto, K.; Tsutsumi, S.; Ihori, T.; Ohta, H. *J. Am. Chem. Soc.* **1990**, *112*, 9614.
81. Kume, Y.; Ohta, H. *Tetrahedron Lett.* **1992**, *33*, 6367.
82. Fuji, I.; Lerner, R. A.; Janda, K. D. *J. Am. Chem. Soc.* **1991**, *113*, 8528.
83. Cavelier, F.; Gomez, R.; Jacquier, R.; Verducci, J. *Tetrahedron: Asymmetry* **1993**, *4*, 2501.
84. Cavelier, F.; Gomez, R.; Jacquier, R.; Verducci, J. *Tetrahedron Lett.* **1994**, *35*, 2891.
85. Reymond, J.-L.; Reber, J.-L.; Lerner, R. A. *Angew. Chem., Int. Ed. Engl.* **1994**, *33*, 475.
86. Novice, M. H.; Seikaly, H. R.; Seiz, A. D.; Tidwell, T. T. *J. Am. Chem. Soc.* **1980**, *102*, 5835.
87. Ishihara, K.; Yamamoto, H. *J. Am. Chem. Soc.* **1994**, *116*, 1561.
88. Ishihara, K.; Miyata, M.; Hattori, K.; Tada, T.; Yamamoto, H. *J. Am. Chem. Soc.* **1994**, *116*, 10520.
89. Kobayashi, S.; Araki, M.; Hachiya, I. *J. Org. Chem.* **1994**, *59*, 3758.
90. Ishihara, K.; Kaneeda, M.; Yamamoto, H. *J. Am. Chem. Soc.* **1994**, *116*, 11179.

8

Silicon(IV) reagents

AKIRA HOSOMI and KATSUKIYO MIURA

1. Introduction

Organosilicon compounds are one of the most widely and frequently used organometallics in organic synthesis. They are easy to handle and synthesize due to their thermal and aerial stability, and exhibit a variety of reactivities derived from the electropositive nature of silicon as well as the steric and electronic effects of the silyl group.[1,2] Silicon reagents play important roles as not only protecting agents and masked nucleophiles but also Lewis acids. The Lewis acidity of alkyl- and alkoxysilanes is much weaker than that of the corresponding group III metal compounds. However, when one of the four ligands on silicon is changed to a soft Lewis base such as triflate (OTf) or iodide (I), the acidity becomes strong enough for synthetic use as Lewis acid catalyst.[3-8] Trimethylsilyl trifluoromethanesulfonate (TMSOTf) and iodotrimethylsilane (TMSI), bearing one Lewis-acidic co-ordination site, interact strongly to various heteroatoms, particularly oxygen, to activate the carbon–heteroatom bond. Although it is also known that five co-ordinated silicates and angle strained silanes have moderate Lewis acidity, these types of Lewis acids are only utilized to intramolecularly assist the addition of silylated nucleophiles as shown in the reactions of allylsilicates[9,10] and enoxysilacyclobutane.[11]

The synthetic reactions mediated by TMSOTf[3-5] and TMSI[6-8] have been extensively developed in the last two decades. The silicon-centred Lewis acids are frequently employed for the reactions of trimethylsilylated nucleophiles (TMSNu) with acetals ($R^1R^2C(OR^3)_2$) and carbonyl compounds ($R^1R^2C=O$), which usually proceed with a catalytic amount of TMSX (X = OTf, I) under mild conditions, forming C–heteroatom, C–C, and C–H bonds in high efficiency (Scheme 8.1). The plausible catalytic cycles A and B consist of three parts: polarization of the substrates by TMSX, nucleophilic attack of TMSNu to the resultant oxonium or carbenium ion intermediates, and dissociation into TMSX and the products. This chapter describes the synthetic utility of the TMSOTf- and TMSI-catalysed reactions.

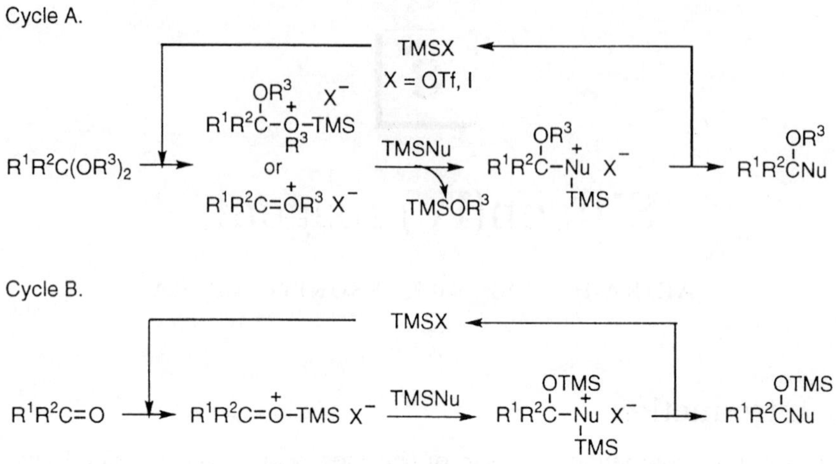

Cycle A.

$$R^1R^2C(OR^3)_2 \rightarrow \begin{matrix} OR^3 \\ R^1R^2C-O-TMS \quad X^- \\ R^3 \\ \text{or} \\ R^1R^2C=\overset{+}{O}R^3 \ X^- \end{matrix} \xrightarrow[\text{TMSOR}^3]{\text{TMSNu}} \begin{matrix} OR^3 \\ R^1R^2\overset{+}{C}-Nu \quad X^- \\ TMS \end{matrix} \rightarrow \begin{matrix} OR^3 \\ R^1R^2CNu \end{matrix}$$

TMSX, X = OTf, I

Cycle B.

$$R^1R^2C=O \rightarrow R^1R^2C=\overset{+}{O}-TMS \ X^- \xrightarrow{\text{TMSNu}} \begin{matrix} OTMS \\ R^1R^2\overset{+}{C}-Nu \quad X^- \\ TMS \end{matrix} \rightarrow \begin{matrix} OTMS \\ R^1R^2CNu \end{matrix}$$

TMSX

Scheme 8.1

2. Carbon–oxygen and carbon–nitrogen bond formation

Aldehydes and ketones readily react with two equivalents of alkyl silyl ethers in the presence of TMSOTf or TMSI to afford dilkyl acetals in good yields (Scheme 8.2).[12,13] The acid-catalysed reaction would initially form silyl alkyl acetals by nucleophilic addition of alkyl silyl ethers *via* the catalytic cycle B, and then, the acetals would turn dialkyl acetals by the successive nucleophilic substitution *via* the catalytic cycle A (Scheme 8.1). The driving force of the reaction is the great stability of hexamethyldisiloxane formed as a by-product. This method is particularly suited for the acetalization of unsaturated aldehydes and ketones without isomerization of the double bond.[14]

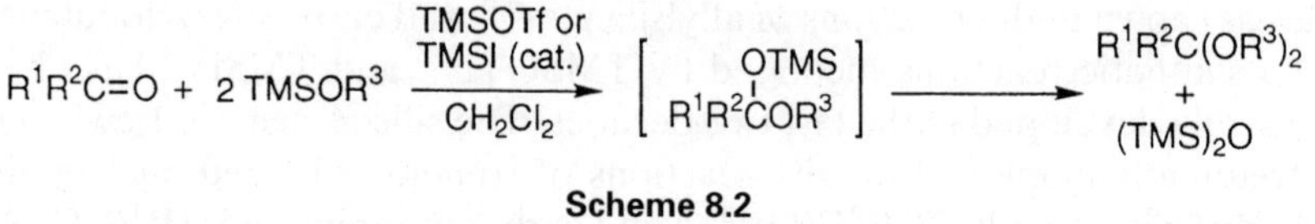

$$R^1R^2C=O + 2\,TMSOR^3 \xrightarrow[\text{CH}_2\text{Cl}_2]{\substack{\text{TMSOTf or} \\ \text{TMSI (cat.)}}} \left[\begin{matrix} OTMS \\ R^1R^2COR^3 \end{matrix} \right] \longrightarrow \begin{matrix} R^1R^2C(OR^3)_2 \\ + \\ (TMS)_2O \end{matrix}$$

Scheme 8.2

Transacetalization with alcohols or alkyl silyl ethers can also be induced by a catalytic or a stoichiometric amount of TMSOTf. The combination of TMSOTf with glycosyl acetates, trichloroacetimides, or phosphites is widely utilized as a tool of *O*-glycosylation for the synthesis of oligosaccharides (Scheme 8.3).[15,16]

The prominent ability of TMSOTf is also exerted in nucleoside synthesis. The oxocarbenium ions generated from glycosyl acetates react with silylated heterocyclic bases to afford nucleosides (Scheme 8.4).[17,18] In this case, an ex-

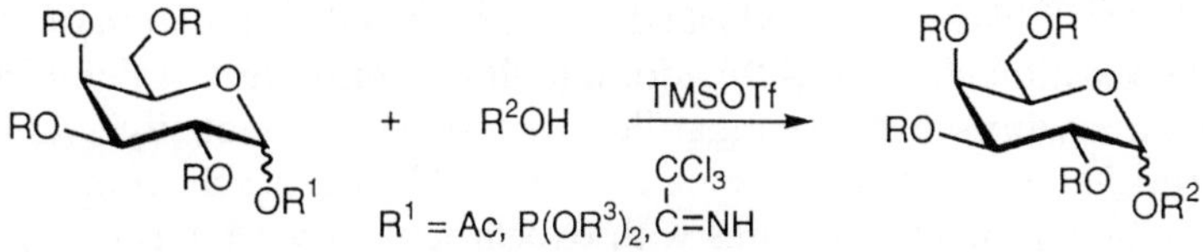

Scheme 8.3

cess amount of TMSOTf is required for high efficiency. The major advantage of TMSOTf compared to SnCl$_4$ or other Lewis acids is to interact reversibly with the heterocycles without their deactivation.

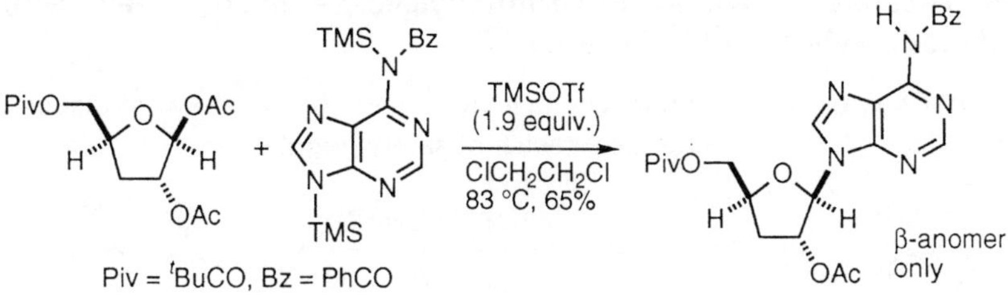

Scheme 8.4

3. Carbon–carbon bond formation

The Mukaiyama cross-aldol reaction is one of the most powerful and selective methods for the construction of carbon–carbon bonds. This well-established transformation involves the reaction of silyl enolates with carbonyl compounds or their derivatives in the presence of an activator such as Lewis acid or fluoride ion.[19,20] TMSOTf and TMSI can effectively promote this process with a catalytic quantity unlike TiCl$_4$, SnCl$_4$, and BF$_3$•Et$_2$O, which are required with a stoichiometric quantity. For instance, acetals and orthoesters react with silyl enolates in the presence of 1–10 mol% of TMSOTf or TMSI in CH$_2$Cl$_2$ at $-78\,°$C to give β-alkoxy carbonyl compounds in high yield (Scheme 8.5 and

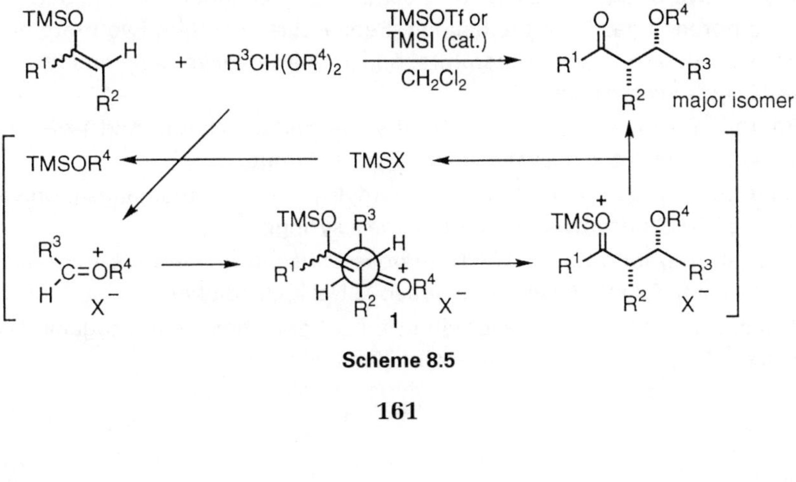

Scheme 8.5

161

Protocol 1).[21-23] The aldol products do not undergo β-elimination of alcohols under the reaction conditions. In addition, this aldol reaction exhibits moderate to high erythroselectivity regardless of the geometry of silyl enolate.[21,22] The stereochemical outcome can be rationalized by the acyclic transition state **1** in the reaction of silyl enolates with oxocarbenium ion intermediates. The aldol reaction of aldehydes and ketones with silyl enolates is also catalysed by TMSOTf,[24] although these electrophiles are less reactive than the corresponding acetals.

Protocol 1.
TMSOTf-catalysed reaction of 1-trimethylsiloxy-1-phenylethene with 1,1-dimethoxybutane[21,22] (Scheme 8.6)

Caution! Carry out all procedures in a well-ventilated hood, and wear disposable vinyl or latex gloves and chemical-resistant safety goggles.

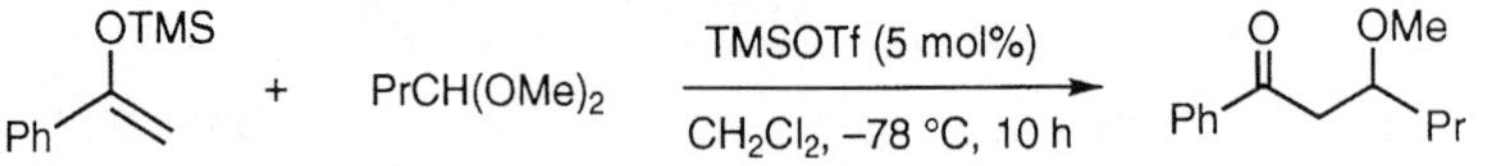

Scheme 8.6

Equipment

- A two-necked, round-bottomed flask (100 mL) fitted with a magnetic stirring bar, rubber septum, and three-way stopcock connected to a vacuum source and a nitrogen balloon

- Dry glass syringes (volume appropriate for quantity of solution to be transferred) with stainless-steel needles
- Dry ice–methanol bath

Materials

- 1,1-Dimethoxybutane[a] (FW 118.2), 1.30 g, 11 mmol — flammable, irritant
- 1-Trimethylsiloxy-1-phenylethene[b] (FW 192.3), 1.92 g, 2.05 mL, 10 mmol — moisture sensitive, irritant
- Trimethylsilyl trifluoromethanesulfonate[c] (FW 222.3), 1.11 g, 0.97 mL, 0.50 mmol — flammable, corrosive
- Freshly distilled (CaH_2) dichloromethane, 30 +5 mL — toxic, irritant

1. Dry the two-necked flask with an electric heat gun under vacuum, and introduce nitrogen gas from the balloon. Repeat this operation two more times. After cooling to room temperature, add 1,1-dimethoxybutane and 30 mL of dry CH_2Cl_2 *via* syringes.

2. Stir the mixture and cool to −78°C (dry ice–methanol bath). Add 1-trimethylsiloxy-1-phenylethene to the solution *via* a syringe.

3. After 10 min, add a solution of trimethylsilyl trifluoromethanesulfonate in 5 mL of CH_2Cl_2 dropwise over 5 min *via* a syringe.

4. After stirring for 10 h at −78°C, remove the septum and pour the reaction mixture into 50 mL of saturated aqueous $NaHCO_3$ solution.

5. Transfer the bilayer to a separating funnel and remove the organic layer. Extract the aqueous layer twice with 30 mL of CH_2Cl_2.

6. Pass the combined organic layer through a short K_2CO_3 column and concentrate *in vacuo*.

7. Purify the resulting oil by column chromatography on silica gel using hexane–ether (10:1). 3-Methoxy-1-phenylhexan-1-one (FW 206.3) is obtained in 75% yield (1.55 g).

[a] Prepared by acid-catalysed reaction of butanal with methanol.[25]
[b] Prepared according to a published procedure.[26] Commercially available from Aldrich.
[c] A colourless fluid liquid boiling at 45–47 °C at 17 mmHg. Commercially available from Aldrich. It can be easily prepared from trimethylsilyl chloride and trifluoromethane sulfonic acid.[27]

The Hosomi–Sakurai reaction, the Lewis acid- or fluoride ion-mediated allylation using allylsilanes, can transfer functionalized allyl groups onto various carbon electrophiles with high regio-, stereo-, and chemoselectivity.[29] This synthetically valuable transformation as well as the Mukaiyama reaction is effectively induced by TMSOTf and TMSI. In the presence of a catalytic amount of TMSOTf or TMSI, acetals react with allylsilanes at the γ-position to give homoallyl ethers in good yields (Scheme 8.7).[30–32] The use of chiral crotylsilane derivatives achieves high levels of acyclic stereocontrol.[33] Aldehydes and ketones can be directly led to homoallyl ethers by *in situ* acetalization with alkyl silyl ethers followed by allylation (Protocol 2).[13] This consecutive reaction is applicable to asymmetric synthesis of homoallyl alcohols using homochiral alkyl silyl ethers.[34]

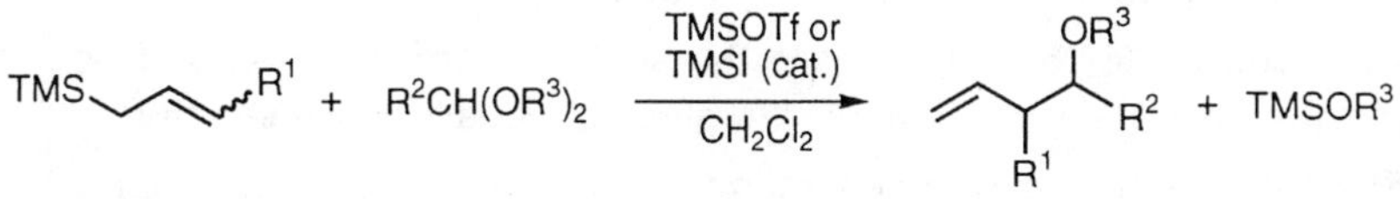

Scheme 8.7

Although TMSOTf and TMSI effect the allylation of aldehydes, their catalytic activities are considerably low. However, TMSB(OTf)$_4$, a supersilylating agent, is an excellent catalyst for the allylation.[35]

Protocol 2.
TMSI-catalysed consecutive acetalization and allylation of hexanal to 4-methoxy-1-nonene[13] (Scheme 8.8)

Caution! Carry out all procedures in a well-ventilated hood, and wear disposable vinyl or latex gloves and chemical-resistant safety goggles.

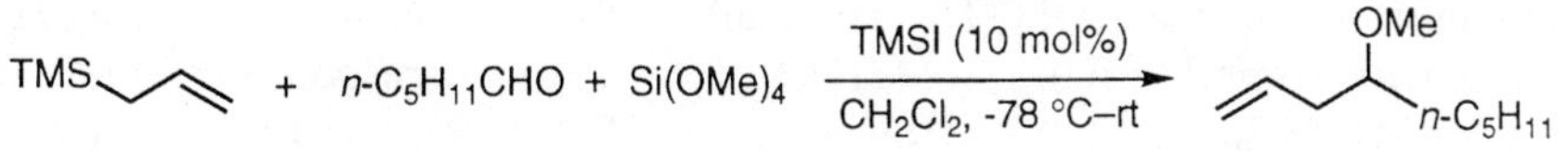

Scheme 8.8

163

Protocol 2. *Continued*

Equipment

- A two-necked, round-bottomed flask (100 mL) fitted with a magnetic stirring bar, rubber septum, and three-way stopcock connected to a vacuum source and a nitrogen balloon
- Dry glass syringes (volume appropriate for quantity of solution to be transferred) with stainless-steel needles
- Dry ice–methanol bath and ice–water bath

Materials

- Freshly distilled (CaCl$_2$) hexanal (FW 100.2), 1.00 g, 1.20 mL, 10 mmol — **flammable, irritant**
- Freshly distilled (CaH$_2$) dichloromethane, 30 mL — **toxic, irritant**
- Tetramethoxysilane (FW 152.2), 1.67 g, 1.62 mL, 11 mmol — **flammable, corrosive**
- Allyltrimethylsilane (FW 114.3), 1.37 g, 1.91 mL, 12 mmol — **flammable, irritant**
- Iodotrimethylsilane[a] (FW 200.1), 0.20 g, 0.14 mL, 1.0 mmol — **flammable, corrosive**
- Pyridine, 0.5 mL — **flammable, toxic**

1. Dry the two-necked flask with an electric heat gun under vacuum, and introduce nitrogen gas from the balloon. Repeat this operation two more times. After cooling to room temperature, add hexanal, dry CH$_2$Cl$_2$ tetramethoxysilane, and allyltrimethylsilane *via* syringes.

2. Stir the mixture and cool to −78 °C (dry ice–methanol bath).

3. Add iodotrimethylsilane to the solution *via* a syringe and stir for 15 min at −78 °C.

4. Then, warm the reaction vessel to 0 °C with an ice–water bath.

5. After stirring for 3 h, add pyridine *via* a syringe.

6. After stirring for 5 min, remove the septum from the flask and pour the reaction mixture into 50 mL of saturated aqueous NaHCO$_3$ solution.

7. Transfer the bilayer to a separating funnel and remove the organic layer. Extract the aqueous layer twice with 30 mL of Et$_2$O.

8. Pass the combined organic layer through a short Na$_2$SO$_4$ column and concentrate *in vacuo*.

9. Purify the resulting oil by column chromatography on silica gel using hexane–ether (20:1). 4-Methoxy-1-nonen (FW 156.3) is obtained in 94% yield (1.47 g).

[a] A colourless liquid boiling at 108 °C at 760 mmHg. Commercially available. It can be easily prepared from hexamethyldisilane and iodine in quantitative yield as shown in Protocol 4.[36]

4. Carbon–hydrogen bond formation

A catalytic amount of TMSOTf successfully induces the reductive cleavage of acetals with trimethylsilane (TMSH) to afford the corresponding ethers in high yields (Scheme 8.9).[37] However, the TMSOTf-and TMSI-catalysed reactions of aldehydes or ketones with TMSH do not give silyl ethers but

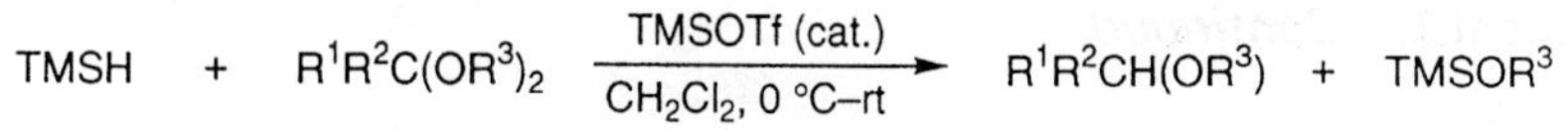
TMSH
+
R1R2C(OR3)2
TMSOTf (cat.)
CH2Cl2, 0 °C–rt
R1R2CH(OR3)
+
TMSOR3

Scheme 8.9

symmetrical ethers (Scheme 8.10).[38] This reductive condensation can be applied for the cyclization of 1,4- and 1,5-diketones to cyclic ethers.[39] The formation of the symmetrical ethers is probably because the intermediary silyl ether attacks the activated C=O bond faster than TMSH. Indeed, under the similar conditions, the reaction of ketones or aldehydes with alkyl silyl ethers results in the selective formation of unsymmetrical ethers (Protocol 3).[38]

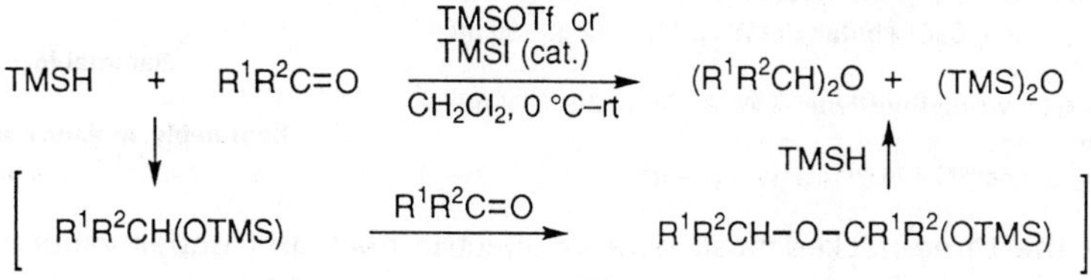
TMSH
+
R1R2C=O
TMSOTf or
TMSI (cat.)
CH2Cl2, 0 °C–rt
(R1R2CH)2O + (TMS)2O
TMSH
R1R2CH(OTMS)
R1R2C=O
R1R2CH–O–CR1R2(OTMS)

Scheme 8.10

The combination of TMSOTf with hydrosilanes is also effective for the reductive cyclization of hydroxyketones (Scheme 8.11).[40]

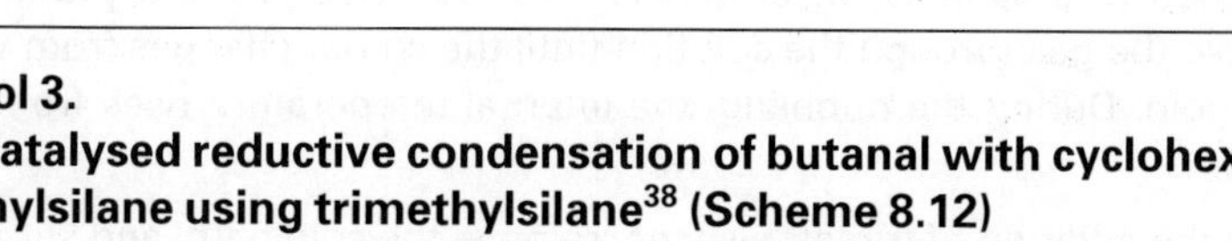
Me
O
O
H
Ph
TMSOTf (1.0 eq.),
Et3SiH (10 eq.)
CH2Cl2, 0 °C, 15 min, 85%
Me
O
Ph

Scheme 8.11

Protocol 3.
TMSI-catalysed reductive condensation of butanal with cyclohexyloxy-trimethylsilane using trimethylsilane[38] (Scheme 8.12)

Caution! Carry out all procedures in a well-ventilated hood, and wear disposable vinyl or latex gloves and chemical-resistant safety goggles.

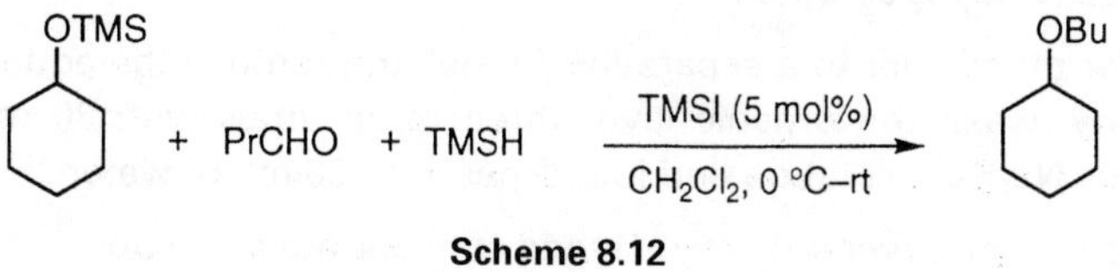
OTMS
+ PrCHO
+ TMSH
TMSI (5 mol%)
CH2Cl2, 0 °C–rt
OBu

Scheme 8.12

Akira Hosomi and Katsukiyo Miura

Protocol 3. *Continued*

Equipment

- A three-necked, round-bottomed flask (100 mL) fitted with a magnetic stirring bar, a rubber septum, thermometer, and three-way stopcock connected to a vacuum source and a nitrogen balloon
- A bubbler containing liquid paraffin
- Tygon tubing attached to a stainless-steel needle
- Dry glass syringes (volume appropriate for quantity of solution to be transferred) with stainless-steel needles
- Ice–water bath

Materials

- Iodine (FW 253.8), 0.13 g, 0.50 mmol — **highly toxic, corrosive**
- Hexamethyldisilane (FW 146.4), 79 mg, 0.11 mL, 0.54 mmol — **flammable, irritant**
- Freshly distilled (CaH$_2$) dichloromethane, 14 + 10 mL — **toxic, irritant**
- Freshly distilled (CaCl$_2$) butanal (FW 72.11), 0.72 g, 0.90 mL, 10 mmol — **flammable, corrosive**
- Cyclohexyloxytrimethylsilane[a] (FW 172.3), 1.72 g, 2.01 mL, 10 mmol — **flammable, moisture sensitive**
- Trimethylsilane[b] (FW 74.2) in a gas cylinder — **flammable**

1. Dry the three-necked flask with an electric heat gun under vacuum, and introduce nitrogen gas from the balloon. Repeat this operation two more times. After cooling to room temperature, remove the septum, quickly add iodine, and attach the septum under a stream of nitrogen.

2. Add hexamethyldisilane and 14 mL of dry CH$_2$Cl$_2$ *via* syringes.

3. Stir the violet solution at room temperature for 10 min, and then, cool to 0°C with the ice–water bath.

4. Add a solution of butanal and cyclohexyloxytrimethylsilane in 10 mL of CH$_2$Cl$_2$ *via* a syringe.

5. After stirring for 10 min, replace the vacuum source with a bubbler and turn the stopcock to the bubbler from the nitrogen source.

6. Add trimethylsilane directly from a gas cylinder by means of Tygon tubing attached to a stainless-steel needle inserted through the septum. Slowly bubble the gas through the solution until the colour changes from violet to red–gold. During the bubbling, the internal temperature rises from 0°C to 15°C.

7. Stop the addition of trimethylsilane, remove the cold bath, and stir at room temperature for 2 h.

8. Remove the septum and pour the reaction mixture into 30 mL of 10% aqueous Na$_2$S$_2$O$_3$ solution.

9. Transfer the bilayer to a separating funnel and remove the aqueous layer. Similarly, wash the organic layer three more times with 30 mL of 10% aqueous Na$_2$S$_2$O$_3$ solution and four times with 30 mL of water.

10. Dry the organic layer over MgSO$_4$ and concentrate *in vacuo*.

166

8: Silicon(IV) reagents

11. Purify the crude product by distillation. Cyclohexyl butyl ether (FW 156.3, b.p. 68–70 °C/0.08 mmHg) is obtained in 93% yield (1.45 g).

[a] Prepared from cyclohexanol and chlorotrimethylsilane in the presence of triethylamine.[41] Commercially available from Shin-Etsu Chemical Co. Ltd.
[b] B.p. 6.7 °C. Commercially available from Strem Chemicals Inc. or Chisso Co. Ltd. It can be prepared from chlorotrimethylsilane and lithium aluminum hydride.[42]

References

1. Patai, S.; Rappoport, Z. *The Chemistry of Organic Silicon Compounds*; Wiley, Chichester, **1989**.
2. Colvin, E. W. *Silicon in Organic Synthesis*; Butterworths, London, **1981**.
3. Simchen, G. *Advances in Silicon Chemistry*. (ed G. L. Larson). JAI Press, Greenwich, **1991**, Vol. 1, p 189.
4. Noyori, R.; Murata, S.; Suzuki, M. *Tetrahedron* **1981**, *37*, 3899.
5. Suzuki, M.; Noyori, R. *J. Syn. Org. Chem. Jpn* **1982**, *40*, 534.
6. Olah, G. A.; Prakash, G. K. S.; Krishnamurti, R. *Advances in Silicon Chemistry* (ed. G. L. Larson). JAI Press, Greenwich, **1991**, Vol. 1, p. 1.
7. Olah, G. A.; Narang, S. C. *Tetrahedron* **1982**, *38*, 2225.
8. Hosomi, A. *J. Syn. Org. Chem., Jpn* **1982**, *40*, 545.
9. Hosomi, A.; Kohra, S.; Tominaga, Y. *J. Chem. Soc., Chem. Commun*, **1987**, 1517.
10. Hosomi, A. *Rev. Heteroatom Chem.* **1992**, *7*, 214.
11. Denmark, S. E.; Griedel, B. D.; Coe, D. M.; Schnute, M. E. *J. Am. Chem. Soc.* **1994**, *116*, 7026.
12. Tsunoda, T.; Suzuki, M.; Noyori, R. *Tetrahedron Lett*, **1980**, *21*, 1357.
13. Sakurai, H.; Sasaki, K.; Hayashi, J.; Hosomi, A. *J. Org. Chem.* **1984**, *49*, 2808.
14. Seyferth, D.; Menzel, H.; Dow, A. W.; Flood, T. C. *J. Organomet. Chem.* **1972**, *44*, 279.
15. Paulsen, H. *Angew. Chem., Int. Ed. Engl.* **1990**, *29*, 823.
16. Lin, C.-C.; Shimazaki, M.; Heck, M.-P.; Aoki, S.; Wang, R.; Kimura, T.; Ritzèn, H.; Takayama, S.; Wu, S.-H.; Weitz-Schmidt, G.; Wong, C.-H. *J. Am. Chem. Soc.* **1996**, *118*, 6826.
17. Vorbrüggen, H.; Krolikiewicz, K.; Bennua, B.; *Chem. Ber.* **1982**, *114*, 1234.
18. McDonald, F. E.; Gleason, M. M. *J. Am. Chem. Soc.* **1996**, *118*, 6648.
19. Mukaiyama, T.; Narasaka, K.; Banno, K. *Chem. Lett.* **1973**, 1011.
20. Gennari, C. *Comprehensive Organic Synthesis* (ed. B. M. Trost, I. Fleming). Pergamon Press, Oxford, **1991**, Vol. 2, p 629.
21. Murata, S.; Suzuki, M.; Noyori, R. *J. Am. Chem. Soc.* **1980**, *102*, 3248.
22. Murata, S.; Suzuki, M.; Noyori, R. *Tetrahedron* **1988**, *44*, 4259.
23. Sakurai, H.; Sasaki, K.; Hosomi, A. *Bull. Chem. Soc. Jpn* **1983**, *56*, 3195.
24. Mezger, F.; Simchen, G.; Fischer, P. *Synthesis* **1991**, 375.
25. Marsmann, H. C.; Horn, H.-G. *Z. Naturforsch., [b]* **1972**, *27*, 1448.
26. House, H. O.; Czuba, L. J.; Gall, M.; Olmstead, H. D. *J. Org. Chem.* **1969**, *34*, 2324.
27. DeGraw, J. I.; Goodman, L.; Baker, B. R. *J. Org. Chem*, **1961**, *26*, 1156.
28. Hosomi, A.; Sakurai, H. *Tetrahedron Lett*, **1976**, 1295.
29. Hosomi, A. *Acc. Chem. Res.* **1988**, *21*, 200.

30. Tsunoda, T.; Suzuki, M.; Noyori, R. *Tetrahedron Lett.* **1980**, *21*, 71.
31. Sakurai, H.; Sasaki, K.; Hosomi, A. *Tetrahedron Lett.* **1981**, *22*, 745.
32. Hosomi, A.; Sakata, Y.; Sakurai, H. *Tetrahedron Lett.* **1984**, *25*, 2383.
33. Panek, J. S.; Yang, M. *J. Org. Chem.* **1991**, *56*, 5755.
34. Tietze, L. F.; Schiemann, K.; Wegner, C. *J. Am. Chem. Soc.* **1995**, *117*, 5851.
35. Davis, A. P.; Jaspars, M. *Angew. Chem. Int. Ed. Engl.* **1992**, *31*, 470.
36. Kumada, M.; Shiina, A.; Yamaguchi, M. *Kogyo Kagaku Zasshi* **1954**, *57*, 230.
37. Tsunoda, T.; Suzuki, M.; Noyori, R. *Tetrahedron Lett.* **1979**, 4679.
38. Sassaman, M. B.; Kotian, K. D.; Prakash, G. K. S.; Olah, G. A. *J. Org. Chem.* **1987**, *52*, 4314.
39. Sassaman, M. B.; Prakash, G. K. S.; Olah, G. A. *Tetrahedron* **1988**, *44*, 3771.
40. Nicolaou, K. C.; Hwang, C.-K., Duggan, M. E.; Nugiel, D. A.; Abe, Y.; Reddy, K. B.; DeFrees, S. A.; Reddy, D. R.; Awartani, R. A.; Conley, S. R.; Rutjes, F. P. J. T.; Theodorakis, E. A. *J. Am. Chem. Soc.* **1995**, *117*, 10227.
41. Corey, E. J.; Snider, B. B. *J. Am. Chem. Soc.* **1972**, *94*, 2549.
42. Steward, O. W.; Pierce, O. R. *J. Am. Chem. Soc.* **1961**, *83*, 1916.

Silver and gold reagents

MASAYA SAWAMURA

1. Introduction

Owing to their weak Lewis-acidity, reagents involving a silver(I) or gold(I) ion have attracted little attention as a Lewis-acid reagent to date. As described in this chapter, however, phosphine-co-ordinated silver(I) and gold(I) complexes have been successfully applied to the aldol-type condensation of isocyanides with aldehydes, where the silver(I) and gold(I) complexes activate the isocyanide through η^1-co-ordination of the isocyano carbon (see Section 2). Although these complexes show rather low affinity to an oxygen functional group such as a carbonyl group, an example of carbonyl group activation can be seen in the silver(I)-catalysed allylation of aldehydes with allyltin reagents as a Lewis acid catalyst (see Section 3).

2. Enantioselective aldol reaction of activated isocyanides and aldehydes

Aldol-type reaction of activated isocyanides with aldehydes are effectively catalysed by a gold(I) or silver(I) complex of a chiral bis(phosphino)ferrocene bearing an ethylenediaminoalkyl pendant, producing optically active oxazolines.[1,2] The activating group on the isocyanides includs an alkoxycarbonyl,[3–20] carbamoyl,[21–23] phosphonyl,[24,25] or sulfonyl group[26] (Scheme 9.1, Tables 9.1 and 9.2). The reaction catalysed by silver(I) complex requires a slow addition of the isocyanide (over 1 h) except for the reaction of tosylmethyl isocyanide.

In most cases, the *trans*-oxazoline is predominantly formed with a high enantioselectivity, whereas the *cis*-oxazoline with low enantiomeric excess is as a minor product. The *trans*-oxazolines can readily be converted into *threo*-β-hydroxy-α-amino acids or their phosphonic acid analogues by acidic hydrolysis (Scheme 9.2). The reactions of α-substituted isocyanoacetate esters give, after acidic hydrolysis, α-alkylated α-amino acids.[7,8] The oxazolines obtained from *N*-methoxy-*N*-methyl-α-isocyanocetamide (α-isocyano Weinreb amide)[23] can be transformed to *N,O*-protected β-hydroxy-α-amino aldehydes and ketones in high yield (Scheme 9.3).

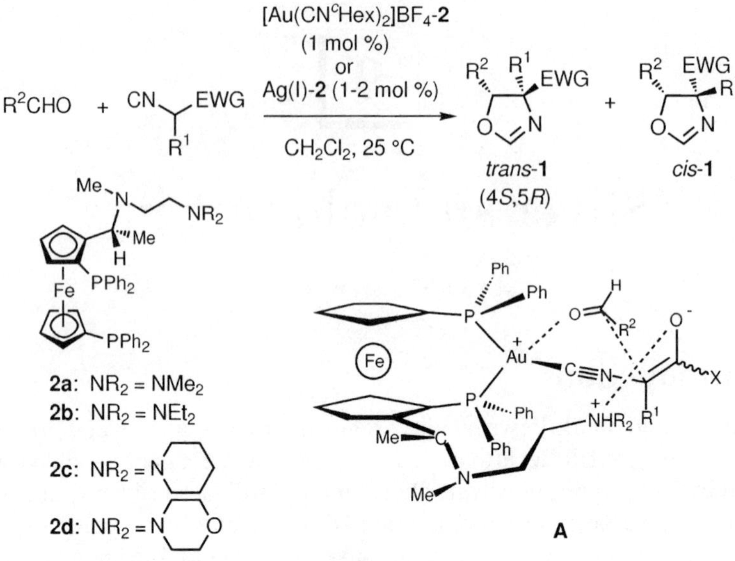

Scheme 9.1

Table 9.1. Gold(I)-catalysed enantioselective aldol reaction of activated isocyanides with aldehydes[a]

EWG	R^1	R^2	Ligand	*trans:cis*	ee (%) (*trans*)
CO_2Me	H	Ph	2b	89:11	93
CO_2Me	H	Ph	2c	94:6	95
CO_2Me	H	Me	2a	78:22	37
CO_2Me	H	Me	2d	89:11	89
CO_2Me	H	*i*-Pr	2c	99:1	94
CO_2Me	H	*t*-Bu	2d	>99:1	97
CO_2Me	H	(*E*)-PrCH=CH	2d	87:13	92
CO_2Me	Me	Ph	2d	93:7	94
CO_2Me	*i*-Pr	Ph	2c	54:46	92
CO_2Me	*i*-Pr	H	2c	—	81
$CONMe_2$	H	$4\text{-BnOC}_6\text{H}_4\text{CH}_2$	2c	>95:5	95
CON(Me)OMe	H	(*E*)-$BnOCH_2CH=CH$	2d	96:4	95
$PO(OPh)_2$	H	Ph	2c	>98:2	96

[a] Yield is generally high.

The terminal amino group of the chiral ligand is essential for the catalyst activity and the stereoselectivities.[3] Structure of the terminal amino group has a substantial effect on the stereoselectivity, six-membered ring amino groups such as piperidino or morpholino groups being generally efficient.[3,4,16] The

Table 9.2. Silver(I)-catalysed enantioselective aldol reaction of activated isocyanides with aldehydes[a]

EWG	R^1	R^2	Ligand	*trans:cis*	ee (%) (*trans*)
CO$_2$Me[b]	H	Ph	2c	96:4	80
CO$_2$Me[b]	H	*i*-Pr	2c	99:1	90
SO$_2$(*p*-Tol)[c]	H	(*E*)-MeCH=CH	2c	97:3	85

[a] Yield is generally high.
[b] Reaction with 2 mol % of AgClO$_4$-2 under slow addition conditions.
[c] Reaction with 1 mol % of AgOTf-2.

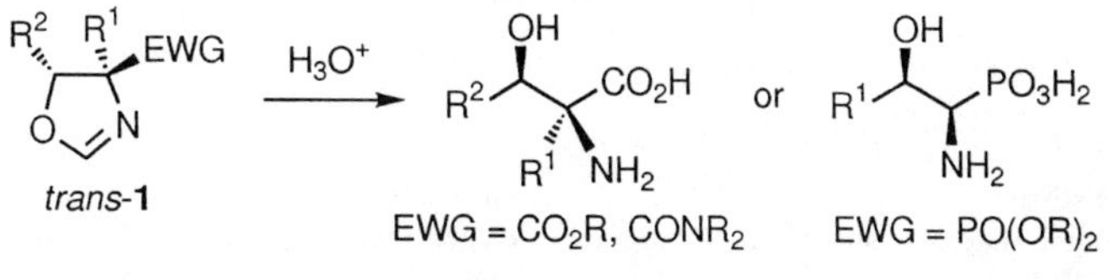

Scheme 9.2

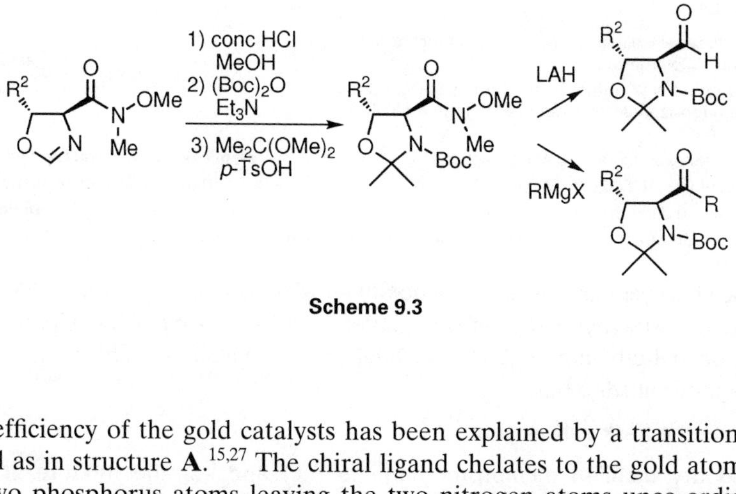

Scheme 9.3

high efficiency of the gold catalysts has been explained by a transition state model as in structure **A**.[15,27] The chiral ligand chelates to the gold atom with the two phosphorus atoms leaving the two nitrogen atoms unco-ordinated. The α-methylene protons of isocyanoacetate are activated through the co-ordination of the isocyano group to the gold atom, and the terminal amino group abstracts one of the activated α-protons, forming an ion pair between enolate anion and ammonium cation. The attractive ligand–substrate inter-action permits a favourable arrangement of the enolate and aldehyde on the gold atom in the stereodifferentiating transition state.

Protocol 1.
Enantioselective aldol reaction of methyl isocyanoacetate with benzaldehyde catalysed by a chiral diamino bisphosphine-gold (I) complex[3] (Scheme 9.4).

Caution! Carry out all procedures in a well-ventilated hood, and wear disposable vinyl or latex gloves and chemical-resistant safety goggles.

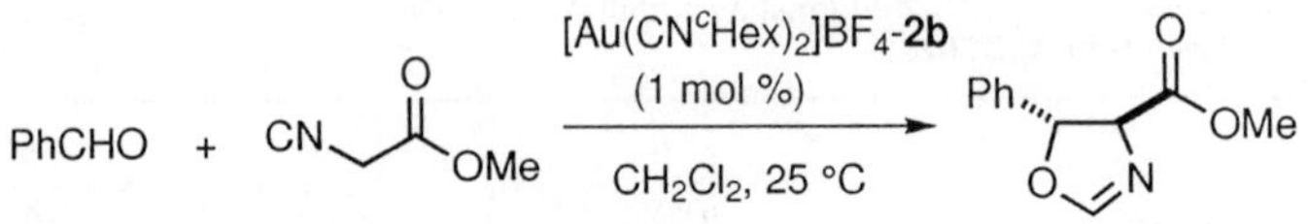

Scheme 9.4

Equipment

- Magnetic stirrer
- Teflon-coated magnetic stirring bar (octagon 15 × 6.5 mm)
- Schlenk flask (30 mL)
- Syringes
- Thermostat
- Vacuum/inert gas source (argon)
- Water bath
- Khugelrohr distillation apparatus

Materials

- Bis(cyclohexyl isocyanide)gold(I) tetrafluoroborate,[28] 25.1 mg, 0.050 mmol   **air-sensitive**
- (R)-N-Methyl-N-[2-(diethylamino)ethyl]-1-[(S)-1′,2-bis (diphenylphosphino)ferrocenyl]ethylamine,[5] 35.5 mg, 0.050 mmol
- Methyl isocyanoacetate,[a] 0.495 g, 5.00 mmol   **stench, irritant, corrosive, lachrymator**
- Benzaldehyde,[a] 0.586 g, 5.50 mmol   **highly toxic, cancer suspect agent**
- Dry, distilled dichloromethane, 5 mL   **toxic, irritant**
- Dichloromethane for the transfer of a reaction mixture, 5 mL   **toxic, irritant**

1. Place bis(cyclohexyl isocyanide)gold(I) tetrafluoroborate (25.1 mg, 0.050 mmol) and (R)-N-methyl-N-[2-(diethylamino)ethyl]-1-[(S)-1′,2-bis(diphenylphosphino)-ferrocenyl]ethylamine (35.5 mg, 0.050 mmol) in a 30 mL Schlenk flask with a magnetic stirring bar.

2. Purge the flask with argon.

3. Add dry distilled dichloromethane (5 mL) and stir the mixture at room temperature for 5 min.

4. Add methyl isocyanoacetate (0.495 g, 5.00 mmol) and immerse the flask into a water bath controlled at 25 °C with a thermostat.

5. Add benzaldehyde (0.586 g, 5.50 mmol) and stir the mixture for 20 h. The completion of reaction can be checked by TLC (silica gel, hexane–ethyl acetate, 2:1).

6. Remove the solvent under reduced pressure.

7. Distil the residual liquid with a Khugelrohr (ca. 100 °C, 1 mmHg) to obtain 1.05 g (98%) of the oxazoline or **3a** as a mixture of *trans*-(4*S*,5*R*) (93% ee) and *cis*-(4*R*,5*R*) (49% ee) isomers in a ratio of 89:11 (¹H NMR).

8. Determine the enantiomeric excesses of both isomers by ¹H NMR analysis with a chiral shift reagent Eu(hfc)₃[b] or by GLC analysis with a chiral stationary phase capillary column (Sumichiral OA-520[c]).

9. The *trans*- and *cis*-isomers can readily be separated by flash column chromatography on silica gel (hexane–ethyl acetate, 5:1).[d]

[a] Purify by distillation before use.
[b] Commercially available.
[c] Sumitomo Chemical Co. (Japan).
[d] Pre-dry the solvent over CaCl₂ and use the supernatant.

---

3. Enantioselective aldehyde allylation with allyltin reagents catalysed by a silver(I)–phosphine complex

The cationic silver(I)-BINAP complex is an efficient catalyst for the enantioselective allylation of aldehydes with allyltin reagents (Scheme 9.5).[29] The catalyst is prepared *in situ* from silver(I) triflate and the BINAP ligand in THF, and typically the reaction is carried out in the same solvent at −20 °C. High enantioselectivity is observed with various aldehydes including aromatic, α,β-unsaturated, and saturated aliphatic aldehydes, whereas the scope for the saturated aliphatic aldehydes are somewhat limited (Table 9.3).

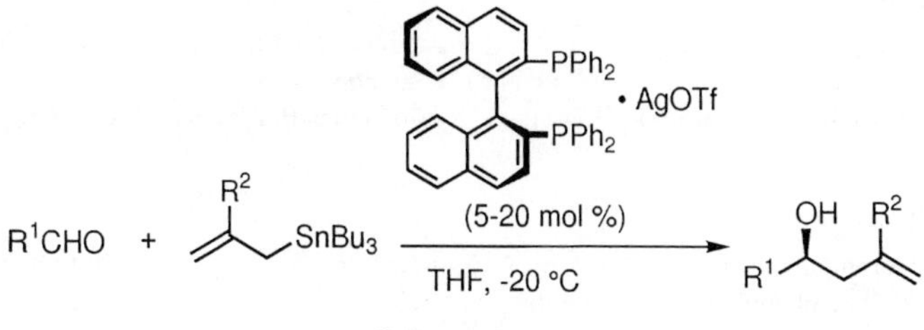

Scheme 9.5

Mechanistic studies suggest that the silver(I)-BINAP complex acts as a Lewis acid catalyst rather than an allylsilver reagent.

Masaya Sawamura

Table 9.3. Enantioselective allylation of aldehydes with allyltin reagents catalysed by a silver(I) complex of (S)-BINAP[a]

R[1]	R[2]	S/C[b]	Yield (%)	ee (%)
Ph	H	20	88	96 (S)
Ph	Me	20	75	92 (R)[c]
2-MeC$_6$H$_4$	H	20	85	97
4-MeC$_6$H$_4$	H	20	59	97
(E)-PhCH=CH	H	6.7	83	88 (S)
PhCH$_2$CH$_2$	H	5[d]	47	88

[a] Allyltin reagent:aldehyde, 1–4:1.
[b] Aldehyde:catalyst.
[c] Reaction with (R)-BINAP.
[d] The reaction was started with 0.1 equiv of the catalyst, and 0.1 equiv of the catalyst was added after 4 h.

Protocol 2.
Asymmetric allylation of benzaldehyde with allyltributyltin catalysed by (S)-BINAP·AgOTf Complex. Preparation of (S)-1-Phenyl-3-butene-1-ol[29] (Scheme 9.6)

Caution! Carry out all procedures in a well-ventilated hood, and wear disposable vinyl or latex gloves and chemical-resistant safety goggles.

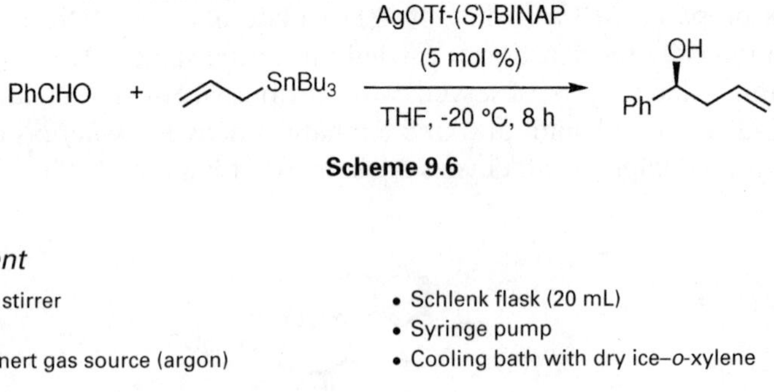

Scheme 9.6

Equipment
- Magnetic stirrer
- Syringes
- Vacuum/inert gas source (argon)
- Schlenk flask (20 mL)
- Syringe pump
- Cooling bath with dry ice–o-xylene

Materials
- Silver trifluoromethanesulfonate,[a] 26.4 mg, 0.103 mmol — **irritant, light-sensitive**
- (S)-(−)-2,2′-Bis(diphenylphosphino)-1,1′-binaphthyl,[a] 66.5 mg, 0.107 mmol
- Benzaldehyde,[b] 208.0 mg, 1.96 mmol — **highly toxic, cancer suspect agent**
- Allyltributyltin,[b] 663.1 mg, 2.00 mmol — **irritant**
- Dry tetrahydrofuran (THF),[c] 6 mL — **flammable, irritant**
- 1 M HCl solution in water, 10 mL — **toxic, corrosive**
- Potassium fluoride, 1 g — **toxic, corrosive**
- Magnesium sulfate — **hygroscopic**

174

1. Ensure that all glass equipment has been dried in an oven before use.

2. Place silver trifluoromethanesulfonate (26.4 mg, 0.103 mmol) and (S)-(−)-2,2′-bis(diphenylphosphino)-1,1′-binaphthyl (66.5 mg, 0.107 mmol) in a 20 mL Schlenk flask with a magnetic stirring bar under argon atmosphere and exclusion of direct light. Add dry tetrahydrofuran (3 mL) and stir the mixture at 20°C for 10 min.

3. After cooling the resulting solution to −20°C, add benzaldehyde (208.0 mg, 1.96 mmol) dropwise. Then add a THF solution (3 mL) of allyltributyltin (663.1 mg, 2.00 mmol) over a period of 4 h with a syringe pump at −20°C.

4. After stirring for 4 h at this temperature, treat the reaction mixture with a mixture of 1 M HCl (10 mL) and solid KF (1 g) at ambient temperature for 30 min.

5. Filter off the resulting precipitate, dry the filtrate over $MgSO_4$ and concentrate *in vacuo*.

6. Purify the crude product by flash column chromatography on silica gel [hexane–ethyl acetate (10:1)] to give the (S)-enriched product (258 mg, 88% yield, 96% ee) as a colourless oil which displays the appropriate ^{1}H NMR (in $CDCl_3$) and IR (neat).

7. Determine the enantiomeric excess by HPLC analysis (Daicel Chiralcel OD-H, hexane–*i*-PrOH, 20:1; flow rate, 0.5 mL/min), t_R = 18.2 min (*R*-isomer), T_R = 19.9 min (*S*-isomer).

[a] Commercially available products can be used without purification. Do not use aged ones.
[b] Purify benzaldehyde and allyltributyltin by distillation before use.
[c] Commercially available anhydrous THF (Aldrich) can be used as received.

References

1. Sawamura, M,; Ito, Y. *Chem. Rev.* **1992**, *92*, 857.
2. Sawamura, M.; Ito, Y. In *Catalytic Asymmetric Synthesis* (ed. I. Ojima). VCH Publishers, New York, **1993**, p 367.
3. Ito, Y.; Sawamura, M.; Hayashi, T. *J. Am. Chem. Soc.* **1986**, *108*, 6405.
4. Ito, Y.; Sawamura, M.; Hayashi, T. *Tetrahedron Lett.* **1987**, *28*, 6215.
5. Hayashi, T.; Sawamura, M.; Ito, Y. *Tetrahedron*, **1992**, *48*, 1999.
6. Ito, Y.; Sawamura, M.; Hayashi, T. *Tetrahedron Lett.* **1988**, *29*, 239.
7. Ito, Y.; Sawamura, M.; Shirakawa, E.; Hayashizaki, K.; Hayashi, T. *Tetrahedron Lett.* **1988**, *29*, 235.
8. Ito, Y.; Sawamura, M.; Shirakawa, E.; Hayashizaki, K.; Hayashi, T. *Tetrahedron* **1988**, *44*, 5253.
9. Soloshonok, V. A.; Hayashi, T. *Tetrahedron Lett.* **1994**, *35*, 2713.
10. Soloshonok, V. A.; Kacharov, A. D.; Hayashi, T. *Tetrahedron* **1996**, *52*, 245.
11. Soloshonok, V. A.; Hayashi, T.; Ishikawa, K.; Nagashima, N. *Tetrahedron Lett.* **1994**, *35*, 1055.

12. Soloshonok, V. A.; Kacharov, A. D.; Avilov, D. V.; Hayashi, T. *Tetrahedron Lett.* **1996**, *37*, 7845.
13. Hayashi, T.; Uozumi, Y.; Yamazaki, A.; Sawamura, M.; Hamashima, H.; Ito, Y. *Tetrahedron Lett.* **1991**, *32*, 2799.
14. Pastor, S. D.; Togni, A. *J. Am. Chem. Soc.* **1989**, *111*, 2333.
15. Togni, A.; Pastor, S. D. *J. Org. Chem.* **1990**, *55*, 1649.
16. Pastor, S. D.; Togni, A. *Helv. Chim. Acta* **1991**, *74*, 905.
17. Togni, A.; Pastor, S. D. *Helv. Chim. Acta* **1989**, *72*, 1038.
18. Togni, A.; Pastor, S. D.; Rihs, G. *Helv. Chim. Acta* **1989**, *72*, 1471.
19. Togni, A.; Häusel, R. Synlett **1990**, 633.
20. Gorla, F.; Togni, A.; Venanzi, L. M.; Albinati, A.; Lianza, F. *Organometallics* **1994**, *13*, 1607.
21. Ito, Y.; Sawamura, M.; Kobayashi, M.; Hayashi, T. *Tetrahedron Lett.* **1988**, *29*, 6321.
22. Soloshonok, V. A.; Hayashi, T. *Tetrahedron Asymmetry* 1994, *5*, 1091.
23. Sawamura, M.; Nakayama. Y.; Kato, T.; Ito. Y. *J. Org. Chem.* **1995**, *60*, 1727.
24. Sawamura, M.; Ito, Y.; Hayashi, T. *Tetrahedron Lett.* **1989**, *30*, 2247.
25. Togni, A.; Pastor, S. D. *Tetrahedron Lett.* **1989**, *30*, 1071.
26. Sawamura, M.; Hamashima, H.; Ito, Y. *J. Org. Chem.* **1990**, *55*, 5935.
27. Sawamura, M.; Ito, Y.; Hayashi, T. *Tetrahedron Lett.* **1990**, *31*, 2723.
28. Bonati, F.; Minghetti, G. *Gazz. Chim. Ital.* **1973**, *103*, 373.
29. Yanagisawa, A.; Nakashima, H.; Ishiba, A.; Yamamoto, H. *J. Am. Chem. Soc.* **1996**, *118*, 4723.

Zr- and Hf-centred Lewis acid in organic synthesis

KEISUKE SUZUKI

1. Introduction

The nature of a given Lewis acid, ML_n, can be modified by changing the metal (M), the ligand (L, chiral or non-chiral) as well as the charge (neutral or cationic), thereby endowing it with useful properties in organic synthesis. Early transition metals are attracting current attention to exploit their hard Lewis acidic characters. Although titanium has been studied most, recent attention is also centred at its group-4 congeners, i.e. zirconium and hafnium. Although they have been extensively used in the polymerization field,[1-3] the use in organic synthesis is still rather limited either as a tailor-made catalyst or as a transient species in the course of the reaction.[4] This chapter describes some unique reactivities of cationic zirconocene or hafnocene complexes, thereby dealing with four modes of functional group activation: (1) C–F bond in glycosyl fluoride, (2) ethers, (3) carbonyl groups, (4) C–C multiple bonds.

2. Activation of C–F bond

Combination of Cp_2HfCl_2 and $AgClO_4$ acts as a powerful activator of glycosyl fluoride, providing a glycosylation method used for the synthesis of complex oligosaccharides or other glycoconjugates (Scheme 10.1).[5-8] The high fluorophilicity of hafnocene perchlorate complex, cationic, or loosely bound covalent species, is invoked to the origin of high activation of C–F bond to generate an oxonium species. Double ligand exchange by Cp_2HfCl_2 and $AgClO_4$ in 1:2

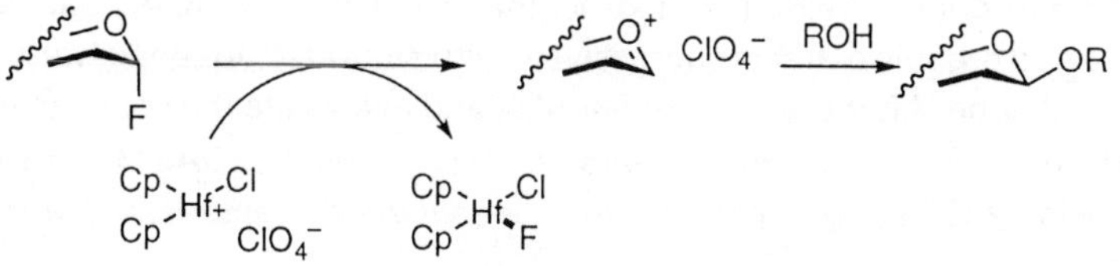

Scheme 10.1

ratio leads to even higher reactivity. Triflate is used as an alternative counter anion considering the potential hazard associated with perchlorate.

Protocol 1.
Metallocene-promoted *O*-glycosylation (Scheme 10.2)

Caution! Carry out all procedures in a well-ventilated hood, and wear disposable vinyl or latex gloves and chemical-resistant safety goggles.

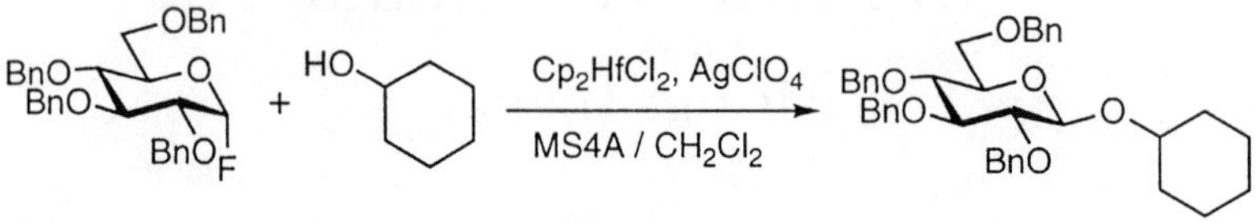

Scheme 10.2

Equipment

- Two-necked, round-bottomed flask (30 mL) containing a magnetic stirring bar, with a side arm bearing a three-way stopcock, and a rubber septum
- Vacuum/inert gas source (gas source may be an argon balloon)
- Dry ice–methanol cooling bath
- All-glass syringe

Materials

- AgClO$_4$ (FW 207), 47.0 mg, 0.227 mmol — **oxidizer, corrosive, explosive**
- Cp$_2$HfCl$_2$ (FW 379), 43.0 mg, 0.113 mmol — **irritant, moisture sensitive**
- dry, distilled CH$_2$Cl$_2$
- Powdered molecular sieves 4A, *c.* 100 mg
- Cyclohexanol (FW 100), 22.6 mg, 0.226 mmol — **irritant, hygroscopic**
- Glycosyl fluoride 1 (FW 543), 61.2 mg, 0.113 mmol

1. Assemble the two-necked flask with a stopcock, a stirring bar, and a rubber septum. Flame dry the flask under vacuum, and, after cooling to ambient temperature, purge with argon through the three-way stopcock. Add Cp$_2$HfCl$_2$ and AgClO$_4$ to the flask.
2. Add dry CH$_2$Cl$_2$ (0.5 mL) via a syringe, and stir the mixture for 10 min to make a slurry.
3. Add a solution of cyclohexanol in CH$_2$Cl$_2$ (0.5 mL). After cooling to −50°C, add glycosyl fluoride 1 in CH$_2$Cl$_2$ (1 mL) via a syringe, and stir the mixture for 1 h.
4. Quench the reaction with saturated aqueous NaHCO$_3$ solution, and filter the mixture through a Celite pad. Extract the products with EtOAc, and wash the combined organic extracts successively with saturated aqueous NaHCO$_3$ solution and brine. After drying over Na$_2$SO$_4$, and evaporate the solvents *in vacuo*.
5. Purify the resulting oil on preparative TLC (hexane–EtOAc, 4:1) to yield *O*-glycoside 2 (65.2 mg, 93%). The product shows characteristic spectroscopic data.

3. Activation of ethers

Cationic hafnocene species are also capable of activating the ether linkages as seen in the unique results in *C*-glycoside synthesis.[9] In the *C*-glycosylation of phenol via *in situ*-formed *O*-glycoside promoted by a Lewis acid (**3** + **4** → [**5**] → **6**: *O* → *C* glycoside rearrangement), the Cp_2HfCl_2–$AgClO_4$ combination is particularly effective in terms of the reactivity in converting **5** into **6** as well as the stereoselectivity. Both of these features have their origin in the effective activation of an ether oxygen by a cationic hafnocene species as illustrated in **A** and **B**[10–13] (Scheme 10.3).

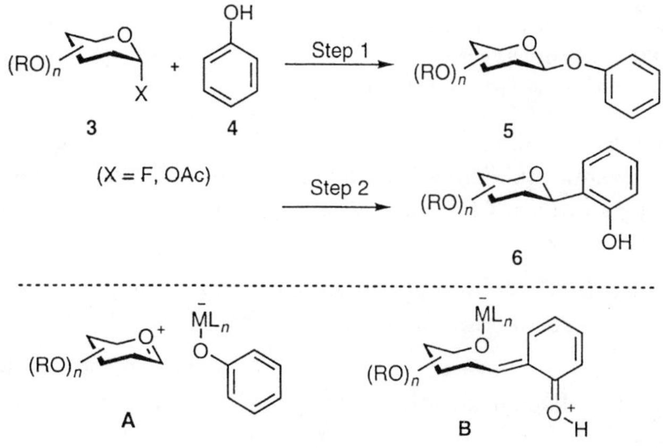

Scheme 10.3

Wipf *et al.*[14–16] reported the activation of epoxide by a cationic zirconocene species to trigger 1,2-rearrangement.

Protocol 2.
O → C-glycoside rearrangement of olivosyl acetate 7 and phenol 8 (Scheme 10.4)

Caution! Carry out all procedures in a well-ventilated hood, and wear disposable vinyl or latex gloves and chemical-resistant safety goggles.

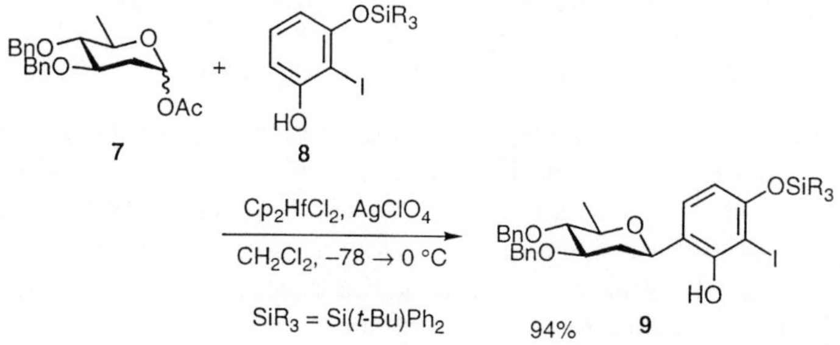

Scheme 10.4

179

Protocol 2. *Continued*

Equipment

- Two-necked, round-bottomed flask (100 mL) fitted with a magnetic stirring bar, a three-way stopcock, and a rubber septum
- Dry ice–methanol cooling bath
- Vacuum/inert gas source (gas source may be an argon balloon)
- All-glass syringe

Materials

- AgClO₄ (FW 207.3), 893 mg, 4.31 mmol — **oxidizer, corrosive, explosive**
- Cp₂HfCl₂ (FW 380), 820 mg, 2.16 mmol — **irritant, moisture sensitive**
- Dry, distilled CH₂Cl₂
- Powdered molecular sieves 4A, ca. 1.2 g
- Phenol **8** (FW 474), 936 mg, 1.97 mmol
- D-olivosyl acetate **7** (FW 370), 666 mg, 1.80 mmol

1. Assemble the 100 mL round-bottomed flask with a stopcock, a stirring bar, and a rubber septum. Flame dry the vessel containing powdered molecular sieves under vacuum. After cooling, purge with argon. Place Cp₂HfCl₂ and AgClO₄ in the flask.

2. Introduce dry CH₂Cl₂ (5 mL) via a syringe through the rubber septum, and form a slurry by stirring for 15 min.

3. After cooling the mixture to −78°C, add a solution of phenol **8** in CH₂Cl₂ (15 mL) and D-olivosyl acetate **7** in CH₂Cl₂ (10 mL) via a syringe to this suspension.

4. Let the mixture to warm to 0°C during 40 min, and stir it for 15 min.

5. Quench the reaction with pH 7 phosphate buffer, and acidify it with 2 M HCl. Filter the mixture through a Celite pad, and extract the products with EtOAc. Wash the combined organic extracts with brine. After drying over Na₂SO₄ and filtration, concentrate the solution *in vacuo*.

6. Purify the resulting oil on silica-gel flash column (Hexane–acetone, 85:15) to yield *C*-glycoside **9** (1.32 g, 93.5%). The product shows characteristic spectroscopic data.

4. Carbonyl activation

Carbonyl groups are also activated by cationic zirconocene species. Alkenyl- or alkyl-zirconocene chlorides, **10** or **12** (Scheme 10.5), readily accessible via hydrozirconation of alkyne or alkene,[17,18] are unreactive nucleophiles, and their Grignard-type carbonyl addition is slow. However, a catalytic amount of AgClO₄ or AgAsF₆ greatly accelerates the reaction, which is ascribed to the carbonyl activation by the cationic species.[19] Recently, Hf(OTf)₄ was prepared and used as a catalyst for Friedel–Crafts reaction, and for Fries rearrangement, which are also ascribed to the carbonyl activation by the hafnium reagent[20,21]

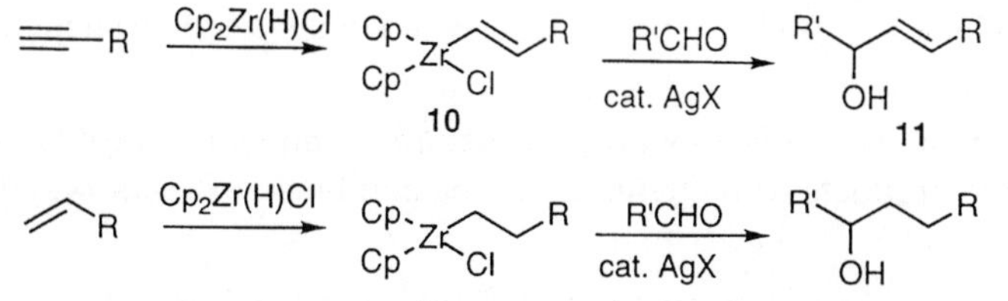

Scheme 10.5

Protocol 3.
AgAsF$_6$ catalysed addition of alkenylzirconocene chloride to aldehyde (Scheme 10.6)

Caution! Carry out all procedures in a well-ventilated hood, and wear disposable vinyl or latex gloves and chemical-resistant safety goggles.

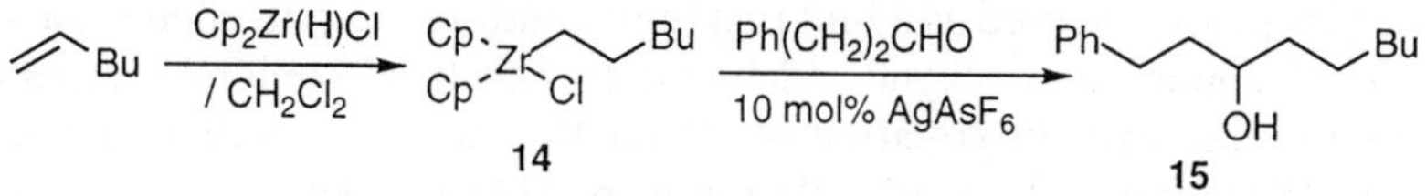

Scheme 10.6

Equipment

- Two-necked, round-bottomed flask (30 mL) fitted with a magnetic stirring bar, a three-way stopcock, and a rubber septum
- Nitrogen gas line
- Vacuum/inert gas source (gas source may be an argon balloon)
- Dry ice–methanol cooling bath
- All-glass syringe

Materials

- Cp$_2$Zr(H)Cl (FW 258), 270 mg, 1.05 mmol   **moisture sensitive, light sensitive**
- AgAsF$_6$ (FW 296), 20.1 mg, 0.0677 mmol   **highly toxic, cancer suspect agent**
- Dry, distilled CH$_2$Cl$_2$
- 1-hexyne (FW 84.2), 93.1 mg, 1.11 mmol   **flammable liquid irritant**
- 3-phenylpropanal (FW 134), 83.1 mg, 0.619 mmol

1. Assemble the 30 mL round-bottomed flask with a stopcock, a stirring bar, and a rubber septum, and flame dry the flask under vacuum. After cooling to room temperature, fill the apparatus with nitrogen.

2. Place Schwartz reagent in the flask, and then rapidly place the rubber septum on the neck of the flask.

3. Add a solution of the 1-hexyne in CH$_2$Cl$_2$ (3 mL) via a syringe to the flask at −78°C, and then allow the mixture to warm to 25°C, and stir the mixture for 10 min.

Protocol 3. *Continued*

4. Add a solution of 3-phenylpropanal in CH_2Cl_2 (3 mL) to the mixture.

5. After 5 min, add $AgAsF_6$ to the mixture, where a deep brown suspension will form.

6. After 10 min, pour the mixture into saturated aqueous $NaHCO_3$ solution, extract the products with EtOAc. Wash the combined extracts with brine and dry over Na_2SO_4.

7. Purify the resulting oil on preparative TLC (Hexane–EtOAc, 4:1) to yield **15** as colourless oil (129 mg, 94.5%). The product shows characteristic spectroscopic data.

---

5. Reductive coupling of allenes and alkynes

It is well documented that cationic zirconocene species plays a key role in Kaminsky polymerization, a recent descendant of the Ziegler–Natta process.[1-3] This process served as a hint to a regio- and stereocontrolled reductive coupling of allenes and alkynes (Scheme 10.7); allylzirconium species **17**, generated by the hydrozirconation of allene **16**, undergoes carbometallation of alkyne **18** promoted by methylaluminoxane (MAO). The overall reaction gives the 'α-internal' coupling product **19** in high selectivity.[22]

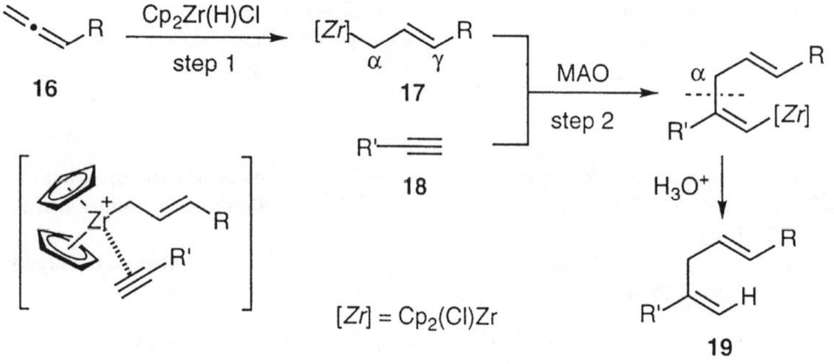

Scheme 10.7

Protocol 4.
MAO-catalysed allylzirconation of 1-alkynes (Scheme 10.8)

Caution! Carry out all procedures in a well-ventilated hood, and wear disposable vinyl or latex gloves and chemical-resistant safety goggles.

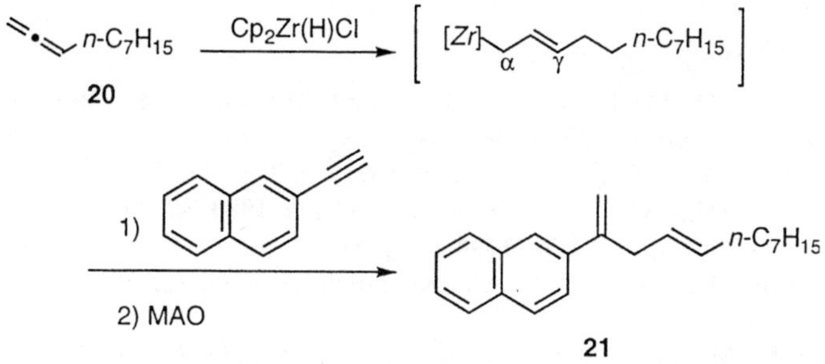

Scheme 10.8

Equipment

- Two-necked, round-bottomed flask (30 mL) fitted with a magnetic stirring bar, a three-way stopcock, and a rubber septum
- Nitrogen gas line

- Vacuum/inert gas source (gas source may be an argon balloon)
- Dry ice–methanol cooling bath
- All-glass syringe

Materials

- $Cp_2Zr(H)Cl$ (FW 258), 190 mg, 0.736 mmol   **moisture sensitive, light sensitive**
- Methylaluminoxane (MAO) (0.93 M solution in toluene), 0.12 mL   **pyrophoric, corrosive**
- Dry, distilled CH_2Cl_2
- Allene **20** (FW 138), 86.3 mg, 0.625 mmol
- 2-Ethynylnaphthalene (FW 152), 55.4 mg, 0.364 mmol

1. Assemble the two-necked flask with a stopcock, stirring bar, and a rubber septum, and flame dry the flask under vacuum. After cooling to room temperature, purge the flask with nitrogen.

2. Transfer Schwartz reagent from Schlenk flask into the flask which is connected to the nitrogen line, and then rapidly equip the neck of the flask with a rubber septum.

3. Add CH_2Cl_2 (0.7 mL) to the flask to form a slurry at $-78\,°C$. Add a solution of allene **20** in CH_2Cl_2 (2.5 mL), warm the mixture to $25\,°C$ over 20 min and stir for a further 0.5 h.

4. Chill the resulting red solution to $-78\,°C$, and add ethynylnaphthalene in CH_2Cl_2 (2.5 mL) via a syringe followed by MAO solution. Warm the mixture to $-20\,°C$ over 15 min, and stir for 20 min at $-20\,°C$. Stop the reaction by carefully adding saturated aqueous K_2CO_3 solution. Stir the mixture for 5 min, and add anhydrous Na_2SO_4.

Protocol 4. *Continued*

5. After filtration through a Celite pad and evaporation, purify the resulting oil on preparative TLC (hexane) to give diene **21** (105 mg, 99%). The product shows characteristic spectroscopic data.

References

1. Brintzinger, H. H.; Fischer, D.; Mülhaupt, R.; Rieger, B.; Waymouth, R. M. *Angew. Chem. Int. Ed. Engl.* **1995**, *34*, 1143.
2. Jordan, R. F. *Adv. Organomet. Chem.* **1991**, *32*, 325.
3. Sinn, H.; Kaminsky, W. *Adv. Organomet. Chem.* **1980**, *18*, 99.
4. Suzuki, K. *Pure Appl. Chem.* **1994**, *66*, 1557.
5. Matsumoto, T.; Maeta, H.; Suzuki, K.; Tsuchihashi, G. *Tetrahedron Lett.* **1988**, *29*, 3567.
6. Suzuki, K.; Maeta, H.; Matsumoto, T.; Tsuchihashi, G. *Tetrahedron Lett.* **1988**, *29*, 3571.
7. Suzuki, K.; Maeta, H.; Matsumoto, T. *Tetrahedron Lett.* **1989**, *30*, 4853.
8. Matsumoto, T.; Maeta, H.; Suzuki, K.; Tsuchihashi, G. *Tetrahedron Lett.* **1988**, *29*, 3575.
9. Suzuki, K.; Matsumoto, T. Total synthesis of aryl *C*-glycoside antibiotics. In *Recent progress in the chemical synthesis of antibiotics and related microbial products*, Vol. 2 (ed. G. Lukacs) Springer Verlag, Berlin, pp 352–403.
10. Matsumoto, T.; Katsuki, M.; Suzuki, K. *Tetrahedron Lett.* **1988**, *29*, 6935.
11. Matsumoto, T.; Hosoya, T.; Suzuki, K. *Synlett* **1991**, 709.
12. Matsumoto, T.; Hosoya, T.; Suzuki, K. *Tetrahedron Lett.* **1988**, *31*, 4629.
13. Matsumoto, T.; Katsuki, M.; Jona, H.; Suzuki, K. *J. Am. Chem. Soc.* **1991**, *113*, 6982.
14. Wipf, P.; Xu, W. *J. Org. Chem.* **1993**, *58*, 825.
15. Wipf, P.; Xu, W. *J. Org. Chem.* **1993**, *58*, 5880.
16. Wipf, P.; Xu, W. *Tetrahedron* **1995**, *51*, 4551.
17. Schwartz, J.; Labinger, J. A. *Angew. Chem., Int. Ed. Engl.* **1976**, *15*, 333.
18. Buchwald, S. L.; LaMaire, S. J.; Nielsen, R. B.; Watson, B. T.; King, S. M. *Org. Synth.* **1992**, *71*, 77.
19. Suzuki, K.; Hasegawa, T.; Imai, T.; Maeta, H.; Ohba, S. *Tetrahedron*, **1995**, *55*, 4483.
20. Hachiya, I.; Moriwaki, M.; Kobayashi, S. *Bull. Chem. Soc. Jpn* **1995**, *68*, 2053.
21. Kobayashi, S.; Moriwaki, M.; Hachiya, I. *Bull. Chem. Soc. Jpn* **1997**, *70*, 267.
22. Yamanoi, S.; Imai, T.; Matsumoto, T.; Suzuki, K. *Tetrahedron Lett.* **1997**, *38*, 3031.

11

Scandium(III) and yttrium(III)

SHIN-ICHI FUKUZAWA

1. Introduction

According to the inorganic chemistry text book written by Cotton and Wilkinson,[1] scandium is not truly a rare earth element but yttrium and lanthanides are rare earth elements. The stable oxidation state of scandium is trivalent and the ionic radius of scandium (III) is significantly smaller (0.89 Å) than those for any of the rare earth elements (1.0–1.17 Å). Chemical behaviour is intermediate between aluminium and that of lanthanides. Yttrium has a trivalent oxidation state similar to scandium and lanthanide elements and the ionic radius of yttrium (III) (1.04 Å) is close to those of erbium (1.03 A) and holmium (1.04 Å). Yttrium resembles lanthanide elements in its chemical properties.

2. Scandium(III) triflate

Among scandium(III) compounds, scandium(III) trifluoromethanesulfonate (triflate) [Sc(OTf)$_3$] is the most attractive reagent and has been intensively studied in organic synthesis because it has a stronger Lewis acidity than that of other scandium (III) compounds.[2] Compared to even yttrium(III) and lanthanide(III) triflates, Sc(OTf)$_3$ has a stronger Lewis acidity and catalyses certain reactions which are not mediated by yttrium(III) and lanthanide(III) triflates. The stronger Lewis acidity of Sc(OTf)$_3$ is probably due to its smaller ionic radius than those of lanthanides. Sc(OTf)$_3$ is commercially available or readily prepared by the reaction of scandium oxide and trifluoromethanesulfonic acid in water.[3] Characteristic features of scandium triflate are as follows. It can be used as a catalyst in the presence of water and catalyses certain reactions and maintains catalytic activity. It can be recovered from the aqueous layer after reaction and reused in further reactions. As it is stable in water and air, it is easy to handle and the catalysed reaction can be carried out by a simple procedure.

2.1 Friedel–Crafts reaction

Sc(OTf)$_3$ catalyses Friedel–Crafts acylation reaction of arenes effectively.[4] It mediates the reaction more efficiently than Yb(OTf)$_3$ and Y(OTf)$_3$. Acylation of benzene and toluene does not proceed, but reaction with more electron-rich arenes gives acylated products. Reaction with anisole, thioanisole, *o*- and *m*-dimethoxybenzene gives the corresponding single acylated product in excellent yields, respectively (Scheme 11.1).

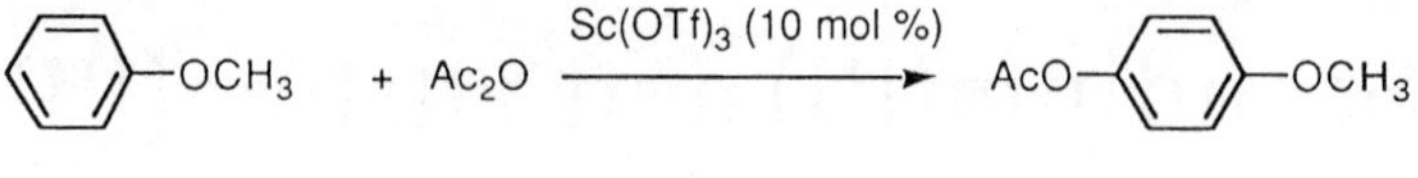

Scheme 11.1

The advantages of the process are as follows. First, a catalytic amount (1 mol %) of Sc(OTf)$_3$ is enough to promote the reaction in contrast to the conventional acylation reaction which requires a stoichiometric amount of Lewis acid. Second, evolution of hydrogen halide during quenching of Lewis acid, such as AlCl$_3$, is avoided. Third, Sc(OTf)$_3$ can be recovered from the aqueous layer and re-used as a catalyst, whereas disposal of a considerable amount of aluminium hydroxide which results from the work-up process is a severe environmental problem in the conventional Friedel–Crafts process.

Sc(OTf)$_3$ has been shown to catalyse *ortho*-selective acylation of phenol and naphthol derivatives with acyl halides to yield the corresponding ketones in high yields.[5] The hydroxynaphthyl ketones (**4**) can also be obtained by Sc(OTf)$_3$-catalysed Fries rearrangement of acyloxy naphthalenes (Scheme 11.2).[6]

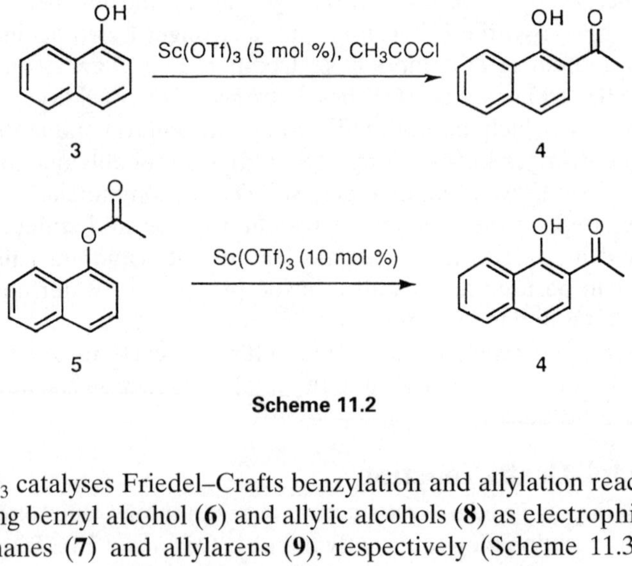

Scheme 11.2

Sc(OTf)$_3$ catalyses Friedel–Crafts benzylation and allylation reactions with arenes using benzyl alcohol (**6**) and allylic alcohols (**8**) as electrophiles to give diarylmethanes (**7**) and allylarens (**9**), respectively (Scheme 11.3).[5] Water,

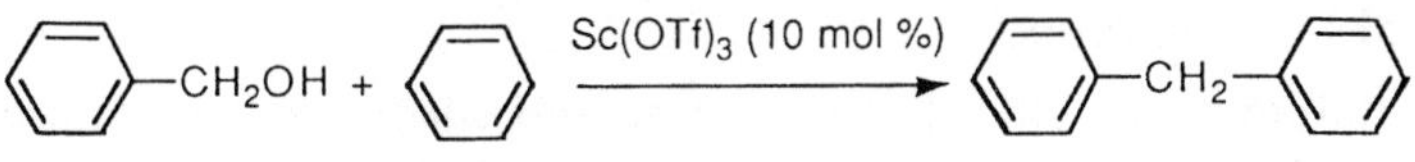

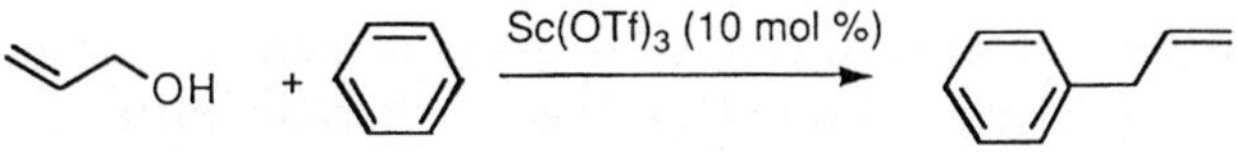

Scheme 11.3

produced during the reaction, does not spoil catalytic activity of $Sc(OTf)_3$ and a catalytic amount (10 mol %) of $Sc(OTf)_3$ is enough for the reaction. In contrast to conventional Friedel–Crafts benzylation and allylation using organic halides as electrophiles which produces troublesome hydrogen halides as byproducts, this process only produces water. Considering the advantages of $Sc(OTf)_3$ in the Friedel–Crafts reaction, it may solve some severe environmental problems induced by conventional Lewis acid-promoted reactions in industry.

Protocol 1.
Preparation of *p*-methoxyacetophenone[4]

Equipment
- Three-necked round-bottomed flask (50 mL)
- Magnetic stirrer
- Thermostatted oil bath
- Reflux condenser
- Three-way stopcock
- Stopcocks

- Thermometer (100°C)
- Drying tube (calcium chloride)
- Separating funnel (300 mL)
- Sintered-glass filter funnel (porosity 3)
- Erlenmeyer flask (200 mL)
- Teflon-coated magnetic stirrer bar (2.0 × 0.7 cm)

Materials
- Distilled (CaH_2 nitromethane (FW 61.04), 10 mL — **flammable**
- Anisole (FW 108.14), 1.08 g, 10.0 mmol — **irritant, hygroscopic**
- Distilled acetic anhydride (FW 102.09), 1.02 g, 10.0 mmol — **corrosive, lachrymator**
- $Sc(OTf)_3$ (FW 492.16), 0.98 g, 2.0 mmol — **irritant, hygroscopic**
- Diethyl ether for extraction, 60 mL — **flammable, toxic**
- Anhydrous magnesium sulfate — **hygroscopic**

1. Place $Sc(OTf)_3$ (0.98 g, 2.0 mmol) and a magnetic stirrer bar in a three-necked round-bottomed flask (50 mL) connected with two stopcocks and a three-way stopcock.
2. Heat a flask in an oil bath at 180°C under vacuum (1 mm Hg) for 1 h.
3. When the flask has cooled to room temperature, connect a reflux condenser, a thermometer and a drying tube.

Shin-ichi Fukuzawa

Protocol 1. *Continued*

4. Add nitromethane (10 mL) to the flask and stir for 10 min, and then add anisole (1.08 g, 10.0 mmol) and acetic anhydride (1.02 g, 10.0 mmol).

5. Heat the solution to 50°C in an oil bath equipped with a temperature controller (bath temperature, 70°C).

6. Keep the temperature at 50°C with stirring and monitor the disappearance of anisole by TLC (visualized with UV). It takes 4 h to complete the reaction.

7. Add water (30 mL) to the solution. Separate the organic layer with the aid of separating funnel (100 mL). Extract the aqueous layer with three portions of diethyl ether (20 mL). Wash the combined organic layers with brine (30 mL). Transfer the solution to an Erlenmeyer flask and dry the solution over anhydrous magnesium sulfate.

8. Filter the dried solution through a sintered-glass filter funnel and remove the solvent on a rotary evaporator to leave a pale-yellow residue. Distil the residue by Kugelrohr at 0.4 mmHg to collect 1.22 g (7.4 mmol, 74%) of the product (oven temperature, 110°C). The compound was characterized by ^{1}H NMR and elemental analysis. The product comprises only *p*-acetoxyanisole; no *o*- and *m*-isomers were detected by capillary GC analysis (capillary column coated with (5%-phenyl)methypolysiloxane).

Protocol 2.
Reaction of benzylalcohol with benzene[7]

Equipment

- Three-necked round-bottomed flask (200 mL)
- Magnetic stirrer
- Thermostatted oil bath
- Reflux condenser
- Three-way stopcock
- Drying tube (calcium chloride)
- Separating funnel (300 mL)
- Sintered-glass filter funnel (porosity 3).
- Erlenmeyer flask (200 mL)
- Teflon-coated magnetic stirrer bar (2.0 × 0.7 cm)

Materials

- Benzene (FW 78.11), 100 mL, 1.11 mol — cancer-suspect agent, flammable
- Distilled benzyl alcohol (FW 108.14) 2.16 g, 20.0 mmol — irritant, hygroscopic
- Sc(OTf)$_3^a$ (FW 492.16), 0.98 g, 2.0 mmol — irritant, hygroscopic
- Diethyl ether for extraction, 50 mL — flammable, irritant
- Anhydrous magnesium sulfate — hygroscopic

1. Assemble a three-neck round-bottomed flask, a magnetic stirrer bar, a reflux condenser, and a drying tube.

2. Add Sc(OTf)$_3$ (0.98 g, 2.0 mmol), benzene (100 mL) and benzyl alcohol (2.16 g, 20.0 mmol) to the flask and heat the solution at refluxing temperature with stirring.

3. After stirring for 8 h under reflux, cool the flask to room temperature.

4. Add water (50 mL) to the solution. Separate the organic layer with the aid of separating funnel (300 mL). Extract the aqueous layer with diethyl ether (50 ml). Wash the combined organic layers with brine (30 ml). Transfer the organic layer to an Erlenmeyer flask and dry the solution over anhydrous magnesium sulfate.

5. Filter the dried solution through a sintered-glass filter funnel and remove the solvent on a rotary evaporator. Distil the residue at 3.0 mmHg to collect 2.7 g (16.2 mmol, 81%) of the product boiling at 97 °C. The compound was characterized by GC/MS, ^{1}H NMR and elemental analysis.

6. Concentrate the aqueous layer with a rotary evaporator and heat the crystal-line residue [Sc(OTf)$_3$] at 180 °C under vacuum for 20 h. The recovered Sc(OTf)$_3$ can be used in the next Friedel–Crafts benzylation and allylation reactions.

[a] Commercial Sc(OTf)$_3$ retains moisture. Dry Sc(OTf)$_3$ by heating in a flask using an oil bath at 180 °C under vacuum (1 mm Hg) for more than 1 h (see Protocol 1).

2.2 Acetalization

Chiral acetals are particularly important precursors for the synthesis of enantio-merically pure compounds.[8] Chiral dioxane and dioxolane are prepared by the direct reaction of aldehydes or ketones with chiral 1,3-diols and 1,2-diols in the presence of a catalytic amount of Sc(OTf)$_3$ at room temperature (Scheme 11.4).[9] Chiral 2,4-pentanediol, 2,3-butanediol, and diethyl tartarate are used as chiral diols. Considering that the chiral acetals derived from diethyl tartarate have only been prepared by trans-acetalization under acidic con-ditions,[10,11] this method is convenient and practical. It is worth noting that removal of water which results from the acetalization is not necessary.

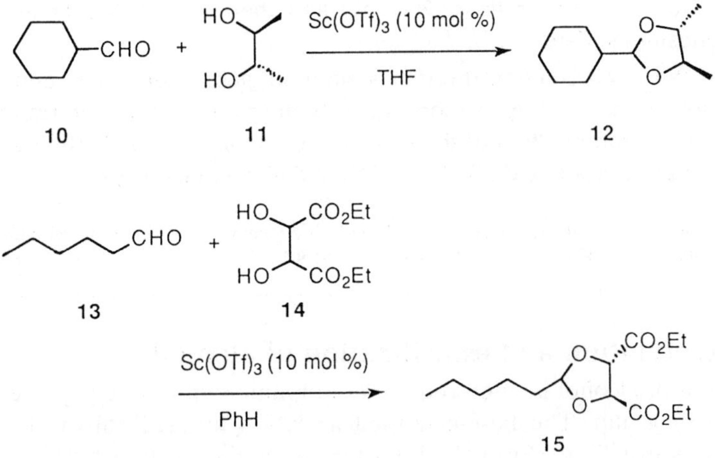

Scheme 11.4

189

Protocol 3.
Scandium(III) triflate catalysed acetalization of aldehyde[9]

Equipment
- Three-necked round-bottomed flask (100 mL)
- Magnetic stirrer
- Three-way stopcock
- Stopcocks
- Drying tube (calcium chloride)
- Separating funnel (300 mL)
- Sintered-glass filter funnel (porosity 3)
- Erlenmeyer flask (200 mL)
- Teflon-coated magnetic stirrer bar (2.0 × 0.7 cm)

Materials
- Dry distilled benzene (P_2O_5) (FW 78.11), 30 mL — **cancer-suspect agent, flammable**
- Distilled hexanal (FW 100.16), 1.00 g, 10.0 mmol — **flammable, irritant**
- L-(+)-Diethyl tartarate (FW 206.19), 2.50 g, 12.1 mmol
- Sc(OTf)$_3$[a] (FW 492.16), 0.98 g, 2.0 mmol — **irritant, hygroscopic**
- Diethyl ether for extraction, 50 mL — **flammable, irritant**
- Anhydrous magnesium sulfate — **hygroscopic**

1. Assemble a three-neck round-bottomed flask (100 mL), a magnetic stirrer bar, a stopcock, a three-way stopcock and a drying tube.
2. Add Sc(OTf)$_3$ (0.98 g, 2.0 mmol) and benzene (15 mL) into a flask and stir the suspension for 10 min at room temperature.
3. Add (1R, 2R)-(+)-diethyl tartarate (2.50 g, 12.1 mmol) and stir the mixture at room temperature for 5 min until the mixture becomes homogeneous. Then, add hexanal (1.0 g, 10.0 mmol) and stir the solution at room temperature.
4. Monitor the production of the acetal by GC. After stirring for 48 h, add water (50 mL) to the solution. Separate the organic layer with the aid of a separating funnel (300 mL). Extract the aqueous layer with diethyl ether (50 mL). Wash the combined organic layers with brine (30 mL). Transfer the organic layer to a Erlenmyer flask (200 mL) and then dry them over anhydrous magnesium sulfate.
5. Filter the dried solution through a sintered-glass filter funnel and remove the solvent on a rotary evaporator. Distil the residue at 0.3 mmHg to collect 1.98 g (6.9 mmol, 69%) of the product boiling at 97–100 °C. The compound was characterized by GC/MS, ^{1}H NMR and elemental analysis.

[a] Commercial Sc(OTf)$_3$ retains moisture. Dry SC(OTf)$_3$ by heating in a flask using oil bath at 180 °C under vacuum (1 mm HG) for more than 1 h (see Protocol 1).

2.3 Acetylation and esterification of alcohol

Acylation of alcohol is often required in organic synthesis, e.g. protection of a hydroxy group. The base-catalysed acylation by acid anhydride or acyl chloride is usually performed with a tertiary amine such as pyridine and its derivatives.[12] Acid-catalysed acylation of alcohols with acid anhydride is an alternative acylation method which requires mild conditions. Sc(OTf)$_3$ is an efficient catalyst for acylation of alcohols by acid anhydrides (Scheme 11.5).[13,14] It also effectively catalyses the esterification of alcohols with carboxylic acid in the presence of *p*-nitrobenzoic anhydride.

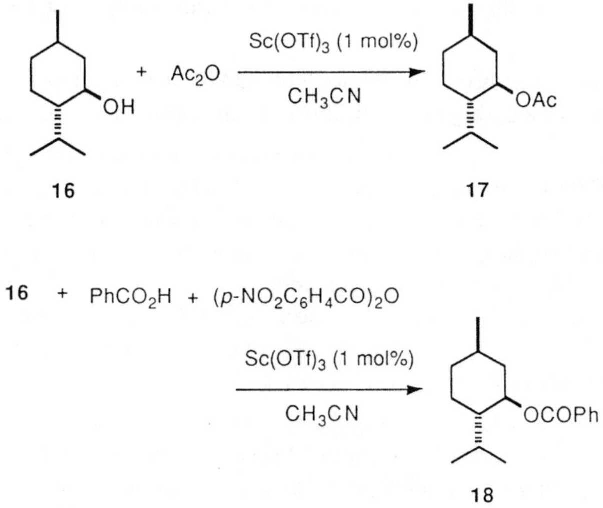

16 + PhCO$_2$H + (*p*-NO$_2$C$_6$H$_4$CO)$_2$O

Scheme 11.5

Protocol 4.
Preparation of L-acetyl menthol (17)[13,14]

Equipment

- Three-necked round-bottomed flask (100 mL)
- Magnetic stirrer
- Three-way stopcock
- Stopcocks
- Drying tube (calcium chloride)
- Separating funnel (300 mL)
- Sintered-glass filter funnel (porosity 3)
- All-glass syringe with a needle-lock luer (volume appropriate for quantity of solution to be transfered)
- Erlenmeyer flask (200 mL)
- Teflon-coated magnetic stirrer bar (2.0 × 0.7 cm)
- Glass column (2.5 cm × 50 cm) packed with silica gel

Materials

- Dry distilled acetonitrile (CaH$_2$) (FW 41.05), 30 mL — flammable, toxic
- Distilled acetic anhydride (FW 102.09), 1.0 mL, 10.6 mmol — corrosive, lachrymator
- L-(−)-Menthol (FW 156.27), 1.56 g, 10.0 mmol — irritant
- Sc(OTf)$_3$[a] (FW 492.16), 0.05 g. 0.1 mmol — irritant, hygroscopic
- Diethyl ether for extraction, 150 mL — flammable, irritant
- *n*-Hexane for column chromatography — flammable, irritant
- Ethyl acetate for column chromatography — irritant
- Anhydrous magnesium sulfate — hygroscopic

Protocol 4. *Continued*

1. Assemble a three-necked round-bottomed flask (100 mL), drying tube, a three-way stopcock, and a stopcock.

2. Add L-menthol (1.56 g, 10.0 mmol), acetic anhydride (1.53 g, 15.0 mmol), and acetonitrile (29 mL) to the flask and stir the solution for 10 min at room temperature.

3. Add dropwise a acetonitrile solution (1.0 mL) of Sc(OTf)$_3$ (49.6 mg, 0.10 mmol) with the aid of a syringe (2.0 ml) through the three-way stopcock.

4. After stirring for 1 h, add a saturated aqueous sodium hydrogen carbonate solution (50 mL) to the solution. Transfer the mixture to a separating funnel (300 mL) and extract the aqueous layer with three portions of diethyl ether (50 mL) and separate the organic layer. Wash the combined organic layers with brine (30 mL), then dry over anhydrous magnesium sulfate. Capillary GC analysis (capillary column coated with 100% dimethylpolysiloxane) of the organic solution shows that L-menthol is converted into the acylated product quantitatively.

5. Filter the dried solution through a sintered-glass filter funnel and remove the solvent on a rotary evaporator. Prepare a column (2.5 cm × 50 cm) for chromatography using silica gel (Merck Silica Gel 60) and *n*-hexane as the eluant. Dissolve the residue in a minimum volume (2 mL) of *n*-hexane and elute the column with *n*-hexane–ethyl acetate (5:1). Evaporate the eluate on a rotary evaporator to yield a colourless residue (2.00 g, 10.0 mmol, 100%). GC/MS and ^{1}H NMR spectra of the products display the suitable structure.

[a] Commercial Sc(OTf)$_3$ retains moisture. Dry Sc(OTf)$_3$ by heating in a flask using an oil bath at 180°C under vacuum (1 mm Hg) for more than 1 h (see Protocol 1).

2.4 Allylation of carbonyl compounds with tetra-allyltin catalysed by scandium(III) triflate

Allylation of carbonyl compounds has been recognized as an important process to produce synthetically useful homoallylic alcohols.[15] Tetra-allyltin is a good allyl transfer agent and itself reacts with ketones and aldehydes in the presence of an acid catalyst under mild conditions.[16] Water-tolerant Sc(OTf)$_3$ is an efficient catalyst for the allylation of carbonyl compounds with tera-allyltin in aqueous media.[17] The allylation reaction of ketones and aldehydes proceeds smoothly in the presence of catalytic amounts of Sc(OTf)$_3$ to produce homoallylic alcohols in good yields under mild conditions. For example, monosaccharides such as D-arabinose (**20**) is easily allylated in aqueous media (Scheme 11.6).

The allylated adducts of saccharides are intermediates for the synthesis of higher saccharides.

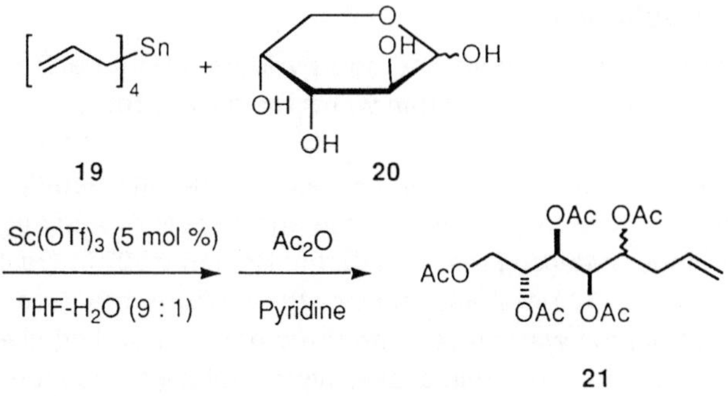

Scheme 11.6

Protocol 5.
Scandium(III) triflate catalysed allylation of D-arabinose[17]

Equipment

- Three-necked round-bottomed flask (200 mL)
- Single-necked round-bottomed flask (300 mL)
- Magnetic stirrer
- Three-way stopcock
- Drying tube (calcium chloride)
- Pressure equalizing dropping funnel (50 mL)
- Separating funnel (300 mL)
- Sintered-glass filter funnel (porosity 3)
- Erlenmeyer flask (200 mL)
- Teflon-coated magnetic stirrer bar (2.0 × 0.7 cm)
- Glass column (2.5 cm × 50 cm) packed with silica gel

Materials

- Dry distilled acetonitrile (CaH$_2$) (FW 41.05), 90 mL — **flammable, toxic**
- Distilled water (FW 18.02), 10 mL
- D-(−)-Arabinose (FW 151.12), 1.51 g, 10.0 mmol
- Tetra-allyltin^a (FW 282.98), 1.41 g, 5.0 mmol — **toxic**
- Sc(OTf)$_3$^{b} (FW 492.16), 0.25 g, 0.5 mmol — **irritant, hygroscopic**
- Distilled pyridine (CaH$_2$) (FW 79.10), 25 mL — **flammable, toxic**
- Acetic anhydride (FW 102.09), 10 mL, 0.13 mmol — **corrosive, lachrymator**
- Diethyl ether for extraction, 150 mL — **flammable, irritant**
- Ethyl acetate for column chromatography — **flammable, irritant**
- Anhydrous magnesium sulfate — **hygroscopic**

1. Assemble a three-necked round-bottomed flask (200 mL), drying tube, a three-way stopcock, and a dropping funnel.

2. Add Sc(OTf)$_3$ (0.25 g, 0.50 mmol) and acetonitrile (40 mL)–water (10 mL) solution of D-arabinose (1.51 g, 10.0 mmol) to the flask.

3. Add acetonitrile (50 mL) solution of tetra-allyltin (1.41 g, 5.0 mmol) dropwise from a dropping funnel at room temperature and stir the solution at the same temperature for 60 h.

4. Transfer the solution into single-necked round-bottomed flask (300 mL) and remove the solvents on a rotary evaporator to leave a white residue.

Protocol 5. *Continued*

5. Cool the flask with a ice-bath and add pyridine (25 mL) and Ac_2O (10 mL, 0.13 mol) with stirring. After 30 min warm the flask to room temperature and stir for 5 h.

6. Transfer the solution into single-necked round-bottomed-flask (200 mL) and remove 70–80% volume of pyridine and excess of Ac_2O using a rotary evaporator under reduced pressure (10 mm Hg, temp 70°C). Treat the residue with cold 1M HCl (100 mL) and transfer the mixture to a separating funnel (300 mL). Extract the water layer with three portions of diethyl ether (50 ml) and wash the combined organic layer with a saturated aqueous solution of $NaHCO_3$ (50 mL). Transfer the organic layer to an erlenmeyer flask and dry the solution over anhydrous magnesium sulfate.

7. Remove the solvent on a rotary evaporator to leave a pale-yellow viscous residue. Prepare a column (2.5 cm × 50 cm) for chromatography using silica gel (Merck Silica Gel 60) and hexane as the eluant. Dissolve the residue in a minimum volume (5 mL) of chloroform and elute the column with ethyl acetate. Evaporate the ethyl acetate eluate on a rotary evaporator to yield a colourless residue (2.78 g, 6.9 mmol, 69%). Capillary GC analysis (Capillary column coated with 100% dimethypolysiloxane) of the residue shows that the product is a mixture of diastereomers (*anti:syn* 5 65:35). ^{1}H NMR spectra of the products display suitable structures.

[a] Tetra-allyltin is prepared by the reaction of allylmagnesium chloride with tin(IV) chloride.[18,19]
[b] Commercial $Sc(OTf)_3$ retains moisture. Dry $Sc(OTf)_3$ by heating in a flask using oil bath at 180°C under vacuum (1 mm Hg) for more than 1 h (see Protocol 1).

2.5 Scandium(III) triflate-catalysed Mukaiyama aldol reaction

Although a wide variety of Lewis acid catalysts has been proposed for the Mukaiyama aldol reaction,[20] $Sc(OTf)_3$ is a promising alternative catalyst for the reaction from the viewpoint that it can be used in aqueous media and recovered and reused (Scheme 11.7).[21] Most Lewis acids usually decompose

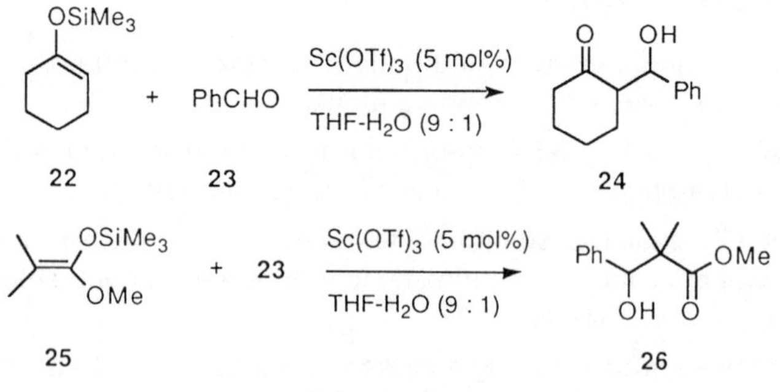

Scheme 11.7

in the presence of even a small amount of water but Sc(OTf)$_3$ works as a catalyst in aqueous media. A catalytic amount of Sc(OTf)$_3$ (5 mol%) effectively catalyses the reaction of silyl enol ethers with aldehyde or acetals to yield the corresponding aldol adducts in H$_2$O–THF (1:9) solvent.

Protocol 6.
Sc(OTf)$_3$-catalysed reaction of 1-trimethylsiloxy-1-cyclohexene (22) with benzaldehyde (23) in aqueous media[21]

Equipment

- Three-necked round-bottomed flask (100 mL)
- Magnetic stirrer
- Three-way stopcock
- Stopcocks
- Septum
- Separating funnel (300 mL)
- Sintered-glass filter funnel (porosity 3)
- All-glass syringe with a needle-lock luer (volume appropriate for quantity of solution to be transferred)
- Erlenmeyer flask (100 mL)
- Teflon-coated magnetic stirrer bar (2.0 × 0.7 cm)
- Glass column (2.5 cm × 50 cm) packed with silica gel

Materials

- Distilled dry tetrahydrofuran (THF) (FW 80.17), 25 mL — **flammable, irritant**
- 1-Trimethylsiloxycyclohexene[a] (FW 170.33), 1.70 g, 10.0 mmol — **moisture sensitive irritant**
- Distilled benzaldehyde (FW 106.12), 1.06 g, 10.0 mmol — **cancer suspect**
- Sc(OTf)$_3$ (FW 492.16),[b] 0.25 g, 0.5 mmol — **irritant, hygroscopic**
- Diethyl ether for extraction, 150 mL — **flammable, irritant**
- n-Hexane for column chromatography — **flammable, irritant**
- Ethyl acetate for column chromatography — **flammable, irritant**
- Anhydrous magnesium sulfate — **hygroscopic**

1. Assemble a three-necked round-bottomed flask (100 mL), a three-way stopcock and stopcocks under argon while the appartus is still hot.

2. Assemble a syringe (20 mL) and needle (30 cm) while hot and allow the assembled syringe to cool to room temperature in a desiccator. Flush the syringe with argon.

3. When the flask has cooled to room temperature, add Sc(OTf)$_3$ (0.246 g, 0.50 mmol). Then, add THF (8.0 mL) and H$_2$O (2.0 mL) to the flask using the syringe through the septum on the three-way stopcock.

4. Add THF (10.0 mL) solution of benzaldehyde (1.06 g, 10.0 mmol) and 1-trimethylsiloxy-1-cyclohexene (1.72 g, 10.0 mmol) using a syringe (20 mL) through the septum on the three-way stopcock at room temperature. Stir the mixture and monitor the reaction by a silica gel TLC (Merck Silica Gel 60 PF$_{254}$).

5. After 7 days, add water (50 mL) to the solution. Transfer the mixture to a separating funnel (300 mL) and extract the aqueous layer with three portions of diethyl ether (50 mL). Wash the combined organic layers with brine (30 mL), then dry over anhydrous magnesium sulfate. Capillary GC analysis

Shin-ichi Fukuzawa

Protocol 6. *Continued*

(capillary column coated with (5%-phenyl)methypolysiloxane) of the organic solution shows the presence of a diastereomeric mixture of the product (*syn:anti* = 75:25).

6. Filter the dried solution through a sintered-glass filter funnel and remove the solvent on a rotary evaporator. Prepare a column (2.5 cm × 50 cm) for chromatography using silica gel (Merck Silica Gel 60) and *n*-hexane as the eluant. Dissolve the residue in a minimum volume (2 mL) of *n*-hexane and elute the column with *n*-hexane–ethyl acetate (5:1). Evaporate the eluate on a rotary evaporator to yield a white solid (1.28 g, 6.3 mmol, 63%). GC/MS and ^{1}H NMR spectra of the products display the suitable structure.

[a] Commercially available.
[b] Commercial Sc(OTf)$_3$ retains moisture. Dry Sc(OTf)$_3$ by heating in a flask using an oil bath at 180°C under vacuum (1 mm Hg) for more than 1 h (see Protocol 1).

2.6 Diels–Alder reaction

Although Lewis acid-catalysed Diels–Alder reactions have been shown to proceed under mild conditions, they are often accompanied by a diene polymerization.[22] They frequently need excess amounts of the catalyst in the carbonyl-containing dienophiles. Sc(OTf)$_3$ is an efficient catalyst for Diels–Alder reactions of carbonyl-containing dienophiles with cyclopentadiene.[23] Thus, 3-crotonoyl-1,3-oxazolidin-2-one (**28**) reacts with cyclopentadiene (**27**) to give a mixture of corresponding adduct (**29**), in the ratio of *endo:exo* = 87:13, in a good yield (Scheme 11.8).

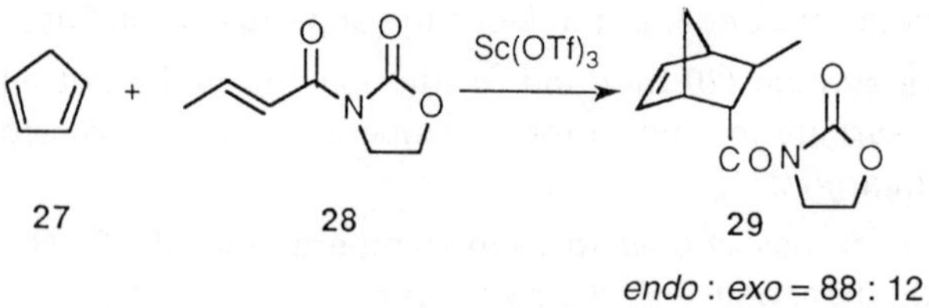

endo : exo = 88 : 12

Scheme 11.8

Asymmetric version of this reaction can be achieved using chiral binaphthol (**30**), *cis*-1,2,3-trimethylpiperidine (**31**), and molecular sieve 4A together with Sc(OTf)$_3$ (Scheme 11.9).[24] The *endo* adduct (*endo:exo* = 86:14–89:11) of the reaction of **28** with **27** is obtained in up to 96% ee. It must be noted that the enantioselectivity of the reaction decreases in accordance with ageing time and temperature of the preparation of the chiral catalyst (**32**).[25] 3-Acetyl-1,3-oxazolidin-2-one is found to be effective for preventing the ageing effect.

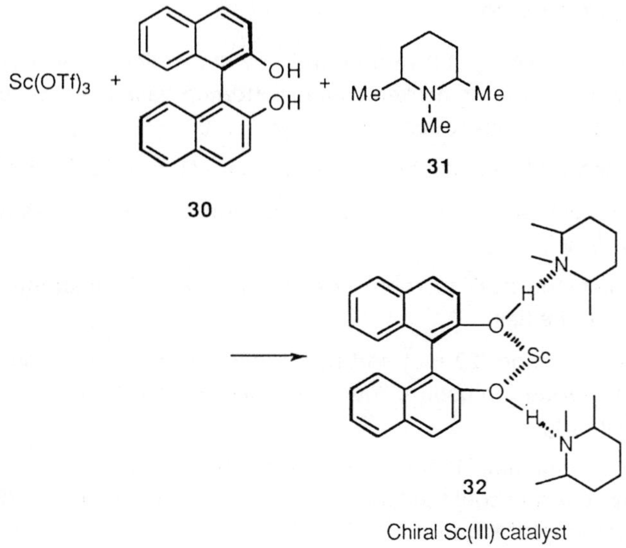

Scheme 11.9

Protocol 7.
Asymmetric Diels–Alder reaction of cyclopentadiene (27) with 3-crotonoyl-1,3-oxazolidin-2-one (28)[24]

Equipment

- Three-necked round-bottomed flask (100 mL)
- Magnetic stirrer
- Three-way stopcock
- Stopcocks
- Septum
- All-glass syringe with a needle-lock luer (volume appropriate for quantity of solution to be transferred)
- Dry ice–methanol bath
- Separating funnel (300 mL)
- Sintered-glass filter funnel (porosity 3)
- Erlenmeyer flask (100 mL)
- Glass column (2.5 cm × 50 cm) packed with silica gel
- Vacuum/inert gas source and inlet

Materials

- Distilled (P_2O_5) dichloromethane (FW 84.93), 25 mL — **toxic, irritant**
- 3-Crotonoyl-1,3-oxazolidin-2-one[a] (FW 155.15), 1.55 g, 10 mmol
- Freshly distilled cyclopentadiene[b] (FW 66.10), 1.98 g, 30.0 mmol — **flammable**
- Powder molecular sieve 4A (activated)[c] 1.0 g
- (R)-Binaphthol[d] (FW 286.33), 0.34 g, 1.18 mmol — **irritant**
- cis-1,2,3-Trimethylpiperidine[e] (FW 127.22), 0.31 g, 2.4 mmol
- Sc(OTf)$_3$ (FW 492.16), 0.49 g, 1.0 mmol — **irritant, hygroscopic**
- Diethyl ether for extraction, 150 mL — **flammable, toxic**
- n-Hexane for column chromatography — **flammable, irritant**
- Ethyl acetate for column chromatography — **flammable, irritant**
- Anhydrous magnesium sulfate — **hygroscopic**

197

Protocol 7. *Continued*

1. Place Sc(OTf)$_3$ (0.49 g, 1.0 mmol) and powder molecular sieves 4A (activated) (1.0 g) in a three-necked round-bottomed flask (100 mL) connected with a three-way stopcock and stopcocks.

2. Heat a flask by oil bath at 190°C under vacuum (1 mm Hg) for 5 h.

3. Flush the flask with argon through several vacuum cycles while the apparatus is still hot.

4. After cooling the flask to room temperature, add (*R*)-binaphthol (0.34 g, 1.2 mmol) to the flask.

5. Assemble a syringe (20 mL) and needle (30 cm) while hot and allow the assembled syringe to cool to room temperature in a desiccator. Flush the syringe with argon.

6. Add dichloromethane (10 mL) using the syringe through the septum on the three-way cock and cool the flask with dry ice–ethanol bath to −78°C under a slight pressure of argon.

7. Add *cis*-1,2,3-trimethylpiperidine (0.31 g, 2.4 mmol) using a syringe (1 mL) through the septum on the three-way stopcock into the flask at −78°C. Stir the mixture at this temperature for 30 min.

8. Add dichloromethane (15.0 mL) solution of 3-crotonoyl-1,3-oxazolidin-2-one (1.55 g, 10 mmol) and freshly distilled cyclopentadiene (1.98 g, 30 mmol) using a syringe (10 mL) through the septum on the three-way stopcock.

9. Stir the mixture at −78°C for 4 h and then warm it slowly to 0°C over a period of 10 h. Monitor the reaction by a silica gel TLC (Merck Silica Gel 60 PF$_{254}$).

10. Add water (50 mL) to the solution and transfer the mixture to a separating funnel (300 mL) and separate the organic layer. Extract the aqueous layer with two portions of diethyl ether (50 mL). Wash the combined organic layers with brine (30 mL), then dry them over anhydrous magnesium sulfate. Capillary GC analysis (capillary column coated with (5%-phenyl)methylpolysiloxane) of the organic solution shows the presence of a diastereomeric mixture of the product (*endo*:*exo* = 88:12).

11. Filter the dried solution through a sintered-glass filter funnel and remove the solvent on a rotary evaporator. Prepare a column (2.5 cm × 50 cm) for chromatography using silica gel (Merck Silica Gel 60) and *n*-hexane as the eluant. Dissolve the residue in a minimum volume (2 mL) of *n*-hexane and elute the column with *n*-hexane–ethyl acetate (10:1). Evaporate the eluate on a rotary evaporator to yield a colourless residue (1.40 g, 7.0 mmol, 70%). GC/MS and ^{1}H NMR spectra of the products display the suitable structure.

12. The enantiomeric excess of the *endo* adduct is determined to be 86% ee by HPLC analysis (Daicel Ciralpak AD; *n*-hexane–*i*-PrOH, 80:20; 0.5 ml/min).

[a] 3-Crotonoyl-1,3-oxazolidin-2-one is prepared by the reaction of crotonoyl chloride with 2-oxazolidione in THF in the presence of *n*-butyl-lithium.[26]
[b] Cyclopentadiene is freshly distilled by thermolysis of commercial dicyclopentadiene at 180 °C just prior to use.
[c] Obtained from Aldrich.
[d] Commercially available.
[e] *cis*-1,2,3-trimethylpiperidine is prepared by *N*-methylation of commercial *cis*-1,3-dimethylpiperidine by formaldehyde and formic acid.[27]

---

3. Cyclopentadienyl yttrium hydride

Since inorganic ytterium (III) compounds are less Lewis acidic than the corresponding scandium(III) compounds, they are seldom used as Lewis acid catalysts. Cyclopentadienyl derivatives of yttrium(III) hydrides have been studied in organic synthesis and found to catalyse hydrosilylation of alkenes[28] and cyclization of dienes.[29] The high regioselectivities and stereoselectivities of these reactions are worthy of note. Cyclopentadienylyttrium hydride $[(\eta^5\text{-}C_5Me_5(_2YH(THF)]$ (**34**) is conveniently prepared *in situ* in the reaction of ciscyclopentadienylalkylyttrium $[(\eta^5\text{-}C_5Me_5)_2YCH_3(THF)]$ (**33**) with hydrogen or phenylsilane and can be used for the catalytic reactions (Scheme 11.10). Thus, $(\eta^5\text{-}C_5Me_5)_2YCH_3(THF)$ catalyses the cyclization/hydrosilylation of dienes with phenylsilane smoothly to afford the cyclized organosilane (**38**) (Scheme 11.11).[30] Cyclization reaction of dienes has been applied to the synthesis of epilupine, one of the lupin alkaloids (**41**) (Scheme 11.12).[31]

$$YCl_3 + 2\ Cp^*Li \xrightarrow{THF} Cp^*_2YCl\ (THF)$$

$$\xrightarrow{MeLi} \underset{\mathbf{33}}{Cp^*_2YMe\ (THF)} \xrightarrow{H_2 \text{ or } PhSiH_3} \underset{\mathbf{34}}{\text{"}Cp^*_2YH\ (THF)\text{"}}$$

Scheme 11.10

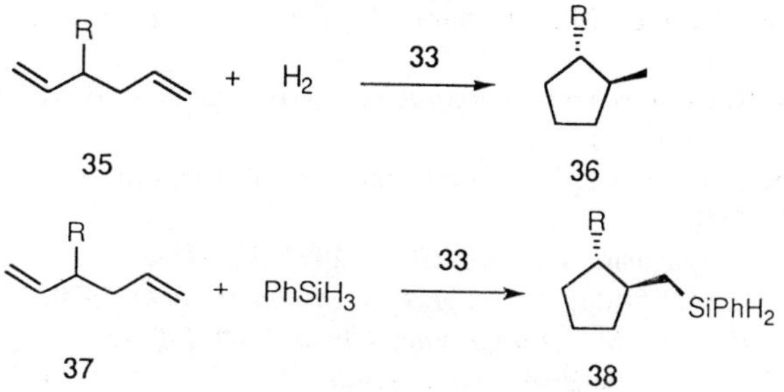

Scheme 11.11

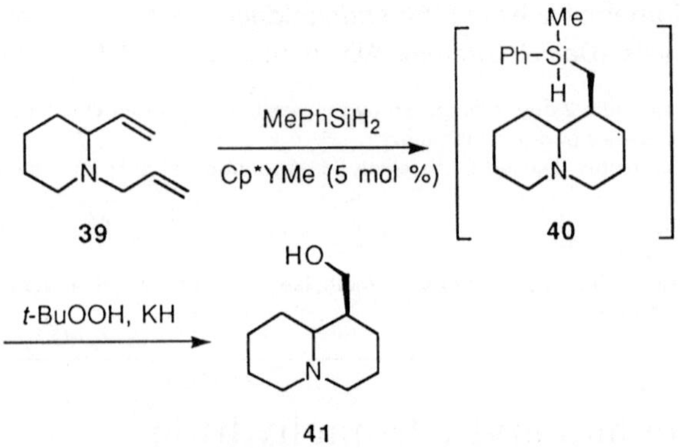

Shin-ichi Fukuzawa

Scheme 11.12

References

1. Cotton, F. A.; Wilkinson, G. *Advanced inorganic chemistry* 5th edn. Wiley, 1988, New York.
2. Kobayashi, S. *Synlett* **1994**, 689.
3. *Synthetic methods of organometallic and inorganic chemistry*, Vol 6; Lanthanides and Actinides (ed. F. T. Edelmann). Thieme, New York, **1997**, pp 36–37.
4. Kawada, A.; Mitamura, S.; Kobayashi, S. *Synlett* **1994**, 545.
5. Kobayashi, S.; Moriwaki, M.; Hachiya, I. *Synlett* **1995**, 1153.
6. Kobayashi, S.; Moriwaki, M.; Hachiya, I. *J. Chem. Soc. Chem. Commun.* **1995**, 1527.
7. Tsuchimoto, T.; Tobita, K.; Hiyama, T.; Fukuzawa, S. *Synlett* **1996**, 557.
8. Alexakis, A.; Mangeney, P. *Tetrahedron Asymmetry* **1990**, *1*, 477.
9. Fukuzawa, S.; Tsuchimoto, T.; Hotaka, T.; Hiyama, T. *Synlett* **1995**, 1077.
10. Fukutani, Y.; Maruoka, K.; Yamamoto, H. *Tetrahedron Lett.* **1984**, *25*, 5911.
11. Fujiwara, J.; Fukutani, Y.; Hasegawa, M.; Maruoka, K.; Yamamoto, H. *J. Am. Chem. Soc.* **1984**, *106*, 5004.
12. Hoefle, G.; Steilich, V.; Vorbrunggen, H. *Angew. Chem., Int. Ed. Engl.* **1978**, *17*, 569.
13. Ishihara, K.; Kubota, M.; Kurihara, H.; Yamamoto, H. *J. Am. Chem. Soc.* **1995**, *117*, 4413.
14. Ishihara, K.; Kubota, M.; Kurihara, H.; Yamamoto, H. *J. Org. Chem.* **1996**, *61*, 4560.
15. Roush, W. R. *Comprehensive organic synthesis*, Vol 2 (ed. B. M. Trost). Pergamon Press, Oxford, **1991**, pp 1–53.
16. Yanagisawa, A.; Inoue, H.; Morodome, M.; Yamamoto, H. *J. Am. Chem. Soc.* **1993**, *115*, 10356.
17. Hachiya, I.; Kobayashi, S. *J. Org. Chem.* **1993**, *58*, 6958.
18. Fishwick, M.; Wallbridge, M. G. H. *J. Organomet. Chem.* **1970**, *25*, 69.
19. Daude, G.; Pereyre, M. *J. Organomet. Chem.* **1980**, *190*, 43.
20. Gennari, C. *Comprehensive organic synthesis*, Vol 2 (ed. B. M. Trost). Pergamon Press, Oxford, **1991**, pp 629–660.

21. Kobayashi, S.; Hachiya, I.; Ishitani, H.; Araki, M. *Synlett* **1993**, 472.
22. Carruthers, W. *Cycloaddition reactions in organic synthesis*, Pergamon Press, Oxford, 1990.
23. Kobayashi, S.; Hachiya, I.; Araki, M.; Ishitani, H. *Tetrahedron Lett.* **1993**, *34*, 3755.
24. Kobayashi, S.; Araki, M.; Hachiya, I. *J. Org. Chem.* **1994**, *59*, 3758.
25. Kobayashi, S.; Ishitani, H.; Araki, M.; Hachiya, I. *Tetrahedron Lett.* **1994**, *35*, 6325.
26. Evans, D. A.; Chapman, K. T.; Bisaha, J. *J. Am. Chem. Soc.* **1988**, *110*, 1238.
27. Icke, R. N.; Wisegarer, B. B.; Alles, G. A. *Organic synthesis*. Wiley, New York, **1955**, Collect. Vol 3, p 723.
28. Molander, G. A.; Julius, M. *J. Org. Chem.* **1992**, *57*, 6347.
29. Molander, G. A.; Hoberg, J. O. *J. Am. Chem. Soc.* **1992**, *114*, 3123.
30. Molander, G. A.; Nichols, P. J. *J. Am. Chem. Soc.* **1995**, *117*, 4415.
31. Molander, G. A.; Nichols, P. J. *J. Org. Chem.* **1996**, *61*, 6040.

12

Lanthanide(III) Reagents

TAKESHI NAKAI and KATSUHIKO TOMOOKA

1 Introduction

The lanthanide group consists of the 15 elements from lanthanum to lutetium in the periodic table. In the past two decades the synthetic uses of a variety of lanthanide complexes, in terms of the kind of central lanthanides, the oxidation state (+2–+4), and the kind of ligands, have been developed.[1] This chapter focuses primarily on the four classes of the most popular trivalent lanthanides, namely, cerium(III) complexes (CeL$_3$), europium(III) complexes (EuL$_3$), ytterbium complexes (YbL$_3$), and lanthanum(III) complexes (LaL$_3$). Note that many synthetic applications have been found not only of divalent samarium reagents such as SmI$_2$[2–4] and tetravalent cerium reagents such as cerium ammonium nitrate (CAN)[5] but also of lanthanide metals such as Ce, Sm, and Yb, which are, however, beyond the scope of this chapter.

The trivalent state is the most common oxidation state for lanthanides. Generally speaking, lanthanide(III) complexes (exept for the alkoxides) are hard Lewis acids and hence they have a strong affinity toward hard bases such as oxygen donor ligands. This strong oxophilicity is one of the most important characteristics of lanthanide(III) reagents and provides the principal basis for their unique synthetic applications.

2. Cerium(III) reagents

Cerium(III) salts, particularly CeCl$_3$, have found many synthetic applications as inexpensive, unique reagents (stoichiometric use). Of the most widespread use are the NaBH$_4$–CeCl$_3$ system that allows the regioselective reductions of α,β-enones (Scheme 12.1)[6,7] and the stereoselective reduction of ketones

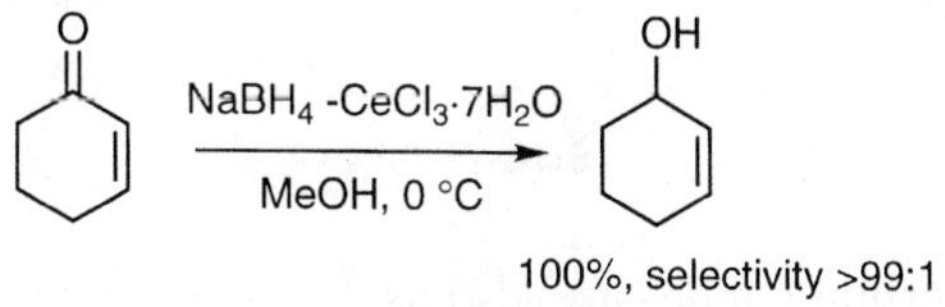

100%, selectivity >99:1

Scheme 12.1

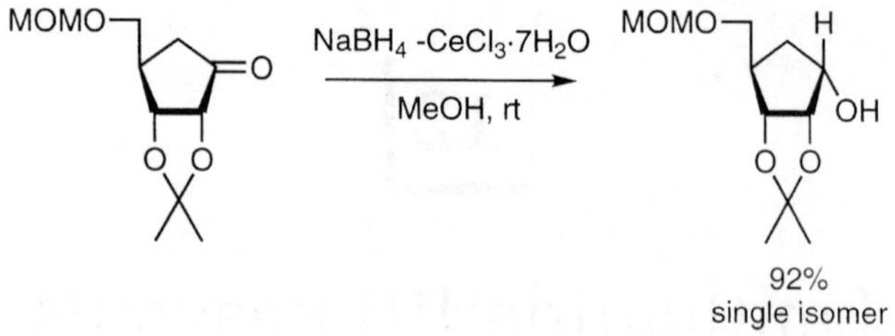

Scheme 12.2

(Scheme 12.2),[8] and the organocerium reagents ('RCeCl$_2$'), generated *in situ* from organolithiums or Grignard reagents and CeCl$_3$, which can smoothly undergo the normal 1,2-addition reactions to carbonyl compounds including α,β-enones (Scheme 12.3).[9] Particularly notable is that the use of CeCl$_3$ as an additive in these reactions significantly supresses unfavourable side reactions (enolization, reduction, conjugate addition, etc.) which are often encountered with the use of organolithiums or Grignard reagents alone. In addition, a higher stereoselectivity is often observed by virtue of the chelating ability of the cerium reagents involved.

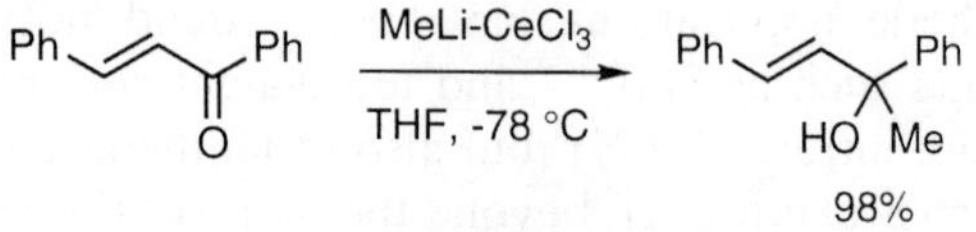

Scheme 12.3

In a similar manner, treatment of lithium enolates with CeCl$_3$ generates the cerium enolates which undergo aldol reactions usually in a higher yield than with lithium enolates alone,[10,11] particularly when sterically hindered and/or easily enolizable carbonyl partners are used (Scheme 12.4).[11] The stereo-selectivity of Ce-enolate reactions is essentially the same as that of Li-enolate reactions, since both reactions proceed through a cyclic transition state.

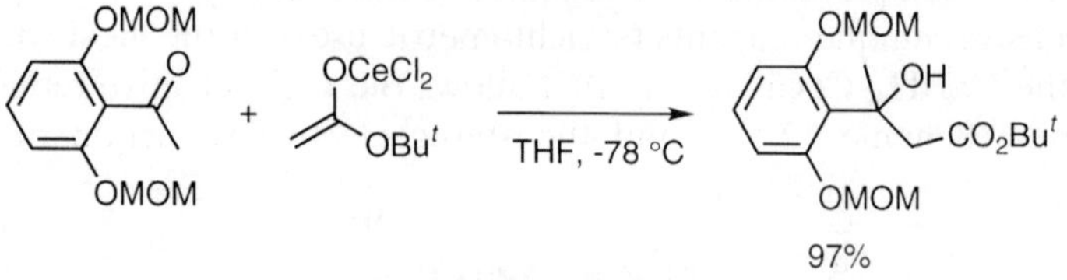

Scheme 12.4

Furthermore, cerium halides are also useful as promoters for the reductive coupling of α-haloketones with carbonyl compounds. The use of CeI$_3$ affords the α,β-unsaturated carbonyl compounds, whereas the $CeCl_3$–NaI or $CeCl_3$–$SnCl_2$ systems produce the aldols.[12]

Protocol 1.
Reduction of 2-cyclopentenone by $NaBH_4$–$CeCl_3$[6,7] (Scheme 12.5).

Caution! Carry out all procedures in a well-ventilated hood, and wear disposable vinyl or latex gloves and chemical-resistant safety goggles.

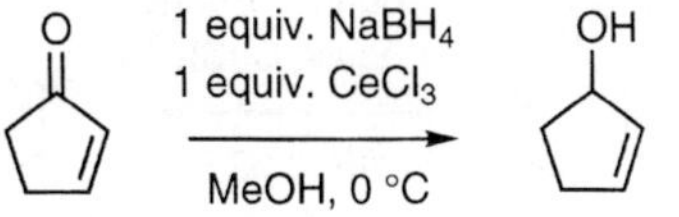

Scheme 12.5

Equipment
- An Erlenmeyer flask (300 mL) equipped with a stirring bar.

Materials
- Cerium chloride heptahydrate (FW 372.6), 19.0 g, 50 mmol   **irritant, hygroscopic**
- 2-Cyclopentenone (FW 82.10), 4.1 g, 50 mmol
- Methanol, 120 mL   **flammable, toxic**
- Sodium borohydride (FW 37.83), 1.9 g, 50 mmol   **flammable, corrosive**

1. Add cerium chloride heptahydrate (50 mmol), 2-cyclopentenone (50 mmol), methanol (120 mL) and stirring bar to the flask. Stir the mixture until the cerium chloride is completely dissolved.

2. Immerse the flask in an ice bath, add sodium borohydride (50 mmol) in portions over 5 min. Hydrogen is evolved vigorously with elevation of temperature to 15–20°C.

3. Stir the mixture for an additional 10 min, and then concentrate the resulting white suspension to about 30 mL under reduced pressure.

4. Add water (100 mL) to dissolve inorganic salts and extract the mixture with ether (3 × 50 mL). Dry the combined extracts over $MgSO_4$ and concentrate to a volume of *c.* 10 mL. Distil the residue under ordinary pressure to collect 2-cyclopenten-1-ol (3.7 g, 88%) with 96% purity.

Protocol 2.

Preparation of organocerium reagents and reaction with carbonyl compounds: reaction of *i*-propylcerium reagent with 2-cyclohexene-1-one[9] (Scheme 12.6)

Caution! Carry out all procedures in a well-ventilated hood, and wear disposable vinyl or latex gloves and chemical-resistant safety goggles.

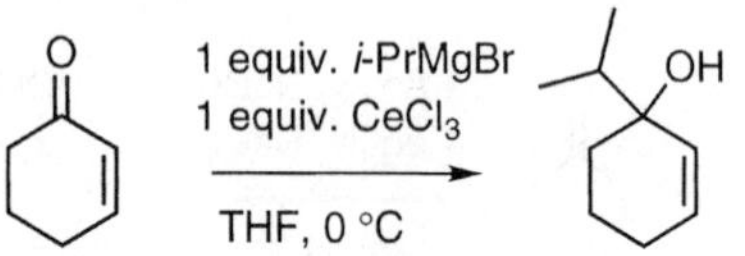

Scheme 12.6

Equipment

- A two-necked, round-bottomed flask (500 mL) fitted with a septum, a three-way stopcock, and a magnetic stirring bar. The three-way stopcock is connected to a vacuum/argon source.
- Vacuum/argon source.
- Syringes
- Mortar
- Oil bath

Materials

- Cerium chloride heptahydrate (FW 372.6), 18.0 g, 48 mmol — **irritant, hygroscopic**
- *i*-Propylmagnesium bromide (FW 147.3), 1.8 M solution in THF, 27 mL, 48 mmol — **flammable, moisture sensitive**
- 2-Cyclohexene-1-one (FW 96.13), 3.85 g, 40 mmol — **highly toxic**
- Dry THF — **flammable corrosive**

1. Quickly finely grind cerium chloride heptahydrate to a powder in a mortar and place it in a two-necked round-bottomed flask. Immerse the flask in an oil bath and heat gradually to 135–140°C with evacuation (<0.5 mmHg). After 1 h at this temperature, completely dry the cerium chloride *in vacuo* by stirring at the same temperature for an additional hour. While the flask is still hot, introduce argon gas and cool the flask in an ice bath. Add dry THF (200 mL) all at once with vigorous stirring. Remove the ice bath and stir the suspension overnight under argon at room temperature.

2. Add a THF (27 mL) solution of *i*-PrMgBr (48 mmol) at 0°C, and stir the mixture at that temperature for 1.5 h.

3. Add a THF (10 mL) solution of 2-cyclohexene-1-one (40 mmol) dropwise over 10 min.

4. After 30 min at 0°C, add an aqueous solution of acetic acid (3%, 150 mL) and extract with ether (3 × 50 mL). Wash the combined extracts with brine, a solution of NaHCO$_3$, and brine. Dry the extract over MgSO$_4$ and concentrate under reduced pressure. Distil the desired product (4.5 g, 80%) as a fraction boiling at 83–85°C at 15 mmHg. This sample may contain 3% of 1,4-addition product as determined by capillary GLC analysis (Silicon OV-17).

Protocol 3.
Aldol reaction with cerium enolate: synthesis of 2-(1-hydroxycyclohexyl)-1-phenyl-1-propanone[10,11] (Scheme 12.7)

Caution! Carry out all procedures in a well-ventilated hood, and wear disposable vinyl or latex gloves and chemical-resistant safety goggles.

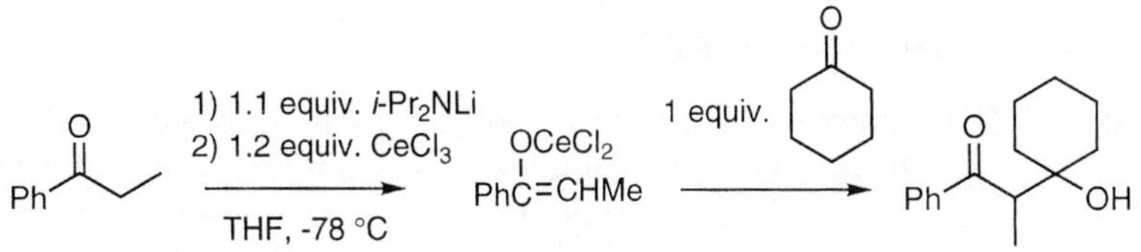

Scheme 12.7

Equipment

- A two-necked, round-bottomed flask (200 mL) fitted with a septum, a three-way stopcock, and a magnetic stirring bar. The three-way stopcock is connected to a vacuum/argon source
- Vacuum/argon source
- Syringes

Materials

- Di-isopropylamine (FW 101.2), 1.5 mL — **flammable, corrosive**
- *n*-BuLi (FW 64.06), 1.53 M solution in hexane, 7.2 mL, 11 mmol — **pyrophric, moisture-sensitive**
- Propiophenone (FW 134.2), 1.34 g, 10 mmol
- Cerium chloride, anhydrous (FW 246.5) suspension in THF (see Protocol 2) — **irritant, hygroscopic**
- Cyclohexanone (FW 98.15), 0.982 g, 11 mmol — **corrosive, toxic**
- Dry THF — **flammable, corrosive**

1. Flame dry the reaction vessel under dry argon. Quickly add a stirring bar while the flask is hot, and then refill the flask with argon through several vacuum cycles. Maintain a slightly possitive argon pressure throughout this reaction.

2. Add THF (15 mL) and di-isopropylamine (1.5 mL) and cool the mixture to −78°C with a dry ice–acetone bath. Add dropwise a hexane solution of *n*-BuLi (11 mmol) with stirring. After stirring for 10 min, add a THF (5 mL) solution of propiophenone (10 mmol).

3. After stirring for 1 h, add a THF (50 mL) suspension of anhydrous cerium chloride (12 mmol) (see Protocol 2) via a syringe and stir the mixture for 30 min at that temperature.

4. Add a THF (2 mL) solution of cyclohexanone (10 mmol).

5. Stirring for 30 min, then pour the mixture into an aqueous solution of acetic acid (3%, 50 mL). Extract with ether (3 × 50 mL), and then wash the combined extracts with brine, a solution of NaHCO₃, and brine. Dry the extract

Protocol 3. *Continued*

over $MgSO_4$ and concentrate under reduced pressure. Purify the crude product by flash chromatography (silica gel, hexane–ethylacetate, 5:1) to give 2-(1-hydroxycyclohexyl)-1-phenyl-1-propane (1.67 g, 72%), $[\alpha]^{20}_D$ −5.3°(*c* 1.2, EtOH).

---

3. Europium(III) reagents

Although Eu(III) complexes such as Eu(fod)$_3$ are familiar to organic chemists as NMR shift agents, the use of Eu complexes as synthetic reagents is a rather recent event. In 1983 Danishefsky *et al.* demonstrated the usefulness of (+)-Eu(hfc)$_3$ as an asymmetric catalyst for the hetero Diels–Alder reaction (Scheme 12.8). Since then, a variety of Eu(III) complexes, including EuCl$_3$ and Eu(dppm)$_3$, have been used as unique catalysts for different synthetic transformations.

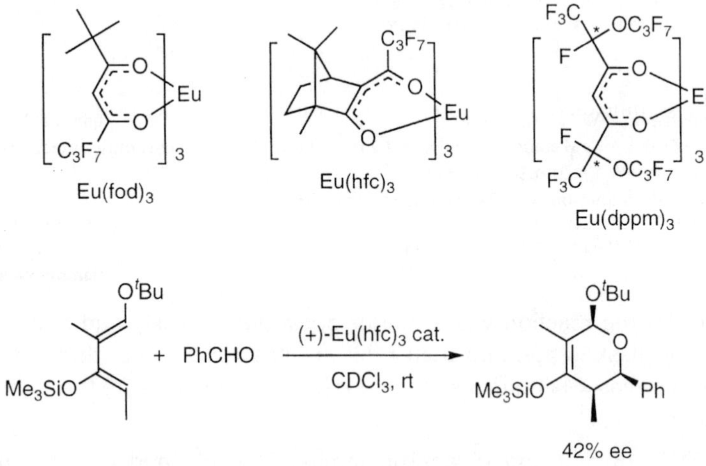

Eu(fod)$_3$ Eu(hfc)$_3$ Eu(dppm)$_3$

Scheme 12.8

To date, europium(III) complexes have found many applications as efficient Lewis acid catalysts for Diels–Alder reactions,[14,15] hetero Diels–Alder reactions,[14,16–18] aldol reactions with ketene silyl acetals (KSA),[19–25] Michael reactions of α,β-enones with KSA,[26] and cyanosilylation of aldehydes with silyl cyanides.[27] Compared with conventional Lewis acids such as TiCl$_4$, SnCl$_4$, and BF$_3$•OEt$_2$ (stoichiometric use), the europium catalyses are often quite unique in chemo- and stereoselectivities. For instance, Eu(dppm)$_3$ is effective as the catalyst for aldol reactions between aldehydes and KSA, but not effective for the reactions between aldehydes and ketone-derived enol silyl ethers.[21]

Scheme 12.9 illustrates the unique ability of Eu(dppm)$_3$ to discriminate the aldehydes, where *p*-nitrobenzaldehyde is less reactive than benzaldehyde. This reversal of reactivity is ascribed to preferential activation of benzaldehyde over *p*-nitrobenzaldehyde by the Eu catalyst. A similar molecular discrimination by Eu(fod)$_3$ is achieved between ketones and α,β-enones in the reactions with KSA in which the Eu catalyst preferentially activates the latter in preference to the former.[22]

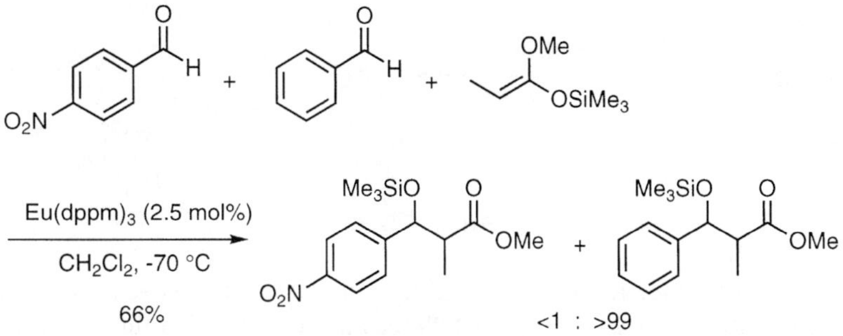

Scheme 12.9

Of more significance is the unique stereocontrolling ability of the Eu-catalysts. For example, the Eu(fod)$_3$-catalysed aldol reactions of chiral α-alkoxy aldehydes with KSA provides high levels of diastereofacial selection and simple diastereoselections, where the stereochemistry of the major product is effectively modulated by changing the protecting group (G) through the molecular recognition (chelation vs. non-chelation) by the Eu catalyst (Scheme 12.10).[23–25] A similar stereomodulation by the Eu catalyst is attained in the hetero Diels–Alder reactions of α-amino and α-alkoxy aldehydes (Scheme 12.11).[28,29] Interestingly, the Michael addition of the 4-siloxy cyclopentenone with the acetate-derived KSA is shown to proceed with the sterically less favourable *syn* diastereofacial selectivity (Scheme 12.12).[30]

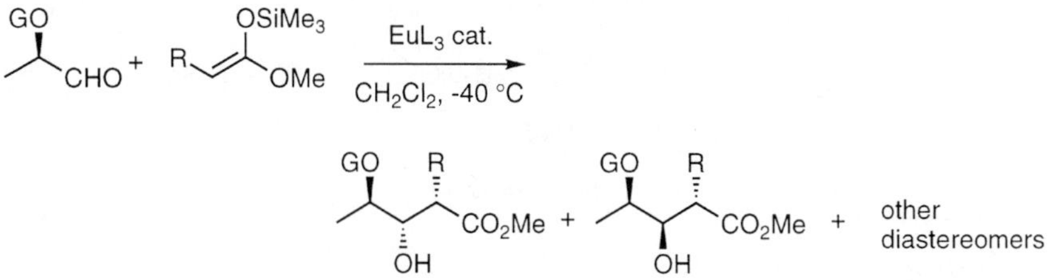

G= TBDMS, R=Me (85% *E*)	94 : 2 : 4
G=Bn, R=OBn (100% *Z*)	2 : 98 : 2

Scheme 12.10

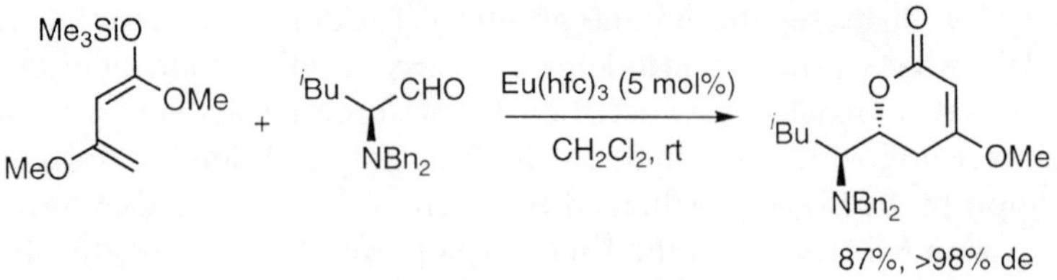

Scheme 12.11

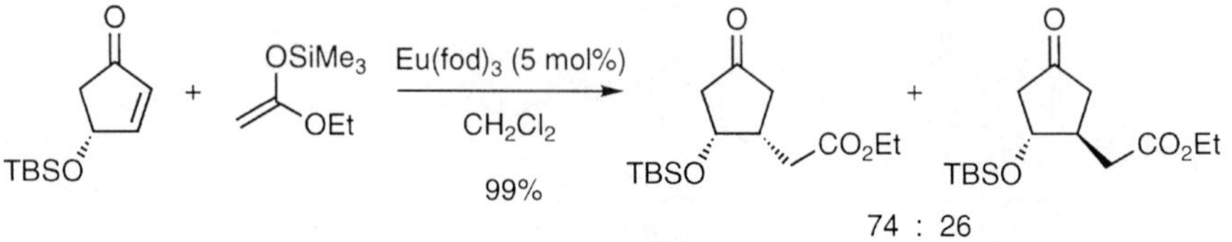

Scheme 12.12

As far as asymmetric catalysis by chiral europium complexes is concerned, modest enantioselectivities (up to 42% ee) are observed in the (+)-Eu(hfc)$_3$ catalysed hetero Diels–Alder reactions (Scheme 12.8),[13] whereas almost no asymmetric inductions (up to 10% ee) are observed in the (−)-Eu(dppm)$_3$-catalysed aldol reactions.[21] More recently, the chiral europium complex, prepared *in situ* from Eu(OTf)$_3$ and the disodium salt of a chiral diamine, has been reported to provide *c.* 40% ee in the aldol reactions of benzaldehyde with KSA.[31] Thus, the development of asymmetric catalysis by chiral Eu complexes remains a challenge.

Protocol 4.
Diels–Alder reaction[17] (Scheme 12.13)

Caution! Carry out all procedures in a well-ventilated hood, and wear disposable vinyl or latex gloves and chemical-resistant safety goggles.

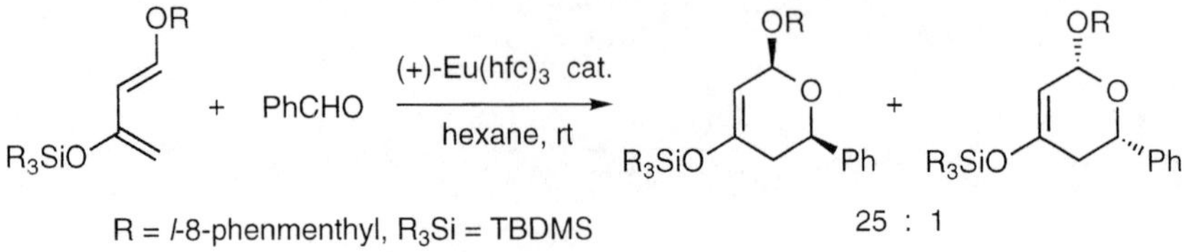

Scheme 12.13

Equipment

- A two-necked, round-bottomed flask fitted with a septum, a three-way stopcock, and a magnetic stirring bar. A three-way stopcock is connected to a vacuum/argon source
- Vacuum/argon source
- Syringes

12: Lanthanide(III) Reagents

Materials

- Benzaldehyde (FW 106.12), 234 mg, 2.20 mmol   **highly toxic, cancer suspect agent**
- *l*-8-Phenmethyoxy diene (FW 414.71), 912 mg, 2.20 mmol
- (+)-Eu(hfc)₃ (FW 1193.73), 131 mg, 0.11 mmol   **hygroscopic**
- Dry Hexane, 12 mL   **flammable, irritant**
- Triethylamine (FW 101.19), 8 mL   **flammable, corrosive**
- Methanol 4 mL   **flammable, toxic**

1. Add a solution of benzaldehyde (2.20 mmol) and *l*-8-phenmethyoxy diene (2.20 mmol) in hexane (12 mL) to the flask. Cool the mixture to −20 °C and add (+)-Eu(hfc)₃ (0.11 mmol). Keep the reaction mixture at between −10 and −20 °C for 60 h.[a]

2. Add triethylamine (8 mL) and methanol (4 mL) and allow the mixture to warm to room temperature. Remove the volatiles *in vacuo*, and pass the crude material through a plug of silica gel. Wash the plug with ethyl acetate. Concentrate the organics under reduced pressure to give enol silyl ether (1.10 g, 95%, dr 25:1). Crystallize this material from ethanol and recrystallize this residue from the mother liquors to give optically pure compound as white crystals (0.69 g, 60%); m.p. 59.5–60.8 °C, [α]²³_D +47.3° (c 1.1, CHCl₃).

[a] After which time NMR analysis showed that no starting aldehyde or diene remained.

Protocol 5.
Eu(fod)₃-catalysed aldol reaction: preparation of optically pure 2-deoxy-D-ribonolactone[23] (Scheme 12.14)

Caution! Carry out all procedures in a well-ventilated hood, and wear disposable vinyl or latex gloves and chemical-resistant safety goggles.

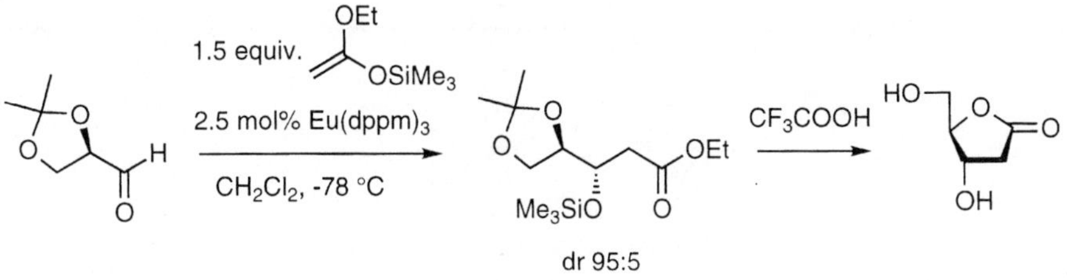

dr 95:5

Scheme 12.14

Equipment

- A two-necked, round-bottomed flask (30 mL) fitted with a septum, a three-way stopcock, and a magnetic stirring bar. The three-way stopcock is connected to a vacuum/argon source
- A round-bottomed flask (30 mL) fitted with a stirring bar
- Vacuum/argon source.
- Syringes

211

Takeshi Nakai and Katsuhiko Tomooka

Protocol 5. *Continued*

Materials

- D-Glyceraldehyde acetonide (FW 130.1), 130 mg, 1.0 mmol
- 1-Ethoxy-1-trimethylsiloxyethene (FW 160.3), 240 mg, 1.5 mmol
- $(-)$-Eu(dppm)$_3$ (FW 2072), 30 w/v% solution in CF$_2$ClCFCl$_2$,[a] 172 μL, 0.025 mmol
- Dry CH$_2$Cl$_2$ **toxic, irritant**

1. Flame dry the reaction vessel under dry argon. Quickly add a stirring bar while the flask is hot, and then refill the flask with argon through several vacuum cycles. Maintain a slightly positive argon pressure throughout this reaction.

2. Add a solution of D-glyceraldehyde acetonide (1 mmol) and 1-ethoxy-1-trimethylsiloxyethene (1.5 mmol) in CH$_2$Cl$_2$ (4 mL) to the flask with the aid of a syringe through the septum and then cool it to −78°C. Add a solution of $(-)$-Eu(dppm)$_3$ in CF$_2$ClCFCl$_2$ (172 μL, 0.025 mmol) and then stir the mixture for 2 h at the same temperature.

3. Add saturated aqueous NH$_4$Cl solution and extract with ethyl acetate (3 × 20 mL). Wash the combined extracts with brine, dry over MgSO$_4$, and concentrate the crude product under reduced pressure (diastereomer ratio; *anti/syn* = 95:5).

4. Dissolve a crude product in THF (2 mL) and to this add trifluoracetic acid (0.5 mL) and one drop of water.

5. After stirring for 1 day, add aqueous NaHCO$_3$ solution and extract with ethyl acetate (5 × 30 mL). Wash the combined extracts with brine and concentrate under reduced pressure. Purify the crude product by flash chromatography (silica gel; hexane–ethylacetate, 5:1) to give optically pure 2-deoxy-D-ribonolactone (0.118 g, 90%), $[\alpha]^{20}_D$ 25.3° (c 1.2, EtOH).

[a] Purchased from Dai-ichi Kagaku Yakuhin.

212

Protocol 6.
(+)-Eu(dppm)₃-catalysed Michael reaction: reaction of a
ketenesilylacetal with 2-cyclopentenone.[26] (Scheme 12.15)

Caution! Carry out all procedures in a well-ventilated hood, and wear disposable vinyl or latex gloves and chemical-resistant safety goggles.

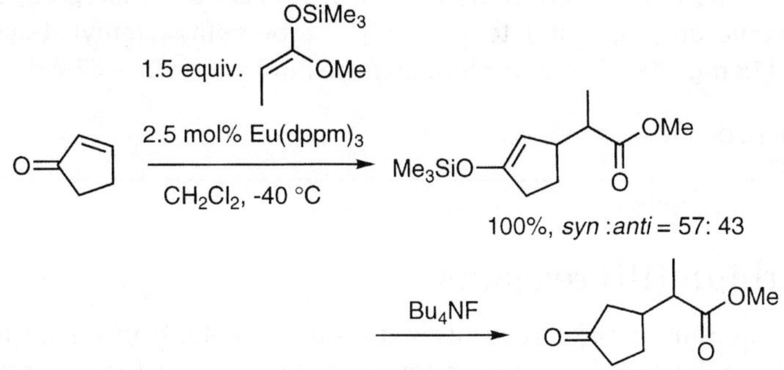

Scheme 12.15

Equipment

- A two-necked, round-bottomed flask (30 mL) fitted with a septum, a three-way stopcock, and a magnetic stirring bar. The three-way stopcock is connected to a vacuum/argon source
- Vacuum/argon source
- Syringes

Materials

- 2-Cyclopentenone (FW 82.10), 41 mg, 0.5 mmol
- (E)-1-[(trimethylsilyl)oxy]-1-methoxy-1-propene
- (+)-Eu(dppm)₃, 30 w/v% solution in CF₂ClCFCl₂[a]
- Dry CH₂Cl₂ — **toxic, irritant**
- Tetrabutylammonium fluoride (FW 261.5), 1.0 M solution in THF — **flammable, irritant**
- Dry THF — **flammable, corrosive**

1. Flame dry the reaction vessel under dry argon. Quickly add a stirring bar while the flask is hot, and then refill the flask with argon through several vacuum cycles. Maintain a slightly positive argon pressure throughout this reaction.

2. Add a solution of 2-cyclopentenone (0.5 mmol) in CH₂Cl₂ (1.5 mL) and a 30 w/v% solution of (+)-Eu(dppm)₃ (0.0125 mmol) in CF₂ClCFCl₂ to the flask with the aid of a syringe through the septum and then cool it to −40 °C. Add (E)-1-[(trimethylsilyl)oxy]-1-methoxy-1-propene (0.75 mmol) to the stirred solution.

3. After stirring for 1 h, pour the mixture into a aqueous saturated solution of NaHCO₃. Extract with ethyl acetate (3 × 30 mL). Wash the combined extracts with brine, dry over MgSO₄, and concentrate the crude product under reduced pressure.

213

Takeshi Nakai and Katsuhiko Tomooka

Protocol 6. *Continued*

4. Dissolve a crude product in THF (1 mL) and to this add a 1.0 M THF solution of Bu₄NF (1 mL) at room temperature. After stirring for several hours and monitor the disappearance of the silyl enol ether by TLC.

5. Pour the mixture into a water (20 mL) and extract with ethyl acetate (5 × 30 mL). Wash the combined extracts with brine and concentrate under reduced pressure. The crude product was purified by flash chromatography (silica gel, hexane–ethyl acetate) to give 3-[(1'-carbomethoxy)ethyl]-1-cyclopentanone (78 mg, 100%) as a diastereomer mixture (*syn:anti* = 57:43).

ᵃ Purchased from Dai-ichi Kagaku Yakuhin.

4. Ytterbium(III) reagents

The most important of these reagents is ytterbium triflate, $Yb(OTf)_3$, which has found wide application as a catalyst for several C–C bond-forming reactions. The most notable is the $Yb(OTf)_3$-catalysed aldol reaction of aldehydes (or acetals) with enol silyl ethers which proceeds smoothly in dichloromethane and, more significantly, even in aqueous media (Scheme 12.16).[32–36] It should be noted that a similar catalytic activity is also observed with $Nd(OTf)_3$, $Gd(OTf)_3$, and $Lu(OTf)_3$.

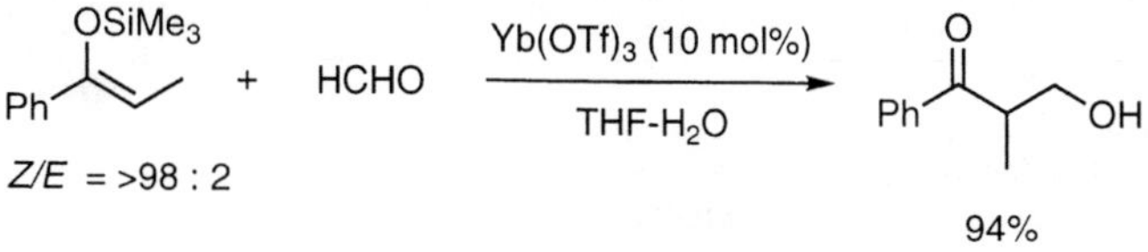

Scheme 12.16

In addition, $Yb(OTf)_3$ is also effective as a catalyst for the Michael reactions of α,β-enones with enol silyl ethers,[37] the Friedel–Crafts acylations with acid anhydrides,[38,39] and the Diels–Alder reactions, wherein $Yb(OTf)_3$ is, however, often less active than $Sc(OTf)_3$. A major merit of the Yb-catalysed reactions is that the catalyst is stable in water and hence can be easily recovered from the aqueous layer and reused.[35]

More significantly, the chiral Yb triflate, prepared *in situ* from $Yb(OTf)_3$, (*R*)-binaphthol, and a tertiary amine, is shown to serve as an efficient asymmetric catalyst for the Diels–Alder reaction (Scheme 12.17).[40] The most striking feature of the asymmetric catalysis is that the chiral Yb catalyst, when combined with 3-phenylacetylacetone as an additive (another achiral ligand), shows the opposite sense of enantioselection to that observed in the cases where the substrate itself or 3-acetyl-1,3-oxazolin-2-one acts as another ligand

214

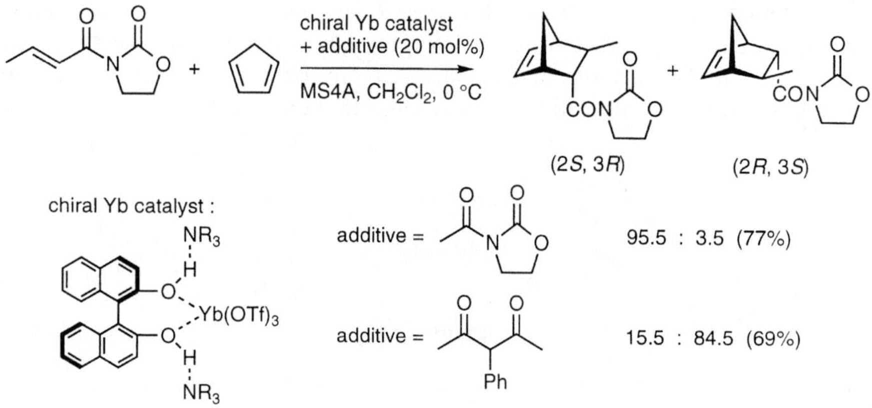

Scheme 12.17

as depicted in Scheme 12.17. The remarkable enantiocontrol by a proper choice of achiral ligands is explained in terms of the significant difference in complexation mode between the chiral catalytic species invoved and the imide dienophile. A similar phenomenon is observed when other lanthanide triflates such as $Lu(OTf)_3$, $Tm(OTf)_3$, and $Ho(OTf)_3$ are used in place of $Yb(OTf)_3$.

Further notable ytterbium reagents are $Yb(fod)_3$ as an efficient catalyst for the ene-type reaction of aldehydes with vinylic ethers,[41] and $Yb(CN)_3$ as an active catalyst for the cyanosilylation of carbonyls, including enolizable ketones and α,β-enones (Scheme 12.18),[42] and for the regioselective ring opening of oxiranes and aziridines with Me_3SiCN (Scheme 12.19).[43,44]

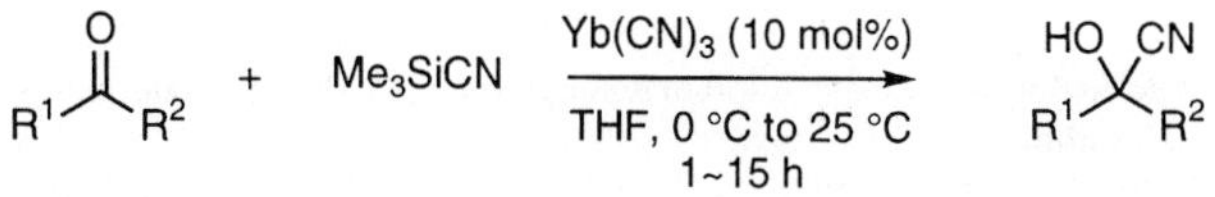

Scheme 12.18

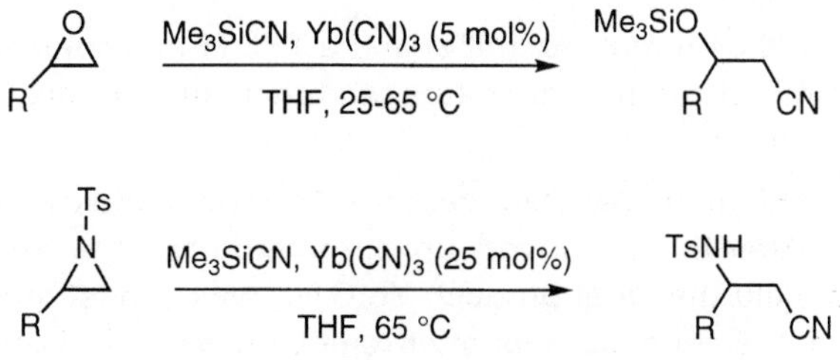

Scheme 12.19

Ytterbium complexes have also found stoichiometric uses in organic synthesis. The alkylytterbium complex, generated *in situ* from $Yb(OTf)_3$ or $YbCl_3$ and an alkyl-lithium or Grignard reagent, provides a unique and high stereoselectivities in the additions to ketones (Scheme 12.20).[45,46]

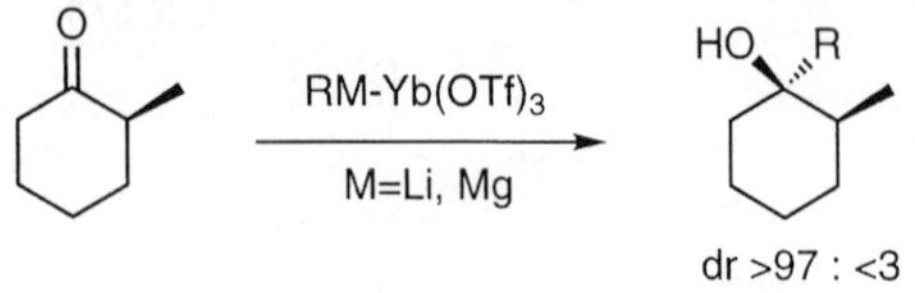

dr >97 : <3

Scheme 12.20

Protocol 7.
$Yb(OTf)_3$-catalysed aldol reaction: reaction of an enol trimethylsilyl ether with formaldehyde. General procedure[36] (Scheme 12.21)

Caution! Carry out all procedures in a well-ventilated hood, and wear disposable vinyl or latex gloves and chemical-resistant safety goggles.

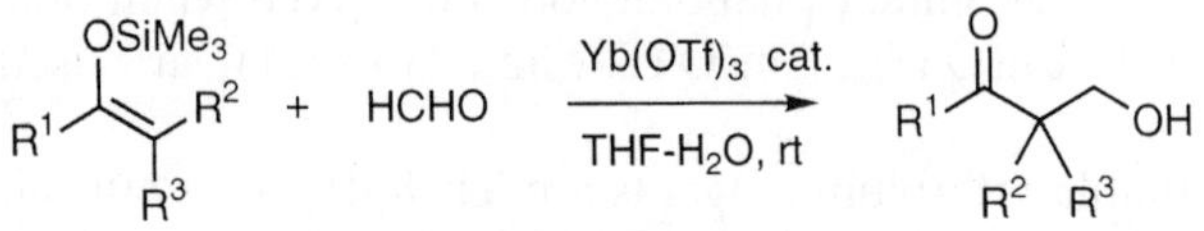

Scheme 12.21

Equipment
• A round-bottomed flask fitted with a stirring bar

Materials
• Formaldehyde (FW 30.03), 40 wt. % solution in water, 1 mL   **highly toxic, cancer-suspect agent**
• Enol silyl ether, 0.4 mmol
• $Yb(OTf)_3$ (FW 620.3), 25 mg, 0.04 mmol   **irritant, hygroscopic**
• THF   **flammable, corrosive**

1. Add commercial formaldehyde solution (40% water solution, 1 mL) and THF (3 mL) to the flask.

2. Add $Yb(OTf)_3$ (0.04 mmol, 10 mol%) and a THF (1 mL) solution of enol silyl ether (0.4 mml) at room temperature, and then stir the mixture for 24 h at this temperature.

3. Remove the THF under reduced pressure. Add water and extract the product with CH_2Cl_2. After the usual work-up, chromatograph the crude product on silica gel to yield the aldol product. $Yb(OTf)_3$ was almost quantitatively recovered from the aqueous layer after removing water and could be reused.

5. Lanthanum(III) reagents

Recently La$_3$(OBu^t)$_9$ has been found to serve as an efficient Lewis base catalyst for C–C bond-forming reactions such as the aldol reaction of α-chloro ketones with aldehydes, the nitroaldol (Henry) reactions of aldehydes with nitroalkanes, and the intramolecular aldol reactions of 1,5-diketones.[47] Based on these findings, Shibasaki's group has developed several (*R*)- or (*S*)-binaphthol (BINOL)-based chiral bimetallic lanthanum complexes, LaM$_3$tris(binaphthoxide) LMB (M = Li, Na, K) (Scheme 12.22), which serve as excellent asymmetric catalysts for the nitroaldol reaction, the hydrophosphonylation of aldehydes and imines, the Diels–Alder reaction, and the Michael reaction.[48]

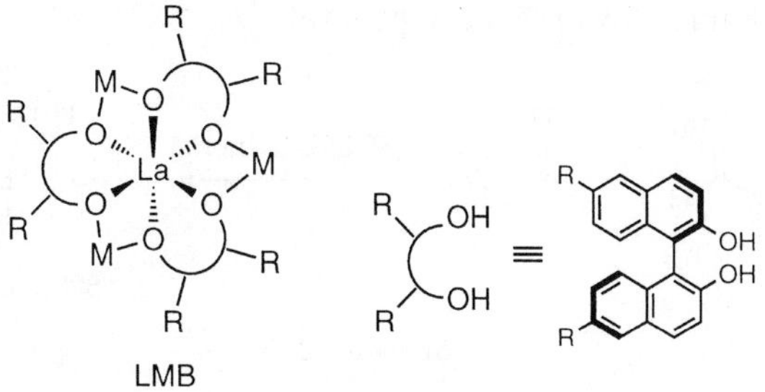

LMB

Scheme 12.22

Of special value is the enantioselective nitroaldol reaction catalysed by (*R*)-LLB, prepared *in situ* from either LaCl$_3$•7H$_2$O, the dilithium salt of (*R*)-BINOL, and NaO*t*Bu or from La(O*i*Pr)$_3$ and (*R*)-BINOL, and *n*-BuLi, and then H$_2$O, which affords the nitroaldol in a high % ee (Scheme 12.23).[49,50] Interestingly, the use of either the second generation LLB-II catalyst (LLB + LiOH) or the 6,6'-disubstituted BINOL-based LLB provides a higher % de and % ee in the reactions leading to a diastereomeric mixture of the nitroaldols.

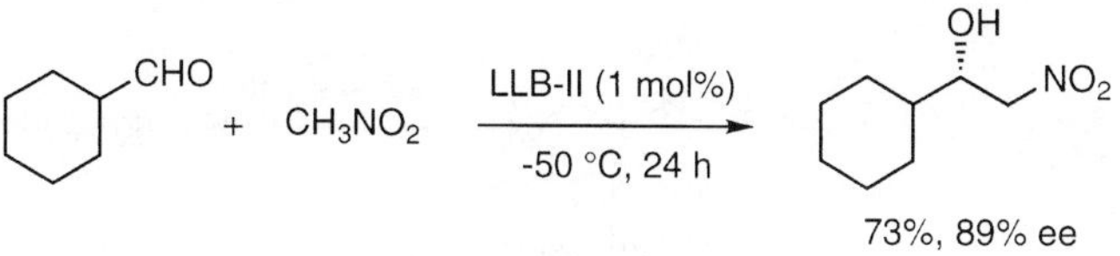

73%, 89% ee

Scheme 12.23

LLB also serves as a an efficient asymmetric catalyst for the Diels–Alder reaction[49] and the hydrophosphonylation of aldehydes,[51–53] but not for the Michael reaction and the hydrophosphonylation of imines. Significantly, however, LSB and LPB are shown to work as an excellent asymmetric catalyst for the Michael reaction (Scheme 12.24)[54,55] and the hydrophosphonylation of

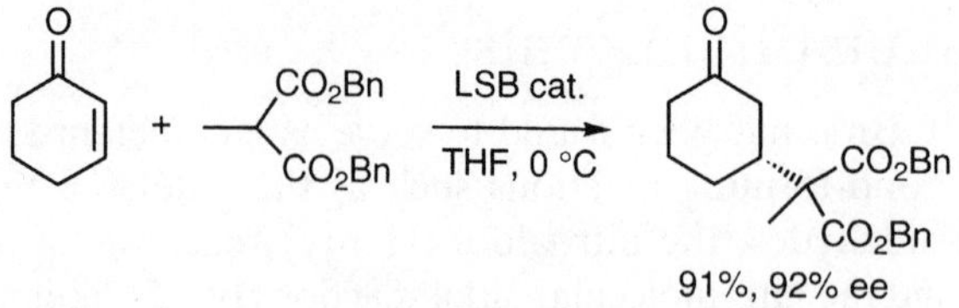

Scheme 12.24

imines (Scheme 12.25),[56] respectively. It should be noted that similar chiral bimetallic complexes derived from other lanthanides such as EuMB, GdMB, and PrMB (M = Li, Na, K) often show a comparably high catalytic activity, but provides a much lower % ee in general.

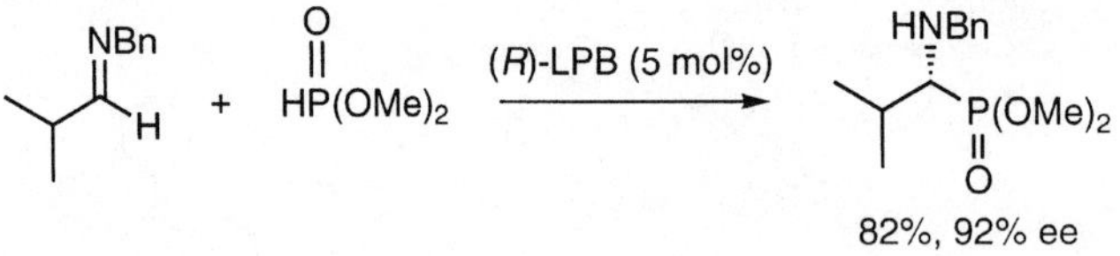

82%, 92% ee

Scheme 12.25

Protocol 8.
Catalytic asymmetric nitroaldol reaction[50]

Caution! Carry out all procedures in a well-ventilated hood, and wear disposable vinyl or latex gloves and chemical-resistant safety goggles.

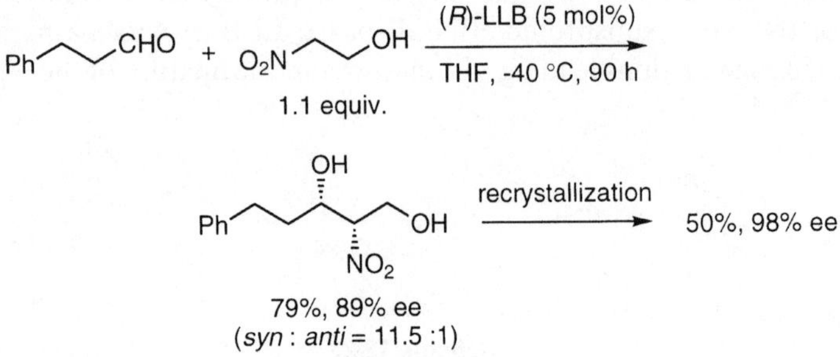

79%, 89% ee
(*syn : anti* = 11.5 :1)

Scheme 12.26

Equipment

- A two-necked, round-bottomed flask fitted with a septum, a three-way stopcock, and a magnetic stirring bar. A three-way stopcock is connected to a vacuum/argon source
- Vacuum/argon source
- Syringes

218

Materials

- (*R*)-(+)-1,1′-Bi-2-naphtol (FW 286.33), 6.5 g, 22.7 mmol — **irritant**
- *n*-Butyllithium (FW 64.06), 1.60 M solution in hexanes, 28.4 mL, 45.4 mmol — **flammable, moisture sensitive**
- $LaCl_3 \cdot 7H_2O$ (FW 371.38), 3.12 g, 8.40 mmol — **irritant, hygroscopic**
- NaO-*t*-Bu (FW 96.11), 242 mg, 2.52 mmol — **flammable, corrosive**
- 2-Nitroethanol (FW 91.07), 19.1 mg, 0.21 mmol — **irritant**
- Hydrocinnamaldehyde (FW 134.18), 25.5 mg, 0.19 mmol — **irritant**
- Dry THF — **flammable, corrosive**

1. Add (*R*)-(+)-1,1′-Bi-2-naphthol (22.7 mmol) to the flask. Dry at 50 °C for 2 h under reduced pressure. Add dry THF (119 mL) under Ar,

2. Add a hexanes solution of *n*-BuLi (1.60 N, 28.4 mL, 45.4 mmol) at 0 °C.

3. Charge a suspension of $LaCl_3 \cdot 7H_2O$ (8.40 mmol) in THF (100 mL) to the different flask. Sonicate for 30 min at room temperature. To this suspension, add the above-prepared solution of dilithium binaphthoxide and a THF solution of NaO-*t*-Bu (0.52 N, 4.85 mL, 2.52 mmol), dropwise at room temperature.

4. Vigorously stir the mixture overnight at room temperature, and then for 48 h at 50 °C. Stand the mixture at room temperature. The resulting supernatant may contain an optically active LLB complex (0.03 M).

5. Charge a THF solution of optically active LLB complex (0.03 M, 209 μL, 0.00627 mmol) to the flask. Add THF (500 μL) and cool the mixture to −40 °C and stir for 30 min.

6. Add a THF (200 μL) solution of 2-nitroethanol (15 μL, 0.21 mmol), and stir the mixture for 30 min at the same temperature.

7. Add hydrocinnamaldehyde (25 μL, 0.19 mmol), and then stir the mixture for 90 h at −40 °C).

8. Add a 1N HCl and then extract with AcOEt (30 mL); wash the organic layer with brine. Dry the extract over Na_2SO_4 and concentrate under reduced pressure. Purify the crude product by chromatography (Lobar Lichropep RI-8 prepacked column, CH_3CN-H_2O 1:1), to give (2*S*,3*S*)-2-nitro-5-phenyl-1,3-pentanediol (70%, *syn:anti* = 11:1, 91% ee). Determine the enantiomeric excess HPLC analysis using DAISEL CHIRALPAK AD (hexane: *i*-PrOH 9:1).

---

6. Other lanthanide(III) reagents

So far, the synthetic utilities of the four important classes of trivalent lanthanide reagents have been described. Outlined below are selected examples of the synthetic uses of other trivalent lanthanide complexes as catalysts.

A series of lanthanide isopropoxides, including $La(OPr^i)_3$ $Ce(OPr^i)_3$, $Sm(OPr^i)_3$, $Gd(OPr^i)_3$, have been found to exhibit an efficient catalytic activity in the Meerwein–Ponndorof–Verley (MPV) reductions. These catalysts

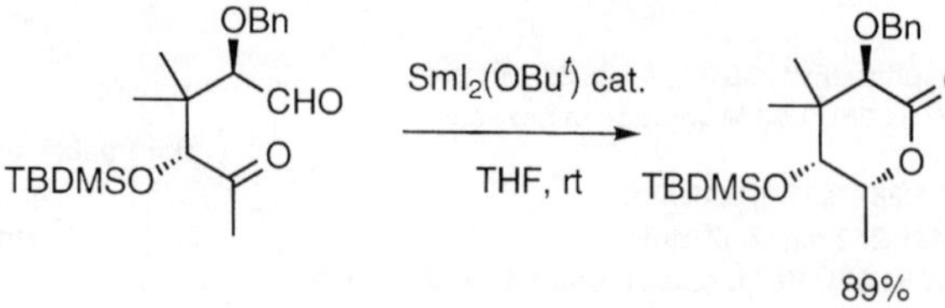

Scheme 12.27

are far more active than the existing catalysts such as $Al(OPr^i)_3$. Interestingly, $Sm(OBu^t)I_2$, prepared *in situ* from SmI_2 and di-*t*-butylperoxide, works well as a catalyst for MPV reduction[57-60] and also for the intramolecular Tishchenko reaction (Scheme 12.27).[61] These lanthanide alkoxides also serve as an efficient catalyst for the transhydrocyanation of acetone cyanohydrin to aldehydes or ketones (Scheme 12.28).[62]

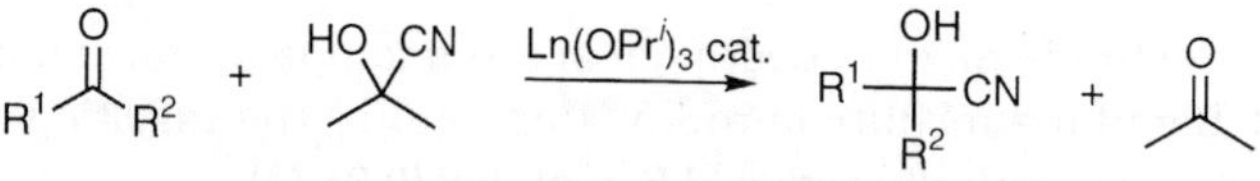

Scheme 12.28

Lu(III) and Sm(III) hydride complexes having pentamethylcyclopentadienyl (Cp') ligands work as an extremely active homogeneous catalyst for the hydrogenation of terminal olefins in particular.[63,64] Of special value is the asymmetric hydrogenation using a chiral samarium catalyst with a chiral auxiliary on the Cp ring (Scheme 12.29).[65]

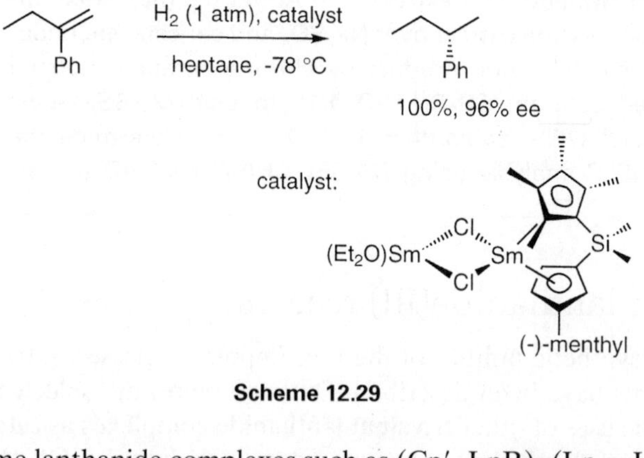

catalyst:

(-)-menthyl

Scheme 12.29

Finally, some lanthanide complexes such as $(Cp'_2 LnR)_2$ (Ln = La, Sm, Yb, Nd, Lu; R = H, Me) have been shown to be extremely active homogeneous catalysts for the polymerization of ethylene (not propylene), methyl methacrylate (MMA), ε-caprolactam, and δ-valerolactone.[66,67] Of interest is the lanthanide-catalyzed polymerization of MMA which proceeds in a living polymerization fashion to produce polymers of high molecular weight ($>5 \times 10^5$) with a narrow polydispersity ($M_w/M_n = 1.02-1.03$) and a high syndiotacticity ($>95\%$).[68,69]

References

1. Imamoto, T. *Lanthanides in Organic Synthesis*. Academic Press; London, **1994**.
2. Kagan, H. B.; Namy, J. L. *Tetrahedron* **1986**, *42*, 6573–6614.
3. Imamoto, T. *J. Synth. Org. Chem.* **1988**, *46*, 540–552.
4. Inanaga, J. *J. Synth. Org. Chem.* **1989**, *47*, 200–211.
5. Ho, T.-L. In *Organic synthesis by oxidation with metal compounds* (ed. M. J. Mijs, C. R. H. I. de Jonge). Plenum, New York, **1986**.
6. Luche, J.-L. *J. Am. Chem. Soc.* **1978**, *100*, 2226–2227.
7. Gemal, A. L.; Luche, J.-L. *J. Am. Chem. Soc.* **1981**, *103*, 5454–5459.
8. Bestmann, H. J.; Roth, D. *Synlett* **1990**, 751–753.
9. Imamoto, T.; Takiyama, N.; Makamura, K.; Hatajima, T.; Kamiya, Y. *J. Am. Chem. Soc.* **1989**, *111*, 4392–4398.
10. Imamoto, T.; Kusumoto, T.; Yokoyama, M. *Tetrahedron Lett.* **1983**, *24*, 5233–5236.
11. Nagasawa, K.; Kanbara, H.; Matsushita, K.; Ito, K. *Tetrahedron Lett.* **1985**, *26*, 6477–6480.
12. Fukuzawa, S.; Tsuruta, T.; Fujinami, T.; Sakai, S. *J. Chem. Soc. Perkin Trans. 1* **1987**, 1473–1477.
13. Bednarski, M.; Maring, C.; Danishefsky, S. *Tetrahedron Lett.* **1983**, *24*, 3451–3454.
14. Morrill, T. C.; Clark, R. A.; Bilobran, D.; Youngs, D. S. *Tetrahedron Lett.* **1975**, 397–400.
15. Ishar, M. P. S.; Wali, A.; Gandhi, R. P. *J. Chem. Soc. Perkin Trans. 1* **1990**, 2185–2192.
16. Bednarski, M.; Danishefsky, S. *J. Am. Chem. Soc.* **1983**, *105*, 3716–3717.
17. Bednarski, M.; Danishefsky, S. *J. Am. Chem. Soc.* **1986**, *108*, 7060–7067.
18. Quimpére, M.; Jankowski, K. *J. Chem. Soc. Chem. Commun.* **1987**, 676–677.
19. Takai, K.; Heathcock, C. H. *J. Org. Chem.* **1985**, *50*, 3247–3251.
20. Vougioukas, A. E.; Kagan H. B. *Tetrahedron Lett.* **1987**, *28*, 5513–5516.
21. Mikami, K.; Terada, M.; Nakai, T. *J. Org. Chem.* **1991**, *56*, 5456–5459.
22. Hanyuda, K.; Hirai, K.; Nakai, T. *Synlett* **1997**, 31–32.
23. Mikami, K.; Terada, M.; Nakai T. *Tetrahedron Asymm.* **1991**, *2*, 993–996.
24. Terada, M.; Gu, J.-H.; Deka, D. C. Mikami, K.; Nakai, T. *Chem. Lett.* **1992**, 29–32.
25. Mikami, K.; Terada, M.; Nakai, T. *J. Chem. Soc. Chem. Commun.* **1993**, 343–345.
26. Terada, M.; Nakai, T.; Mikami, K. *Inorg. Chim. Acta* **1994**, *222*, 377–380.
27. Gu, J.-H.; Okamoto, M.; Terada, M.; Mikami, K.; Nakai, T. *Chem. Lett.* **1992**, 1169–1172.
28. Midland, M. M.; Afonso, M. M. *J. Am. Chem. Soc.* **1989**. *111*, 4368–4371.
29. Midland, M. M.; Koops, R. W. *J. Org. Chem.* **1990**, *55*, 4647–4650, 5058–5065.
30. Otera, J.; Fujita, Y.; Fukuzumi, S.; Hirai, K.; Gu, J.-H.; Nakai, T. *Tetrahedron Lett.* **1995**, *36*, 95–98.

31. Uotsu, K.; Sasai, H.; Shibasaki, M. *Tetrahedron Asymm.* **1995**, *6*, 71–74.
32. Kobayashi, S. *Chem. Lett.* **1991**, 2187–2190.
33. Kobayashi, S.; Hachiya, I. *Tetrahedron Lett.* **1992**, *33*, 1625–1628.
34. Kobayashi, S.; Hachiya, I.; Takahori, T. Synthesis **1993**, 371–373.
35. Kobayashi, S. *Synlett* **1994**, 689–701.
36. Kobayashi, S.; Hachiya, I. *J. Org. Chem.* **1994**, *59*, 3590–3596.
37. Kobayashi, S.; Hachiya, I.; Takahori, T.; Araki, M.; Ishitani, H. *Tetrahedron Lett.* **1992**, *33*, 6815–6818.
38. Kawada, A.; Mitsumura, S.; Kobayashi, S. *J. Chem. Soc., Chem. Commun.* **1993**, 1157–1158.
39. Mikami, K.; Kotera, O.; Motoyama, Y.; Sakaguchi, H.; Maruta, M. Synlett **1996**, 171–172.
40. Kobayashi, S.; Ishitani, H. *J. Am. Chem. Soc.* **1994**, *116*, 4083–4084.
41. Deaton, M. V.; Ciufolini, M. A. *Tetrahedron Lett.* **1993**, *34*, 2409–2412.
42. Matsubara, S.; Takai, T.; Utimoto, K. *Chem. Lett.* **1991**, 1447–1450.
43. Matsubara, S.; Onishi, H.; Utimoto, K. *Tetrahedron Lett.* **1990**, *31*, 6209–6212.
44. Matsubara, S.; Kodama, T.; Utimoto, K. *Tetrahedron Lett.* **1990**, *31*, 6379–6380.
45. Molander, G. A.; Burkhardt, E. R.; Weinig, P. *J. Org. Chem.* **1990**, *55*, 4990–4991.
46. Utimoto, K.; Nakamura, A.; Matsubara, S. *J. Am. Chem. Soc.* **1990**, *112*, 8189–8190.
47. Sasai, H.; Suzuki, T.; Arai, S.; Arai, T.; Shibasaki, M. *J. Am. Chem. Soc.* **1992**, *114*, 4418–4420.
48. Shibasaki, M.; Sasai, H.; Arai, T. *Angew. Chem. Int. Ed. Engl.* **1997**, *36*, 1236–1256.
49. Arai, T.; Yamada, Y. M. A.; Yamamoto, N.; Sasai, H.; Shibasaki, M. *Chem. Eur. J.* **1996**, *2*, 1368–1372.
50. Sasai, H.; Watanabe, S.; Shibasaki, M. *Enantiomer* **1988**, *2*, 267–271.
51. Yokomatsu, T.; Yamagishi, T.; Shibuya, S. *Tetrahedron Asymm.* **1993**, *4*, 1783–1784.
52. Rath, N. P.; Spilling, C. D. *Tetrahedron Lett.* **1994**, *35*, 227–230.
53. Sasai, H.; Bougauchi, M.; Arai, T.; Shibasaki, M. *Tetrahedron Lett.* **1997**, *38*, 2717–2720.
54. Sasai, H.; Arai, T.; Shibasaki, M. *J. Am. Chem. Soc.* **1994**, *116*, 1571–1572.
55. Sasai, H.; Arai, T.; Satow, Y.; Houk, K. N.; Shibasaki, M. *J. Am. Chem. Soc.* **1995**, *117*, 6194–6198.
56. Sasai, H.; Arai, S.; Tahara, Y.; Shibasaki, M. *J. Org. Chem.* **1995**, *60*, 6656–6657.
57. Namy, J. L.; Souppe, J.; Collin, J.; Kagan, H. B. *J. Org. Chem.* **1984**, *49*, 2045–2049.
58. Collin, J.; Namy, J. L.; Kagan, H. B. *Nouv. J. Chim.* **1986**, *10*, 229–232.
59. Lebrun, A.; Namy, J. L.; Kagan, H. B. *Tetrahedron Lett.* **1991**, *32*, 2355–2358.
60. Okano, T.; Matsuoka, M.; Konishi, H.; Kiji, J. *Chem. Lett.* **1987**, 181–184.
61. Uenishi, J.; Masuda, S.; Wakabayashi, S. *Tetrahedron Lett.* **1991**, *32*, 5097–5100.
62. Ohno, H.; Mori, A.; Inoue, S. *Chem. Lett.* **1993**, 375–378.
63. Mauermann, H.; Swepston, P. N.; Marks, T. J. *Organometallics* **1985**, *4*, 200–202.
64. Jeske, G.; Lauke, H.; Mauermann, H.; Schumann, H.; Marks, T. J. *J. Am. Chem. Soc.* **1985**, *107*, 8111–8118.
65. Conticello, V. P.; Brard, L.; Giardello, M. A.; Tsuji, Y.; Sabat, M.; Stern, C. L.; Marks, T. J. *J. Am. Chem. Soc.* **1992**, *114*, 2761–2762.

66. Jeske, G.; Lauke, H.; Mauermann, H.; Swepston, P. N.; Schumann, H.; Marks, T. J. *J. Am. Chem. Soc.* **1985**, *107*, 8091–8103.
67. Watson, P. L.; Parshall, G. W. *Acc. Chem. Res.* **1985**, *18*, 51–56.
68. Yasuda, H.; Yamamoto, H.; Yokota, K.; Miyake, S.; Nakamura, A. *J. Am. Chem. Soc.* **1992**, *114*, 4908–4910.
69. Yasuda, H.; Furo, M.; Yamamoto, H.; Nakamura, A.; Miyake, S.; Kibino, N. *Macromolecules* **1992**, *25*, 5115–5116.

13

Other transition metal reagents: chiral transition-metal Lewis acid catalysis for asymmetric organic synthesis

HISAO NISHIYAMA and YUKIHIRO MOTOYAMA

1. Introduction

Transition-metal Lewis acid catalysts can supply one or occasionally two vacant sites as Lewis acids for substrates possibly to meet the 18-electron rule for transition metals, and moreover they need not change their valency during the reactions. The unique vacant sites prepared by releasing one or two ligands from the characteristic configurations ML_n (n = 4–6, M = metal, L = ligand) consists of different kinds of steric circumstances. Therefore, design of their ligands with consideration not only of stereochemistry but also of valency of metals has been a remarkably attractive subject to develop new Lewis acid catalysts especially for asymmetric synthesis.

The complexes of the early transition metals such as the family of Sc, Ti, V, Cr, or Mn (where d^n, $n \leq 6$), being electron-poor, can easily supply vacant sites as Lewis acids as mentioned in the early chapters. In contrast, those of the late transition metals such as Fe, Co, Ni, or Cu etc., being electron-rich, should be converted into their cationic species to act as acids, such as $[ML_{n-1}]^+$ (n = 4–6), accompanied by certain conjugate bases as counter anions.

2. Activation of haloalkanes and carbonyl compounds

Simple haloalkanes have been found to bind to transition-metal complexes via the halogen lone pair. The co-ordinating iodomethane can act as a methylating reagent which methylated several nucleophiles faster than free iodomethane. For example, Crabtree *et al.* reported that the reaction of the iodomethane complex of Cp-Ru **1** exhibited methylation of the cyclohexanone enamine **2** to selectivity give a 2-methylcyclohexanone **3** in high yield by C-alkylation (Scheme 13.1),[1,2] although free enamines accompany *N*-alkylation products. Upon co-ordination to the cationic ruthenium(II), the methyl carbon atom of MeI increases electrophilicity attributed to activation by the cationic ruthenium Lewis acid as **4**.

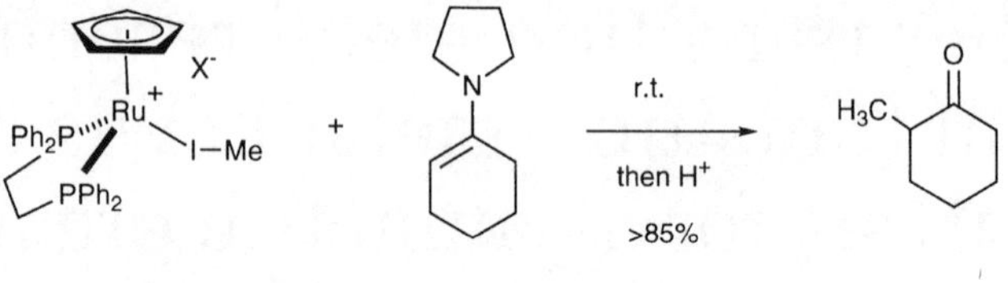

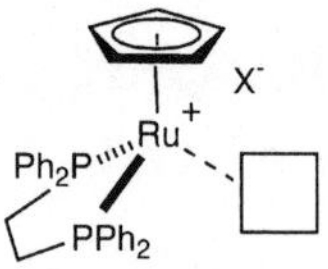

Scheme 13.1

Similar haloalkane complexes of Cp-rhenium have also been reported.[3,4] The cationic complex [CpRe(NO)(PPh$_3$)(IMe)] $^+$BF$_4$$^-$ can methylate triphenylphosphine at room temperature to produce a phosphonium salt in *c.* 90% yield.[5] The CpRe(NO)(PPh$_3$)(vacant)$^+$ species **5** (Scheme 13.2) can act as a Lewis acid and activate the carbon–halogen bonds. Moreover, the Cp-Re species **5** maintain the chirality-at-metal during several reactions.[5] In the sense of chirality-at-metal, a similar observation was disclosed by Brunner in 1974 that the manganese complex CpMn(NO)(COR)(PAr$_3$) exhibits a dissociative substitution by PAr$'_3$ with retention of the configuration via the intermediate **6**.[6–8]

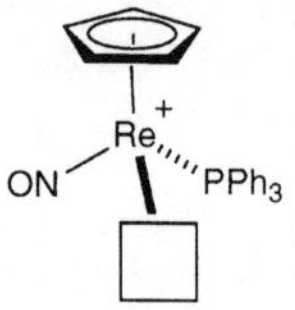

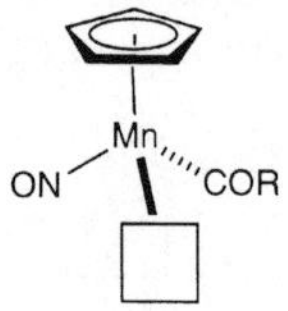

Scheme 13.2

The σ-complex of acetophenone–rhenium was prepared from the methyl complex **7** by treatment with HPF$_6$ followed by addition of acetophenone via the unsaturated species **5** (Scheme 13.3).[9] The acetophenone complex **8** when analysed by X-ray showed that the rhenium atom preferentially binds *syn* to the smaller methyl substituent. The methyl is also directed towards the smallest NO ligand, where the *re*-face of the ketone is opened. Therefore, hydride reduction of the complex (+)-(*R*)-**8** was converted into the corresponding alkoxide complex **9** followed by acidification to produce the optically active 1-phenylethanol **10** in 99% ee (*S*) (Scheme 13.3).

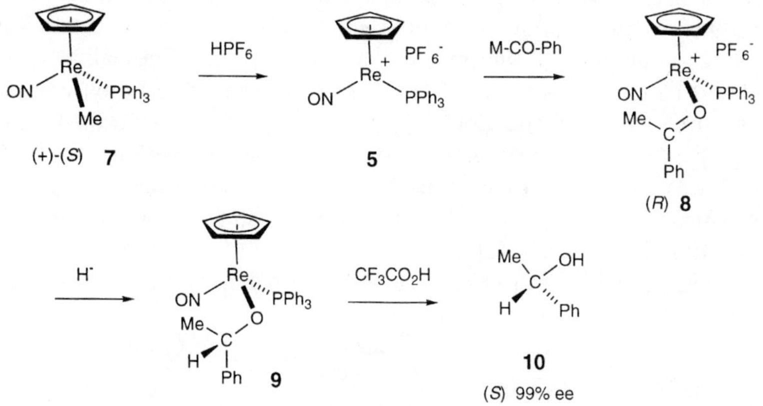

Scheme 13.3

Similarly, the α,β-unsaturated ketone moiety of the chiral rhenium–enone complex (+)-(*R*)-**11** was attacked by dimethyl cuprate followed by acidification to give β-methyl cyclohexanone **13** in 85% ee via the adduct **12** (Scheme 13.4).[10]

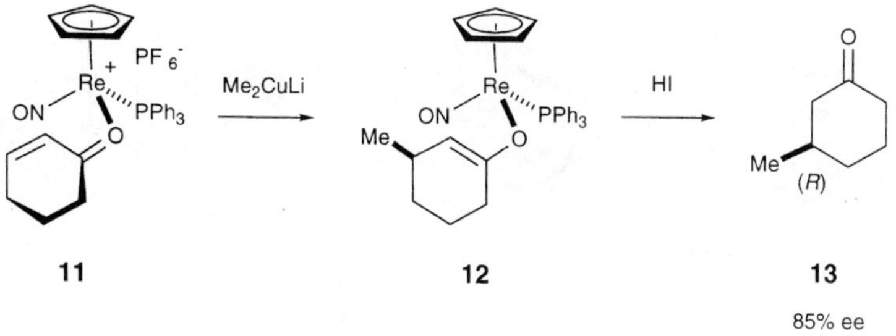

Scheme 13.4

Thus, stoichiometric modes of activation of haloalkanes and carbonyl compounds have been realized in both non-asymmetric and asymmetric reactions by use of the transition-metal complexes as Lewis acids. Many other examples are referred to in the literatures.[5–10]

3. Catalytic asymmetric Diels–Alder reactions

3.1 Achiral transition-metal catalysts

Low oxidation state transition-metal Lewis acids were first reported as catalysts of the Diels–Alder reactions by Hersh et al.[11,12] They originally found the tungsten nitrosyl complex $[W(PMe_3)(CO)_3(NO)]^+(\mu\text{-F-SbF}_5)^-$ **14** (Scheme 13.5) as a catalyst of butadiene or cyclopentadiene and α,β-unsaturated enones at 0°C. They suggested that the mode of catalysis is likely due to activation of the α,β-unsaturated enone by simple $\eta^1(\sigma)$-carbonyl co-ordination on the basis of X-ray analysis of the tungsten–acrolein adduct. They also compared the related cationic fragments of the corresponding η^1-carbonyl complexes, such as $Cp(CO)_2Fe^+$ **15** and $Cp(CO)_3Mo^+$ **16**, to well-known transition-metal Lewis acids.[12] The catalytic activity of these reagents for the Diels–Alder reactions was compared; **14**>**15**>**16**. Furthermore, the order of Lewis acidity was measured on the basis of ^1H NMR chemical shifts of crotonaldehyde on co-ordination; Lewis acids (relative power), BBr_3 (1.00) > $AlCl_3$ (0.82) > BF_3 (0.77) > $TiCl_4$ (0.66) > **14** (0.62) > **16** (0.47) > $AlEt_3$ (0.44) > **15** (0.36).

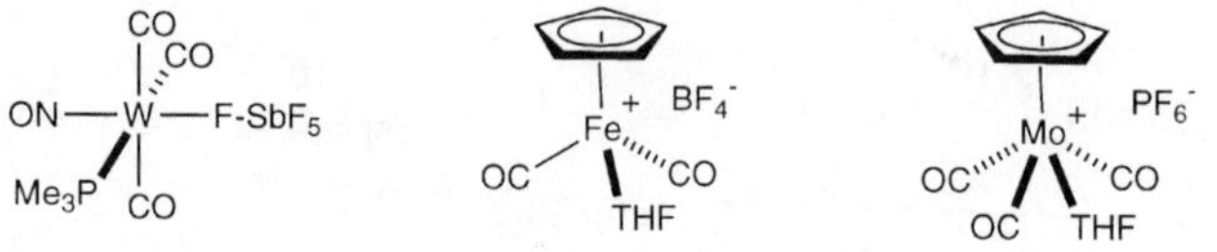

Scheme 13.5

Kelly et al.[13] reported ferrocenium ion, **17** (Scheme 13.6) acting as a catalyst for the Diels–Alder reactions of cyclopentadiene and α,β-unsaturated enones

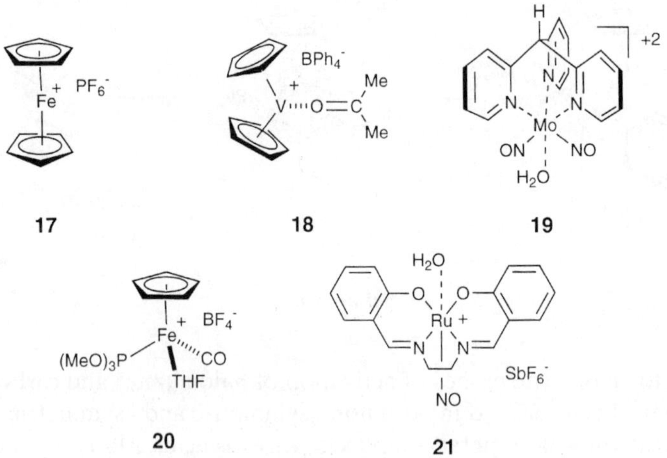

Scheme 13.6

in good yields at 0–20°C. The possibility that metallocenes may serve as Lewis acids can be inferred from the vanadocenium complex of aceteone $(Cp)_2V(\eta^1\text{-}O=CMe_2)^+$, **18**. However, other metallocenes, such as zirconocene dichloride, titanocene dichloride, cobaltocenium ion, and ferrocene, did not exhibit the catalytic activity of the Diels–Alder reaction as Lewis acids.

Other transition-metal Lewis acids for the Diels–Alder reactions have been reported: the doubly charged complex $[HC(py)_3Mo(NO)_2(\text{solvent})]^{+2}(SbF_6)_2$, **19**,[14] $(Cp)(CO)[P(OMe)_3]Fe^+$, **20**,[15] and $[Ru(salen)(NO)(H_2O)]SbF_6$, **21**.[16,17]

3.2 Chiral transition-metal catalysts

Many of transition-metal complexes, as described above, can sufficiently activate dienophiles by co-ordination of the carbonyl groups. The reactions proceed smoothly under mild conditions to provide high diastereoselectivities and enanthioselectivities. Therefore, much effort has been devoted to creating a new asymmetric Diels–Alder reaction with the combination of transition-metals and chiral auxiliaries. Many original chiral auxiliaries have been developed for this purpose.[18,19]

Faller *et al.*[20] found that the Cp–Ru cationic complex $[CpRu(PPh_3)_2(C_2H_4)]PF_6$, **22** (5 mol%) catalysed the hetero-Diels–Alder reaction of Danishefsky's diene **23** and benzaldehyde at room temperature to give the adduct **24** in 78% yield (Scheme 13.7). The catalytically active species generated by dissociation of ethylene from **22**, activate benzaldehyde reacting easily with the diene **23**. The optically active bidentate phosphines, Diop and Chiraphos, also entered to this reaction. However, the enantioselectivities were not high as expected, 16% ee with the Ru–Diop catalyst, **25**, and 25% ee with the Ru–Chiralphos catalyst, **26**. More recently, Ghosh *et al.*[21] improved

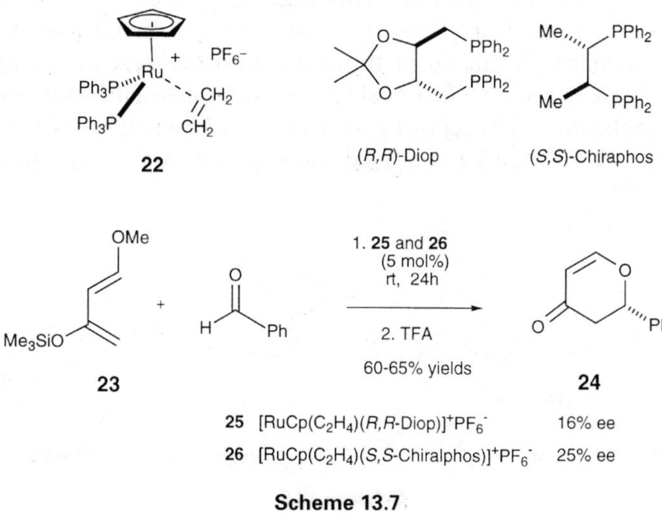

Scheme 13.7

229

the hetero-Diels–Alder reaction of diene **23** with glyoxylates in the presence of chiral copper–bisoxazoline catalysts to attain 70% ee.

Optically active oxovanadium(IV) complexes with camphor-derived diketonato ligands have been investigated for the asymmetric hetero-Diels–Alder reaction.[22] The oxovanadium complex, **27**, (5 mol%) effectively catalysed the reaction of the diene **28** and benzaldehyde at −78°C for *c.* 1 day to give the adduct **29** in 90% yield and in 85% ee (Scheme 13.8).

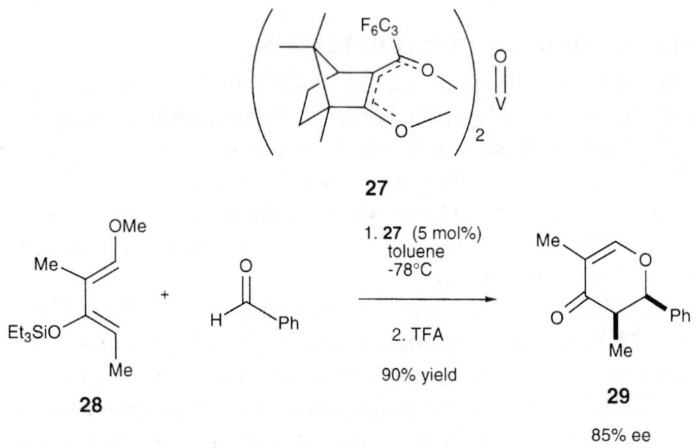

27

28

29

85% ee

Scheme 13.8

As a bidentate dienophile, the acylimide **30** was introduced to the Diels–Alder reaction by Narasaka in 1986 (Scheme 13.9).[23] To capture this component, the reaction may preferentially need an acid catalyst providing two vacant sites. The octahedral structure of transition-metal complexes can supply this demand, as shown in the chiral iron–bisoxazoline complex, **32**, developed by Corey *et al.*[24] (Protocol 1) and the iron–bissulfoxide complexes, **33** and **34** (Scheme 13.10) reported by Khiar *et al.*[25] (Table 13.1). The iron–bisoxazoline, **32** (10 mol%) catalysed the reaction of the acylimide **30** and

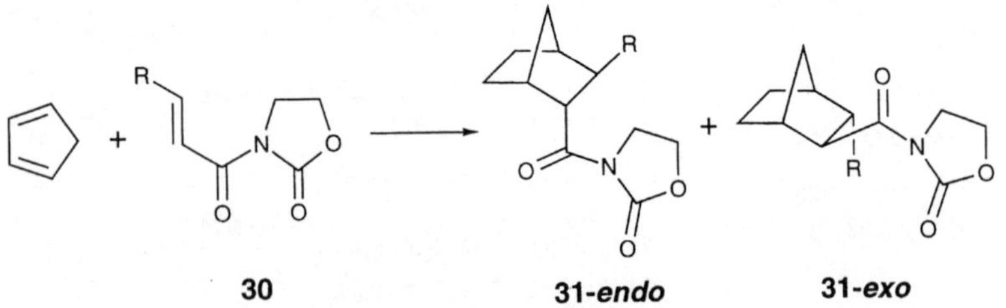

30 + **31-endo** + **31-exo**

Scheme 13.9

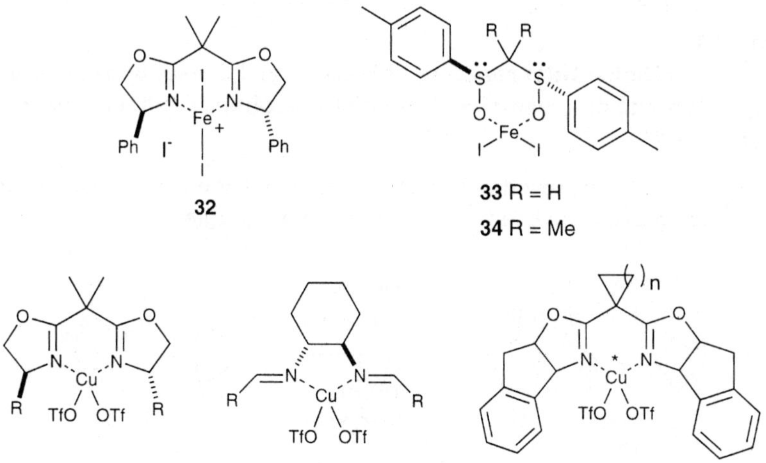

33 R = H
34 R = Me

32

35 R = *tert*-Bu
36 R = *iso*-Pr

37 R = 2,6-Cl₂Ph
38 R = Ph

39 n = 1-4
* bite angle

Scheme 13.10

Table 13.1. Asymmetric Diels–Alder reaction with the acylimide **30** and cyclopentadiene with the catalysts **32–39**

Catalyst	mol%	R=	°C, h	yield(%)	ratio of endo:exo	% ee of endo
32	10	H	−50, 15	–	99:1	86
33	10	H	−50, 5	74	68:32	36
34	10	H	−50, 5	78	78:22	56
35	10	H	−78, 18	86	98:2	98
36	10	H	−78–50, 4	93	96:4	58
35	10	Me	−15, 30	85	96:4	97
37	9	H	−78, 36	87	80:20	92
38	9	H	−78, 36	–	80:20	85
37	9	Me	−10, 30	90	65:35	83
38 n=1	10	H	−70, –	90	96:1	98
39 n=4	10	H	−50, –	90	26:1	83

cyclopentadiene at −50°C to give an *endo:exo* ratio of 99:1 and 86% ee for the *endo* isomer of **31**. It was postulated for both reactions that the iron catalysts **32–34** may supply two vacant sites of the equatorial–equatorial or the equatorial–axial.

Protocol 1.
Asymmetric Diels–Alder reaction of 3-acryloyl-1,3-oxazolin-2-one with cyclopentadiene catalysed by chiral bis(oxazoline)-Fe(III) complex[24] (Scheme 13.11)

Caution! Carry out all procedures in a well-ventilated hood, and wear disposable vinyl or latex gloves and chemical-resistant safety goggles.

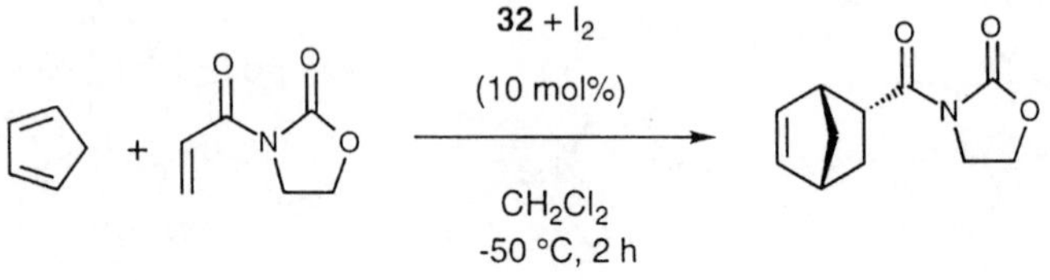

Scheme 13.11

Equipment
- Magnetic stirrer
- Syringes
- Oil bath
- Schlenk flask (10 mL)
- Vacuum/inert gas source (argon)
- Cooling bath with dry ice–acetone

Materials
- Cyclopentadiene, 0.68 mL, 8.19 mmol — flammable, toxic
- 3-Acryloyl-1,3-oxazolin-2-one, 385 mg, 2.73 mmol — flammable, toxic
- 2,2-Bis[2-[4(S)-phenyl-1,3-oxazolinyl]]propane, 109 mg, 0.326 mmol
- Powdered iron, 15.2 mg, 0.272 mmol — moisture-sensitive
- Iodine, 173.1 mg, 0.680 mmol — toxic, corrosive
- Acetonitrile, 4 mL — flammable, lachrymator
- Dichloromethane, 7 mL — harmful by inhalation
- Triethylamine, 0.1 mL — flammable, corrosive
- Ether — flammable, irritant
- Pentane — flammable, irritant
- Sodium sulfite — moisture-sensitive, irritant
- Copper acetate — irritant, hygroscopic

1. Prepare a catalyst solution of powdered iron (15.2 mg, 0.272 mmol) and iodine (104 mg, 0.408 mmol) in acetonitrile (2 mL) under argon atmosphere. After stirring for 1 h at 40 °C, add a solution of 2,2-bis[2-[4(S)-phenyl-1,3-oxazolinyl]]propane (109 mg, 0.326 mmol) in acetonitrile (2 mL) at 23 °C and stir for 1 h at 40 °C. Evaporated under reduced pressure to afford a dark viscous oil, and add dichloromethane (7 mL).

2. After cooling the catalyst solution to −78 °C, add iodine (69.1 mg, 0.282 mmol), 3-acryloyl-1,3-oxazolin-2-one (385 mg, 2.73 mmol) and cyclopentadiene (0.68 mL, 8.19 mmol).

3. After stirring for 2 h at −50 °C, add triethylamine (0.1 mL) and dilute with ether–pentane to the reaction mixture.

4. Wash the dichloromethane solution with aqueous sodium sulfite and copper acetate. Dry the organic layer and then remove the organic solvent using a rotary evaporator.

5. Apply the crude product to a flash silica gel column with 1:1 mixture of *n*-hexane and ethyl acetate to obtain pure product in 95% yield (539 mg, 96% *endo*, 82.2% ee).

6. Determine the enantiomeric excess by HPLC analysis (Daicel CHIRALCEL OD, 10% *i*-PrOH in hexane, flow rate 1 mL/min), t_R = 21.4 min (2*S*), 23.5 min (2*R*).

The most remarkable progress came from the use of the catalysts, **35** and **36** derived from copper(II) triflate and chiral bisoxazoline.[26] The reaction of the acylimide **30** and cyclopentadiene in the presence of the catalyst **35** (10 mol%) proceeds at −78°C for 5 h to give the adduct **31** in 86% with 98:2 ratio of the *endo:exo* isomers and 98% ee for the *endo* isomer. Similarly, the copper(II)– bisimine catalyst, **37** and **38**, exhibited high activity at −78°C to give 92% ee (Protocol 2). The sense of asymmetric induction with these copper catalysts can be interpreted by assuming the reaction via a square-planar Cu(II)(chiral ligand)(acylimide) complex, rather than a tetragonal one, which is attacked on the least hindered side by the diene. In a similar system to the copper catalysts, the copper catalysts of new bisoxazolines, **39**, were synthesized to show that the larger the bite angle of the bisoxazolines, the higher the enantioselectivity up to 98 from 82.[28]

Protocol 2.
Asymmetric Diels–Alder reaction of 3-acryloyl-1,3-oxazolin-2-one with cyclopentadiene catalysed by chiral bis(imine)-Cu(II) complex[27] (Scheme 13.12)

Caution! Carry out all procedures in a well-ventilated hood, and wear disposable vinyl or latex gloves and chemical-resistant safety goggles.

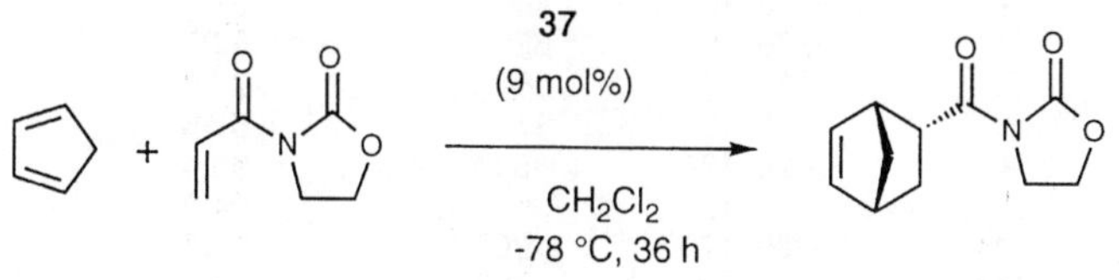

Scheme 13.12

Equipment

- Magnetic stirrer
- Syringes
- Cooling bath with dry ice–acetone
- Schlenk flask (10 mL)
- Vacuum/inert gas source (argon)

Hisao Nishiyama and Yukihiro Motoyama

Protocol 2. *Continued*

Materials

- Cyclopentadiene, 0.68 mL, 8.19 mmol — flammable, toxic
- 3-Acryloyl-1,3-oxazolin-2-one, 385 mg, 2.73 mmol — flammable, toxic
- Bis(imine), 43 mg, 0.1 mmol
- Copper(II) triflate, 32.5 mg, 0.09 mmol — corrosive, moisture-sensitive
- Dichloromethane, 2 mL — harmful by inhalation
- Ether — flammable, irritant

1. Prepare a catalyst solution of bis(imine) (43 mg, 0.1 mmol) and cooper(II) triflate (32.5 mg, 0.09 mmol) in dichloromethane (2 mL) at room temperature under nitrogen atmosphere and stir for 5 h.
2. After cooling of the catalyst solution to −78°C, add 3-acryloyl-1,3-oxazolin-2-one (141 mg, 1 mmol) and cyclopentadiene (0.83, 10 mmol).
3. After stirring for 36 h at −78°C, dilute the reaction mixture with ether and filter through a small plug (1 cm × 2 cm) of silica gel. Wash the plug three times with ether (totally 30 mL) : 87% yield, 80% *endo*, 92% ee.

The non-asymmetric Diels–Alder reaction of α,β-unsaturated aldehydes have also been intensively examined with the transition-metal Lewis acids **14–21** as described above (Scheme 13.13). The first asymmetric reaction with chiral transition-metal Lewis acids was realized by Kündig et al. in 1994.[29] They demonstrated that the chiral iron–bisphosphite catalyst, **42**, (5 mol%), which itself has one acrolein molecule, exhibits catalytic activity for the reaction of acrolein derivatives and cyclopentadiene at −30°C giving in 84–99% ee (Table 13.2). The reaction of methacrolein with **42** as a catalysts is described in Protocol 3 (Scheme 13.14).[29]

Table 13.2. Asymmetric Diels–Alder reaction with acrolein derivatives and cyclopentadiene with the catalysts **42–44**

Catalyst	R=	°C, h	yield(%)	ratio of endo:exo	% ee of major
42	H	−30, 16	46	62:38	84
42	Me	−20, 20	62	3:97	90
42	Et	−20, 20	55	2:98	94
42	Br	−40, 16	87	5:95	95
43a[a]	H	−20, 120	>95[b]	96:4	85
43b[a]	H	−20, 18	>95[b]	94:6	85
43a[a]	Me	−20, 120	>95[b]	4:96	85
43b[a]	Me	−40, 8	>95[b]	3:97	92
43a[a]	Br	−40, 60	>95[b]	3:97	87
43b[a]	Br	−78, 12	>95[b]	2:98	96
44	Me	−78, 24	60	2:98	60
44	Br	−78, 24	93	2:98	68

[a] **43** (5 mol%).
[b] Conversion.

234

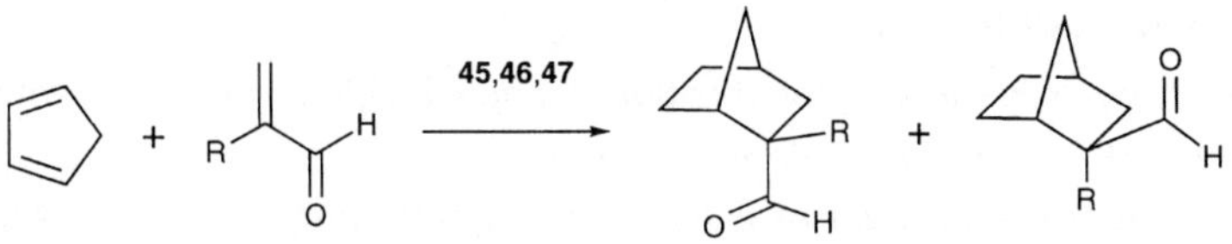

40   **41-*endo***   **41-*exo***

Scheme 13.13

Protocol 3.
Asymmetric Diels–Alder reaction of methacrolein with cyclopentadiene catalysed by chiral phosphite-Fe(II) complex[29] (Scheme 3.14)

Caution! Carry out all procedures in a well-ventilated hood, and wear disposable vinyl or latex gloves and chemical-resistant safety goggles.

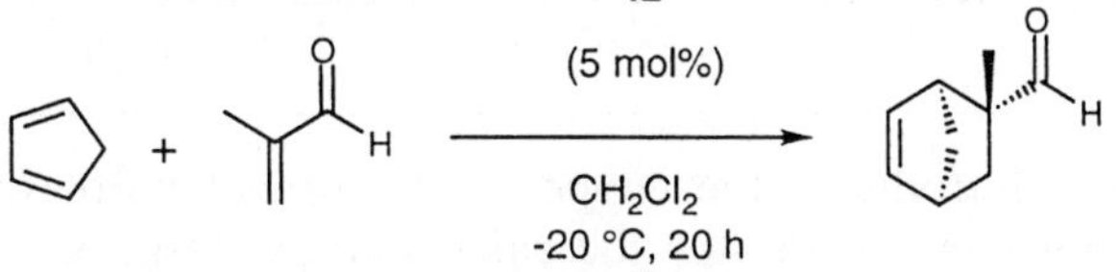

42

(5 mol%)

CH_2Cl_2

-20 °C, 20 h

Scheme 13.14

Equipment
- Magnetic stirrer
- Syringes
- Cooling bath with dry ice–acetone

- Schlenk flask (10 mL)
- Vaccum/inert gas source (argon)

Materials
- Phosphite-Fe-acrolein complex, 110 mg, 0.10 mmol   **flammable, toxic**
- Cyclopentadiene, 170 μL, 2.0 mmol   **flammable, toxic**
- Methacrolein, 170 μL, 2.0 mmol   **flammable, toxic**
- 2,6-Di-*tert*-butylpyridine, 12 μL, 0.05 mmol   **irritant**
- Dichloromethane, 2 mL   **harmful by inhalation**
- *n*-Hexane   **flammable, irritant**
- Ether   **flammable, irritant**

1. Prepare a catalyst solution of phosphite-Fe–acrolein complex (110 mg, 0.10 mmol) and 2,6-di-*tert*-butylpyridine (12 μL, 0.05 mmol) in dichloromethane (2 mL) at −40 °C under argon atmosphere.

2. Add methacrolein (170 μL, 2.0 mmol) and stir for 15 min, then add cyclopentadiene (170 mL, 2.0 mmol) slowly.

3. After stirring for 20 h at −20 °C, add hexane to precipitate the catalyst and wash three times with the same solvent.

4. Expose the combined organic phases to air for removing traces of residual

Protocol 3. *Continued*

catalyst by oxidation. Filter through a pad of Celite and then remove the organic solvent with a rotary evaporator.

5. Apply the crude product to a flash silica gel column with *n*-hexane–ether (10:1) to obtain pure product in 62% yield (174 mg, 97% *exo*, 90% ee).

6. Determine the *exo/endo* ratio by ¹H NMR; δ 9.40 (*endo*), 9.70 (*exo*).[a]

7. Determine the enantiomeric excess by capillary GC using PEG-25M, column temperature 80°C; t_R = 36.4 (*exo*-2*R*), 38.1 min (*exo*-2*S*), 25.5, 33.1 min (*endo*), after converting the adduct to the corresponding (*R,R*)-2,4-dimethyl-1,3-dioxane. A mixture of the Diels–Alder adduct (10–20 mg), (*R,R*)-2,4-pentanediol (1.5 equiv), triethyl orthoformate (1.1 equiv), and *p*-toluene-sulfonic acid (1–2 mg) in dry benzene (1 mL) was stirred at room temperature for several hours (TLC check).[a]

[a] Furuta, K.; Shimizu, S.; Miwa, Y.; Yamamoto, H. *J. Org. Chem.* **1989**, *54*, 1481–1483. Determine the *exo/endo* ratio by capillary GC using PEG-20M, column temperature 100°C; t_R = 8.7 min (*exo*), 10.3 min (*endo*).

Evans *et al.*[30] introduced the copper(II)-bis(oxazolinyl)pyridine (pybox) catalysts, **43**, (Scheme 13.15) for the same purpose, especially to find the importance of the counterion effect. They anticipated that the Cu–pybox has only one accessible co-ordination site for the carbonyl dienophiles, such as acrolein derivatives, because pybox is a tridentate ligand.[31,32] The addition of methacrolein and cyclopentadiene with the Cu–pybox-*tb*, **43a**, (X = OTf) (5 mol%) at −20°C for 120 h gave the adducts in the ratio of the *exo:endo*, 96:4 and 85% ee for the *exo* isomer (Table 13.2). However, the reaction with the Cu–pybox-*tb*, **43b** (X = SbF₆) proceeds only for 8 h to give 95% yield of

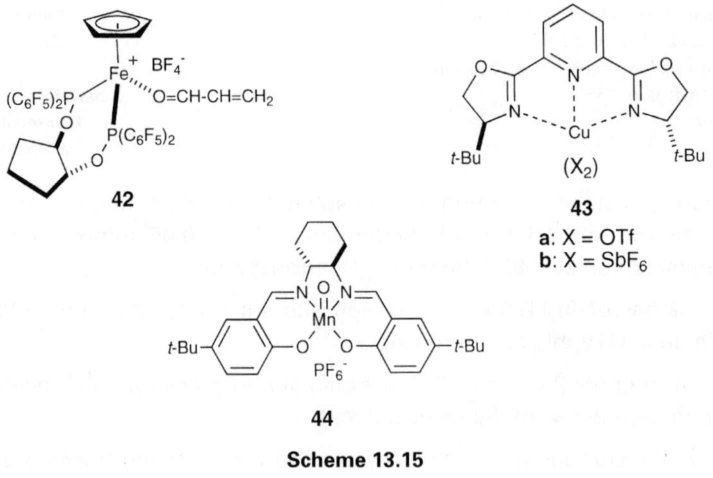

42

43

a: X = OTf
b: X = SbF₆

44

Scheme 13.15

236

the *exo*-isomer with 92% ee (Protocol 4) (Scheme 13.16). The same tendency was observed for bromoacrolein. The same counterion acceleration effect for the Diels–Alder reaction of the acylimide **30** with the Cu–bisoxazoline **35** (X = OTf => SbF$_6$) as also been shown.[30] The SbF$_6$ complex of **35** gave high yields of 96% for the less reactive β-substituted acylimides **30** (R = Ph and Cl) with 95–96% ees.

Protocol 4.
Asymmetric Diels–Alder reaction of methacrolein with cyclopentadiene catalysed by chiral bis(oxazolinyl)pyridine-Cu(II) complex[30] (Scheme 13.16)

Caution! Carry out all procedures in a well-ventilated hood, and wear disposable vinyl or latex gloves and chemical-resistant safety goggles.

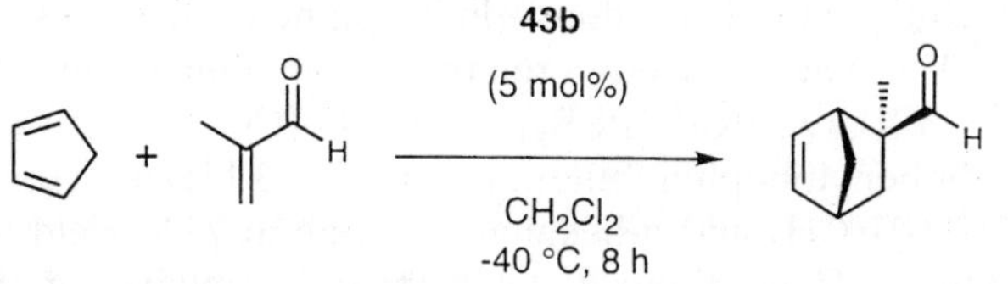

Scheme 13.16

Equipment
- Magnetic stirrer
- Syringes
- Cooling bath with dry ice–acetone
- Schlenk flask (10 mL)
- Vacuum/inert gas source (argon)

Materials
- Cyclopentadiene, 158 mg, 2.4 mmol — flammable, toxic
- Methacrolein, 140 mg, 2.0 mmol — flammable, corrosive
- *tert*-Butyl[pyridine-bis(oxazoline)], 33 mg, 0.10 mmol
- Cupper bromide, 22 mg, 0.10 mmol — irritant, hygroscopic
- Silver hexafluoroantimonate, 69 mg, 0.20 mmol — corrosive, hygroscopic
- Dichloromethane, 4 mL — harmful by inhalation

1. Prepare a catalyst solution of powdered copper bromide (22 mg, 0.10 mmol), silver hexafluoroantimonate (69 mg, 0.20 mmol) and *tert*-butyl[pyridine-bis (oxazoline)] (33 mg, 0.10 mmol) in dichloromethane (4 mL). After stirring for 6 h, filter through a plug of cotton to give a clear blue–green solution.

2. After cooling of the catalyst solution to −78°C, add methacrolein (140 mg, 2.0 mmol) and cyclopentadiene (158 mg, 2.4 mmol): >95% yield, 97% *exo*, 92% ee.

Protocol 4. *Continued*

3. Determine the *exo/endo* ratio by capillary GC (DB-1701, 110 °C, 5 psi), t_R = 5.40 (*exo*), 6.01 min (*endo*). Determine the enantiomeric excess by capillary GC (DB-1701, 110 °C, 5 psi) after the adduct was converted into the corresponding (*R,R*)-2,4-dimethyl-1,3-dioxane, t_R = 29.89 (2*S*), 30.05 min (2*R*).

Very interestingly, optically active oxo(salen)manganese(V) complex **44** serves as a Lewis acid to catalyse the Diels–Alder reaction of bromoacrolein and cyclopentadiene in 93% yield (98:2 = *exo:endo*) with 68% ee for the *exo* isomer.[33]

4. Asymmetric aldol condensations

A crossed aldol reaction of trimethylsilyl enol ethers and certain aldehydes catalysed by achiral cationic rhodium–diphosphine catalyst was first reported by Sato *et al.*[34] The reaction needs a relatively high temperature, 100 °C for 15 h, but 2 mol% of $[(COD)Rh(DPPB)]^+ClO_4^-$ **45** (COD = 1,5-cyclooctadiene and DPPB = diphenylphophinobutane) (Scheme 13.17) catalyses the reaction of $Me_3SiOC(CH_3)=CH_2$ and *n*-hexanal resulting in 74% yield of the corresponding adduct. They showed no catalytic activity of $HRh(PPh_3)_n$ (*n* = 3 and 4), $HClRu(PPh_3)_3$, and $PdCl_2(CH_3CN)_2$ for that purpose, and they proposed a rhodium enolate as an intermediary.

Reetz *et al.*[35] reported similar results of the aldol reaction with more active cationic rhodium–diphosphine complex, $[(solvent)_nRh(DPPB)]^+$, **46**, which was generated by treating the corresponding $[(COD)Rh(DPPB)]^+$ complex under a hydrogen atmosphere. The catalysed reaction of a ketene silylacetal, $Me_3SiOC(OMe)=CMe_2$, **48**, and benzaldehyde with the rhodium catalysts (5 mol%) resulted in 81% yield of the adduct **49** for only 2 h at 22 °C. It was pointed out that the mechanism of the aldol reaction may involve co-ordination of aldehyde molecules at the metal leading to activation of the aldehydes, but they did not deny the metal–enolate intermediate. They also applied the aldol reaction to an asymmetric version. The chiral Rh-Norphos cationic complex **47** (5 mol%) catalyses the aldol condensation of the ketene silylacetal **48** and benzaldehyde in good yields (>75%) but in a low enantioselectivity (Scheme 13.17).

Bosnich and colleagues has developed the achiral cationic ruthenium-based complex, $[Ru(salen)(NO)(H_2O)]^+SbF_6^-$, **21** as a powerful transition-metal Lewis acid catalyst for the crossed-aldol reaction at 25 °C in nitromethane solution, as well as the Diels–Alder reaction with the same catalyst described in the preceding section.[17,36] The reaction of the ketene silylacetal **48** and benzaldehyde with **21** (1 mol%) for only 3 min resulted in over 90% reaction (Protocol 5) (Scheme 13.18).

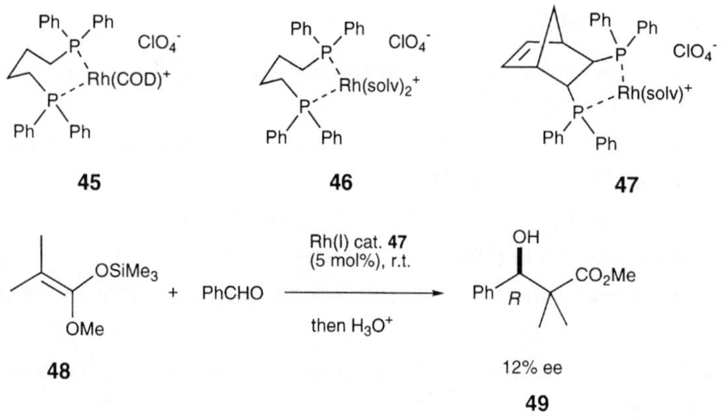

13: Other transition metal reagents

45

46

47

48

Rh(I) cat. **47**
(5 mol%), r.t.

then H₃O⁺

49

12% ee

Scheme 13.17

Protocol 5.
Mukaiyama crossed-aldol reaction of 1-phenyl-1-[(trimethylsilyl)oxy]ethylene with benzaldehyde catalysed by [Ru(salen)(NO)H₂O]SbF₆ Complex[36] (Scheme 13.18)

Caution! Carry out all procedures in a well-ventilated hood, and wear disposable vinyl or latex gloves and chemical-resistant safety goggles.

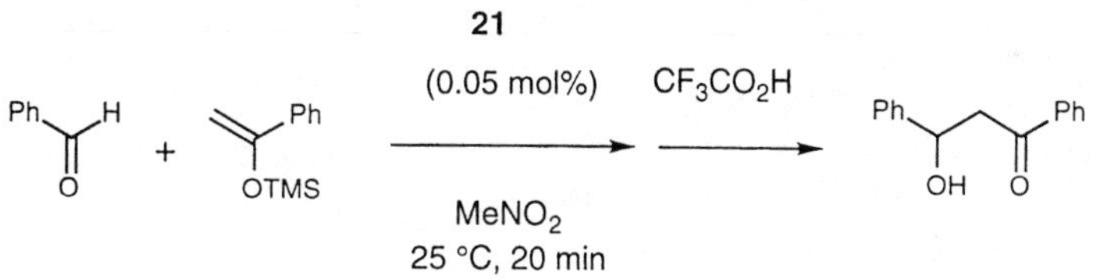

21

(0.05 mol%)   CF₃CO₂H

MeNO₂
25 °C, 20 min

Scheme 13.18

Equipment

- Magnetic stirrer
- Syringes
- Cooling bath with dry ie–acetone
- Schlenk flask (10 mL)
- Vacuum/inert gas source (argon)

Materials

- [Ru(salen)(NO)H₂O]SbF₆, 0.31 mg, 0.48 μmol
- Benzaldehyde, 104 mL, 1 mmol   **toxic, cancer-suspect agent**
- 1-Phenyl-1-[(trimethylsilyl)oxy]ethylene, 204 μL, 1 mmol   **moisture-sensitive, irritant**
- Trifluoroacetic acid, 100 μL, 0.59 mmol   **corrosive, toxic**
- Nitromethane, 700 μL, 2 mL   **flammable**
- Water, 3 mL
- Dichloromethane   **toxic, irritant**

239

Protocol 5. *Continued*

1. Prepare a catalyst solution of [Ru(salen)(NO)H$_2$O]SbF$_6$],[16] (0.31 mg, 0.48 µmol) in nitromethane (700 µL).

2. Add benzaldehyde (104 µL, 1 mmol) and 1-phenyl-1-[(trimethylsilyl) oxy]ethylene (204 µL, 1 mmol).

3. After stirring for 20 min at 25°C, dilute with nitromethane (2 mL) and stir with trifluoroacetic acid (100 µL) for 15 min.

4. Add water (3 mL) and stir vigorously for 15 min, then dilute with dichloromethane.

5. Extract the organic layer with dichloromethane and concentrate under reduced pressure.

6. Take up in dichloromethane and pass through a short column of Florisil, and remove the eluent solvent under reduced pressure to give the product, PhCOCH$_2$CH(OH)Ph, (0.21 g, 92% yield) as an oil, pure by ^{1}H NMR.

The catalytic asymmetric aldol reaction with chiral palladium catalysts was reported by Shibasaki.[37,38] The cationic complex PdCl[(R)-Binap]$^+$, **50** (Scheme 13.19) in DMF-H$_2$O catalysed the condensation of the acetophenone silyl enol ether **51** and benzaldehyde at 23°C to give 96% yield of the adduct with 71–73% ee. However, they concluded that the reaction does not involve a palladium Lewis acid catalyst, and that the reaction is the first example of the asymmetric aldol reaction via Pd(II) enolate species.

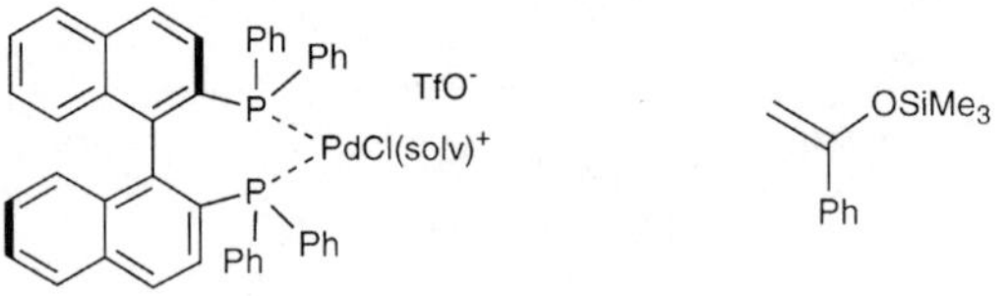

50   **51**

Scheme 13.19

A remarkable system was discovered using copper(II) complexes as Lewis acid catalysts, which activates α-alkoxy aldehydes through bidentate coordination.[39,40] Evans *et al.* applied the cationic complexes of the copper(II)– bisoxazoline, **35** (OTf and SbF$_6$) and the copper(II)–pybox-*ph*, **52**, as catalysts (Scheme 13.20). The α-(benzyloxy)acetaldehyde, **53**, and the silylketene acetal, **54**, reacted in the presence of Cu(II)(OTf)$_2$–bisoxazoline-*tb*-(S,S), **35** (5 mol%) at −78°C to give the (R)-adduct **55-R** in 91% ee (Scheme 13.20(1)). In contrast, the combination catalyst **53** of Cu(II)(SbF$_6$)$_2$ (5 mol%) and pybox-*ph*-(S,S) gave the (S)-adduct **55-S** in 99% ee (Protocol 6) (Scheme

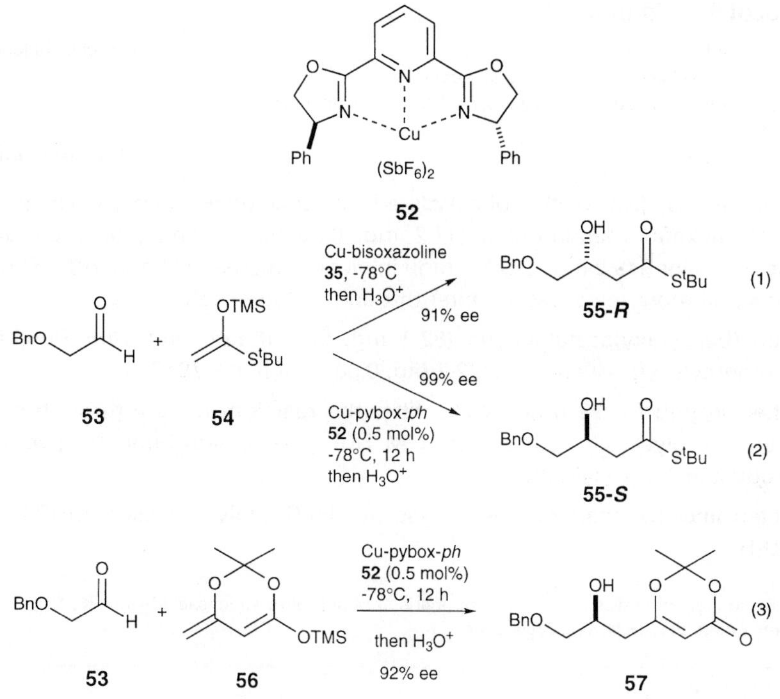

Scheme 13.20

13.20(2). Upon optimization, 0.5 mol% of the catalyst **53** catalysed the reaction for 12 h giving more than 94% yield and 92–98% ees for the several silyl enolethers, for example, from **56** to **57** in 92% ee (Scheme 13.20(3)).

Protocol 6.
Asymmetric mukaiyama-aldol reaction of (benzyloxy)acetaldehyde with 1-*tert*-butylthio-[(trimethylsilyl)oxy]ethene catalysed by chiral bis(oxazolinyl)pyridine-Cu(II) complex (Scheme 13.20[39,40]

Caution! Carry out all procedures in a well-ventilated hood, and wear disposable vinyl or latex gloves and chemical-resistant safety goggles.

Equipment
- Magnetic stirrer
- Syringes
- Cooling bath with dry ice–acetone
- Schlenk flask (10 mL)
- Vacuum/inert gas source (argon)

Materials
- Phenyl[pyridine-bis(oxazoline)], 9.2 mg, 0.025 mmol
- Copper chloride, 3.4 mg, 0.025 mmol — **irritant, hygroscopic**
- Silver hexafluoroantimonate, 17.2 mg, 0.05 mmol — **corrosive, hygroscopic**

241

Protocol 6. *Continued*

• Dichloromethane, 4 mL   harmful by inhalation
• (Benzyloxy)acetaldehyde, 82.1 mg, 0.50 mmol
• 1-*tert*-Butylthio-[(trimethylsilyl)oxy]ethene, 122.7 mg, 0.60 mmol
• 1 M HCL   toxic
• Tetrahydrofuran   flammable, irritant

1. Prepare a catalyst solution of powdered copper chloride (3.4 mg, 0.025 mmol), silver hexafluoroantimonate (17.2 mg, 0.05 mmol) and phenyl[pyridine-bis(oxazoline)] (9.2 mg, 0.025 mmol) in dichloromethane (4 mL). After stirring for 4 h at room temperature, filter through a plug of cotton.

2. Add (benzyloxy)acetaldehyde (82.1 mg, 0.50 mmol) and 1-*tert*-butylthio-[(trimethylsilyl)oxy]ethene (122.7 mg, 0.60 mmol) at −78°C.

3. After stirring for 12 h at −78°C, filter the reaction mixture through silica, then hydrolyse silyl ether with 1 M HCl in tetrahydrofuran to give pure product in 100% yield (99% ee).

4. Determine the enantiomeric excess by HPLC analysis (Daicel CHIRALCEL OD-H).

Absolute configuration was assigned by comparison of optical rotation, see; Mikami, K.; Matsukawa, S. *J. Am. Chem. Soc.* **1994**, *116*, 2363–2364.

References

1. Kulawiec, R. J.; Crabtree, R. H. *Organometallics* **1988**, *7*, 1891.
2. Burk, M. J.; Segmuller, B.; Crabtree, R. H. *Organometallics* **1987**, *6*, 2241.
3. Winter, C. H.; Arif, A. M.; Gradysz, J. A. *J. Am. Chem. Soc.* 1987, *109*, 7560.
4. Winter, C. H.; Veal, W. R.; Garner, C. M.; Arif, M. A.; Gradysz, J. A. *J. Am. Chem. Soc.* **1989**, *111*, 4766.
5. Fernández, J. M.; Gladysz, J. A. *Organometallics* **1989**, *8*, 207.
6. Brunner, H.; Aclasis, J.; Langer, M.; Steger, W. *Angew. Chem., Int. Ed. Engl.* **1974**, *13*, 810.
7. Brunner, H. *Adv. Organomet. Chem.* **1980**, *18*, 151.
8. Boone, B. J.; Klein, D. P.; Seyler, J. W.; Méndez, N. Q.; Arif, A. M.; Gladysz, J. A. *J. Am. Chem. Soc.* **1996**, *118*, 2411.
9. Fernández, J. M.; Emerson, K.; Larsen, R. D.; Gradysz, J. A. *J. Chem. Soc. Chem. Commun.* **1988**, 37.
10. Wang, Y.; Gladysz, J. A. *J. Org. Chem.* **1995**, *60*, 903.
11. Honeychuck, R. V.; Bonnesen, P. V.; Farahi, J.; Hersh, W. H. *J. Org. Chem.* **1987**, *52*, 5293.
12. Bonnesen, P. V.; Ruckett, C. L.; Honeychuck, R. V.; Hersh, W. H. *J. Am. Chem. Soc.* **1989**, *111*, 6070.
13. Kelly, T. R.; Maity, S. K.; Meghani, P.; Chandrakumar, N. S. *Tetrahedron Lett.* **1989**, *30*, 1357.
14. Faller, J. W.; Ma, Y. *J. Am. Chem. Soc.* **1991**, *113*, 1579.

15. Olson, A. S.; Seitz, W. J.; Hossain, M. M. *Tetrahedron Lett.* **1991**, *32*, 5299.
16. Odenkirk, W.; Reingold, A. L.; Bosnich, B. *J. Am. Chem. Soc.* **1992**, *114*, 6392.
17. Hollis, T. K.; Odenkirk, W.; Robinson, N. P.; Whela, J.; Bosnich, B. *Tetrahedron* **1993**, *49*, 5415.
18. Kagan, H. B.; Riant, O. *Chem. Rev.* **1992**, *92*, 1007.
19. Maruoka, K.; Yamamoto, H. In *Catalytic asymmetric synthesis* (ed. I. Ojima). VCH, New York, 1993; Chapter 9, p 413.
20. Faller, J. W.; Smart, C. J. *Tetrahedron Lett.* **1989**, *30*, 1189.
21. Ghosh, A. K.; Mathivanan, P.; Cappiello, J.; Krishnan, K. *Tetrahedron Asymm.* **1996**, *7*, 2165.
22. Togni, A. *Organometallics* **1990**, *9*, 3106.
23. Narasaka, K.; Inoue, M.; Okada, N. *Chem. Lett.* **1986**, 1109.
24. Corey, E. J.; Imai, N.; Zhang, H.-Y. *J. Am. Chem. Soc.* **1991**, *113*, 728.
25. Khiar, N.; Fernández, I.; Alcudta, F. *Tetrahedron Lett.* **193**, *34*, 123.
26. Evans, D. A.; Miller, S. J.; Lectka, T. *J. Am. Chem. Soc.* **1993**, *115*, 6460.
27. Evans, D. A.; Lectka, T.; Miller, S. J. *Tetrahedron Lett.* **1993**, *34*, 7027.
28. Davies, I. W.; Gerena, L.; Castonguay, L.; Senanayake, C. H.; Larsen, R. D.; Verhoeven, T. R.; Reider, P. J. *J. Chem. Soc., Chem. Commun.* **1996**, 1753.
29. Kündig, E. P.; Bourdin, B.; Bernardinelli, G. *Angew, Chem. Int. Ed. Engl.* **1994**, *33*, 1856.
30. Evans, D. A.; Murry, J. A.; von Matt, P.; Norcross, R. D.; Miller, S. J. *Angew. Chem. Int. Ed. Engl.* **1995**, *34*, 798.
31. Nishiyama, H.; Kondo, M.; Nakamura, T.; Itoh, K. *Organometallics* **1991**, *10*, 500.
32. Nishiyama, H.; Itoh, Y.; Sugawara, Y.; Matsumoto, H.; Aoki, K.; Itoh, K. *Bull. Chem. Soc. Jpn.* **1995**, *68*, 1247.
33. Yamashita, Y.; Katsuki, T. *Synlett.* **1995**, 829.
34. Sato, S.; Matsuda, I.; Izumi, Y. *Tetrahedron Lett.* **1986**, *27*, 5517.
35. Reetz, M. T.; Vouggioukas, A. E. *Tetrahedron Lett.* **1987**, *28*, 793.
36. Odenkirk, W.; Whelan, J.; Bosnich, B. *Tetrahedron Lett.* **1992**, *33*, 5729.
37. Sodeoka, M.; Ohrai, K.; Shibasaki, M. *J. Org. Chem.* **1995**, *60*, 2648.
38. Kiyooka, S.; Tsutsumi, T.; Maeda, H.; Kaneko, Y.; Isobe, K. *Tetrahedron Lett.* **1995**, *36*, 6531.
39. Evans, D. A.; Murry, J. A.; Kozlowski, M. C. *J. Am. Chem. Soc.* **1996**, *118*, 5814.
40. Evans, D. A.; Kozlowski, M.; Tedrow, J. S. *Tetrahedron Lett.* **1996**, *37*, 7481.

Lewis acid-assisted anionic polymerizations for synthesis of polymers with controlled molecular weights

TAKUZO AIDA and DAISUKE TAKEUCHI

1. Introduction

Unlike some naturally occurring macromolecules such as enzymes, synthetic polymers, except those obtained by stepwise approach, are of broad molecular weight distributions (MWD). In other words, they are mixtures of macromolecules with different molecular weights. Since 'molecular weight' is an essential factor affecting fundamental properties of polymer materials, it is important to develop a method for the synthesis of a polymer with a desired molecular weight with a narrow MWD. MWD is caused by the heterogeneity of the growth of polymer chains. Scheme 14.1 shows a schematic diagram of

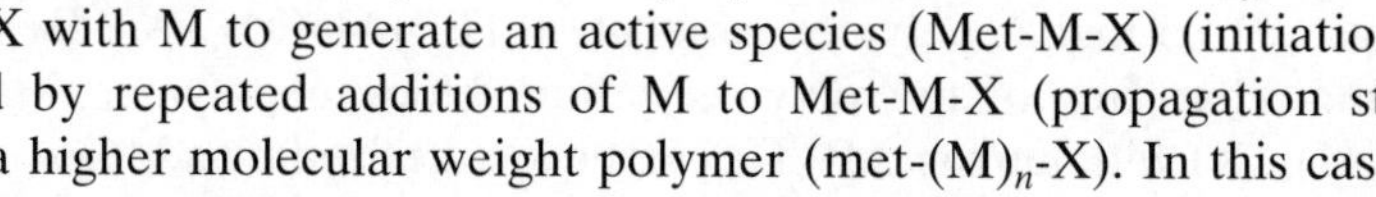

$$\text{Met—X} \xrightarrow{\text{M}} \text{Met—M—X} \xrightarrow{(n-1)\,\text{M}} \text{Met}\left(\text{M}\right)_n\text{—X}$$

Scheme 14.1

addition polymerization of an unsaturated monomer (M) with Met-X as initiator, where the chain growth of a polymer molecule starts by the reaction of Met-X with M to generate an active species (Met-M-X) (initiation step), followed by repeated additions of M to Met-M-X (propagation step), to furnish a higher molecular weight polymer (met-$(M)_n$-X). In this case, if (1) the initiation is much faster than the propagation, and (2) the propagation proceeds uniformly with respect to all growing polymer molecules (Met-$(M)_n$-X), a polymer with a uniform molecular weight should be formed. However, polymerization is generally accompanied by side reactions such as termination and chain transfer reactions, which lead to irreversible deactivation of the growing species. Since these side reactions interfere with the uniform growth of polymer chains, broadening of polymer MWD results.

In 1956, the first example of the formation of a narrow MWD polymer was discovered by Szwarc.[1] In his basic study on electron transfer reactions from organometallic compounds to unsaturated compounds, he noticed that mixing of styrene with sodium naphthalide resulted in the formation of a polymer, which is of very narrow MWD, as indicated by the ratio of weight- to number-average molecular weights (M_w/M_n) close to unity.[2] Upon addition of a fresh feed of styrene to this system after the complete consumption of the first feed, second-stage polymerization ensues, resulting in further growth of all growing polymer molecules. Polymerization with this character is named 'living polymerization', since the growth pattern of polymer can be viewed as analogous to the growth of a biological organism.

(TPP)Al-X (1)

In order to achieve 'living polymerization', it is essential to develop well-behaved initiators, since the initiator affects the relative rate of initiation to propagation, and the potential for side reactions during chain growth. Aida and Inoue *et al.*[3] have discovered that some metalloporphyrin complexes such as aluminium porphyrins ((TPP)Al–X, **1**; TPP = 5, 10, 15, 20-tetraphenyl-porphinato) serve as excellent initiators for living anionic polymerization, where the most characteristic feature is their exceptionally wide applicability for a variety of monomers (Scheme 14.2).[4] With **1** as initiator, the polymerization proceeds by the insertion of a monomer (M) into the Al-X bond in **1** to give a (TPP)Al-M-X (initiation step), which reacts with further monomer molecules (M) to grow to a higher polymer molecule ((TPP)Al–(M)$_n$–X). By virtue of the 'living' character of the polymerization, block copolymers ((TPP)Al–(M')$_m$–(M)$_n$–X) with narrow MWDs can be synthesized by addition of different monomers (M') to (TPP)Al–(M)$_n$–X.[5-10]

The polymerization with **1** is strongly affected by the structure of the porphyrin ligand of **1**, since the nucleophilic growing species always carries a (porphinato)aluminium, derived from **1**, as the counter species.[11-14] In the course of this study, we have discovered that bulky Lewis acids such as **2** dramatically accelerate the polymerization without spoiling the living character, and the polymerization is called 'Lewis acid-assisted high-speed living

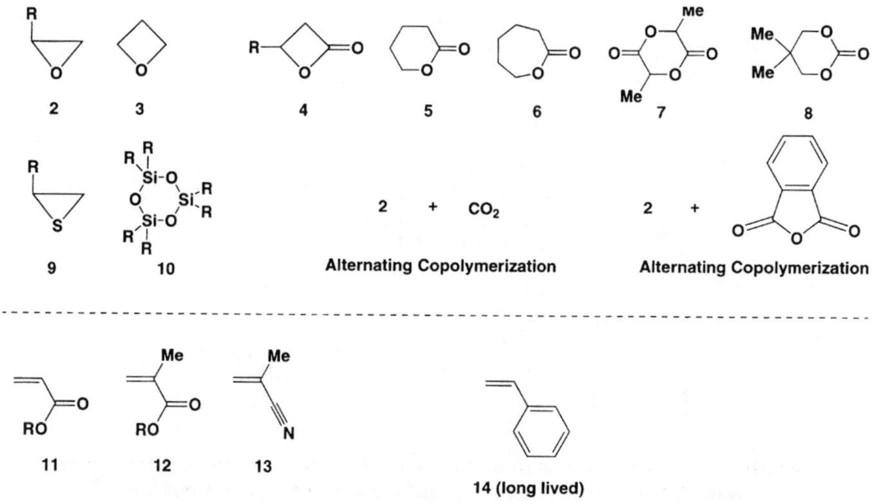

Scheme 14.2. Monomers of living polymerization with metallorphyrins[3].

anionic polymerization'.[15] Such bulky organoaluminium phenolates have been pioneered by Yamamoto *et al.* in organic syntheses.[16–17]

2. Principle, scope, and limitations of 'Lewis acid-assisted high-speed living anionic polymerization'

Detailed studies on the above-mentioned accelerated polymerization have shown that nucleophiles (initiators or active polymer terminals) attached to aluminium porphyrins and bulky Lewis acids do not directly interact (react) with each other but can coexist due to a steric repulsion between them.[16,17] Therefore, the Lewis acids can retain their inherently high ability to coordinate with and activate Lewis basic monomers for nucleophilic attack. This is the basic principle of the 'Lewis acid-assisted high-speed living anionic polymerization'.[18]

As shown in Scheme 14.3, representative Lewis acids for the high-speed living anionic polymerization involve methylaluminium diphenoxides having *ortho*-substituents at the phenyl rings.[18,19] In sharp contrast, with simple trialkylaluminium compounds and methylaluminium diphenoxides without *ortho*-substituents at the phenyl rings, the polymerization is terminated before completion due to undesired direct reactions between the nucleophilic growing species and the Lewis acids, resulting in the formation of a polymer with a broad MWD. An exceptional case is the polymerization with 'hindered' aluminium tetraphenylporphyrins as initiators, where simple trialkylaluminium compounds without any steric protection are usable as accelerators. For

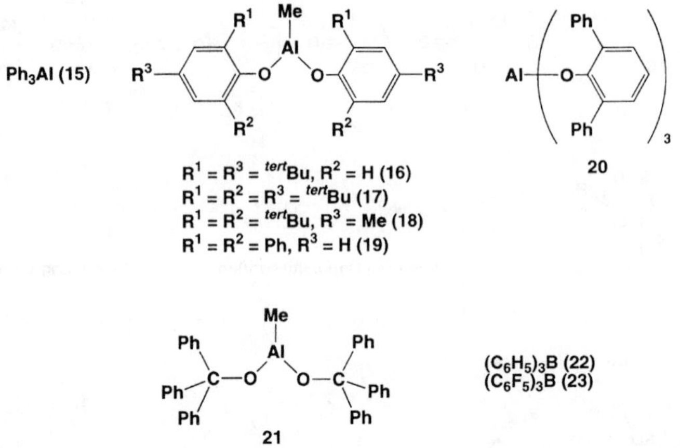

R¹ = R³ = ^{tert}Bu, R² = H (16)
R¹ = R² = R³ = ^{tert}Bu (17)
R¹ = R² = ^{tert}Bu, R³ = Me (18)
R¹ = R² = Ph, R³ = H (19)

$(C_6H_5)_3$B (22)
$(C_6F_5)_3$B (23)

Scheme 14.3. Lewis acids as monomer activators for high-speed living anionic polymerization with aluminium porphyrins as nucleophilic initiators[18].

example, the polymerization of methyl methacrylate initiated with methyl-aluminium tetramesitylporphyrin is accelerated by the addition of tri-isobutyl-aluminium, and proceeds to attain 100% monomer conversion, affording a polymer with a narrow MWD.[20] Triphenylaluminium is also usable, but the activity is lower than the hindered aluminium phenoxides. Trialkylboron compounds, in sharp contrast with trialkylaluminiums, do not terminate nor accelerate the polymerization. On the other hand, triarylborons such as tri-phenylboron and tris(pentafluorophenyl)boron serve as accelerators for the living polymerization.[21]

In addition to the acceleration of anionic polymerization, combined use of bulky nucleophiles and Lewis acids also allows controlled polymerization of monomers with cationic polymerizability: Although cyclic ethers such as oxiranes are polymerizable both anionically and cationically, oxetanes are only cationically polymerizable due to their strong basicity, and no example had been reported for the controlled synthesis of polyoxetanes because of inherently high potentials of cationic processes for side reactions. In contrast, by using an aluminium porphyrin as nucleophilic initiator in conjunction with a bulky Lewis acid, the polymerization of non-substituted oxetane proceeds in a living anionic fashion to give a polymer with a narrow MWD.[22] A sequential high-speed living anionic polymerization of methyl methacrylate and oxetane with this amphiphilic initiating system results in the formation of a narrow MWD block copolymer. Without this method, such a block copolymer could not be available, since methacrylic esters are polymerizable anionically and radically but not cationically.

In place of aluminium porphyrins, Schiff base and phthalocyanine com-plexes of aluminium can be used in conjunction with bulky Lewis acids for the controlled polymerization of cyclic ethers.[23] Later, this method has been extended to much simpler systems composed of quaternary onium salts and bulky Lewis acids, which allow high-speed living anionic polymerization of oxiranes and oxetanes.[24] In this case, bulky ate complexes are formed from onium salts and Lewis acids, which attack activated monomers through co-ordination with excess Lewis acids. Alcohol/bulky Lewis acid systems are also effective for the polymerization of lactones and cyclic carbonates.[25] More recently, organolithium compounds coupled with bulky Lewis acids have been found to initiate the living stereospecific polymerization of methacryl-ates at a low temperature to give narrow MWD polymers rich in heterotactic sequence.[26]

Protocol 1.
Accelerated synthesis of a narrow MWD poly(methyl methacrylate).
Polymerization of methyl methacrylate (12, = Me) initiated with
methylaluminium 5, 10, 15, 20-tetraphenylporphine (1, X = Me) in the
presence of methylaluminium bis(2, 6-di-*tert*-butyl-4-
methylphenoxide) (18).[28]

The procedure consists of three steps; (1) preparation of methylaluminium 5, 10, 15, 20-tetraphenylporphine (TPP)AlMe) (**1**, X = Me), (2) preparation of methylaluminium bis(2, 6-di-*tert*-butyl-4-methylphenoxide) (**18**) and (3) poly-merization of methyl methacrylate (MMA; **12** R x5 Me).

Step 1. Preparation of methylaluminium 5, 10, 15, 20-tetraphenylporphine ((TPP)AlMe) (**1**, X = Me) (Scheme 14.4)

Caution! Carry out all procedures in a well-ventilated hood, and wear disposable vinyl or latex gloves and chemical-resident safety goggles.

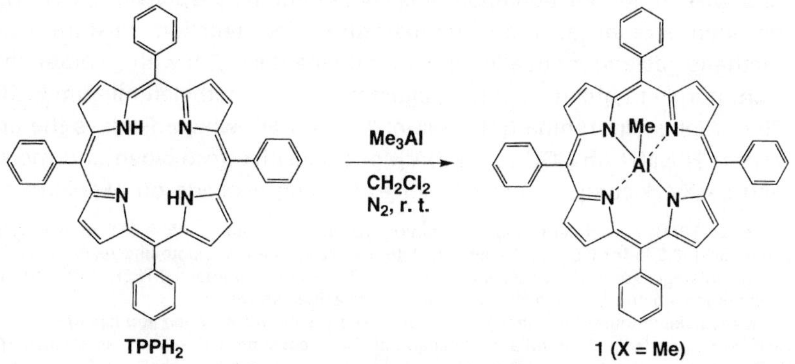

TPPH₂   Me₃Al, CH₂Cl₂, N₂, r. t.   1 (X = Me)

Scheme 14.4

Takuzo Aida and Daisuke Takeuchi

Protocol 1. Continued

Equipment

- Magnetic stirrer
- One-necked, round-bottomed flask (100 mL)
- Three-way stopcock
- Teflon-coated magnetic stirring bar (1 × 0.4 cm)
- All-glass syringes (0.25 and 50 mL) with a needle-lock luer
- Needles
- Source of dry nitrogen (preferably from a nitrogen line)

Materials

- 5, 10, 15, 20-tetraphenylporphine (TPPH$_2$)^a (FW 614.7), 0.615 g, 1 mmol — **light-sensitive**
- Trimethylaluminium^b (FW 72.1), 0.096 mL, 1 mmol — **pyrophoric, air- and moisture-sensitive**
- Dichloromethane (FW 84.9),^{c} 40 mL — **irritant, toxic**

1. Clean all glassware, syringes, needles, and stirring bar and dry for at least 5 h in a 110°C electric oven before use.

2. Put purified TPPH$_2$ (0.614 g, 1 mmol) and a Teflon-coated magnetic stirring bar in a one-necked, round-bottomed flask, and equip the neck of the flask with a three-way stopcock using a Demnum grease (Daikin).

3. Support the apparatus using a clamp and a stand with a heavy base, and connect the apparatus to a vacuum/nitrogen line via the three-way stopcock.

4. Dry the above apparatus with a hair dryer under vacuum (10^{-2} mm Hg) for 1 h, then back-fill the apparatus with nitrogen. Repeat to a total of three times. Do not use an electric heat gun for drying, since overheating leads to decomposition of TPPH$_2$.

5. Charge the apparatus containing TPPH$_2$ with distilled dichloromethane (40 mL), using a syringe through the three-way stopcock in a nitrogen stream, to give a purple suspension of TPPH$_2$.

6. Support a flask, attached to a three-way stopcock, containing distilled trimethylaluminium under nitrogen, using a clamp and a stand with a heavy base, and connect the flask to a nitrogen line via the three-way stopcock.

7. Fill a syringe with trimethylaluminium (0.096 mL, 1 mmol) from the flask through the three-way stopcock in a nitrogen stream, and add the reagent dropwise to the reaction apparatus containing the suspension of TPPH$_2$ in a nitrogen stream at room temperature. The reaction mixture evolves methane gas and gradually turns homogeneous in 2 h with a colour change from purple to greenish purple, characteristic of methylaluminium 5, 10, 15, 20-tetraphenylporphine ((TPP)AlMe; **1**, X = Me), which displays the appropriate ^{1}H NMR (in CDCl$_3$). (TPP)AlMe is sensitive to oxygen and moisture, and easily decomposes to the corresponding alkoxide on exposure to air.

^{a} Synthesize TPPH$_2$ by condensation of benzaldehyde (0.8 mol) and pyrrole (0.8 mol) in refluxing propionic acid (2.5 L) for 0.5 h under air.[27] Isolate the crystalline precipitate and recrystallize it from chloroform/methanol after washing with hot water (~20% yield). Commercial TPPH$_2$ (Aldrich) can also be used but must be purified in a similar manner as described above.
^{b} Distil trimethylaluminium fractionally under reduced pressure in a nitrogen atmosphere.
^{c} Wash commercial dichloromethane subsequently with concentrated sulfuric acid/water/aqueous NaHCO$_3$/water in order to remove stabilization agents, and distill fractionally after refluxing over calcium hydride under nitrogen.

Step 2. Preparation of methylaluminium bis(2, 6-di-*tert*-butyl-4-methylphenolate) (**18**) (Scheme 14.5)

Caution! Carry out all procedures in a well-ventilated hood, and wear disposable vinyl or latex gloves and chemical-resident safety goggles.

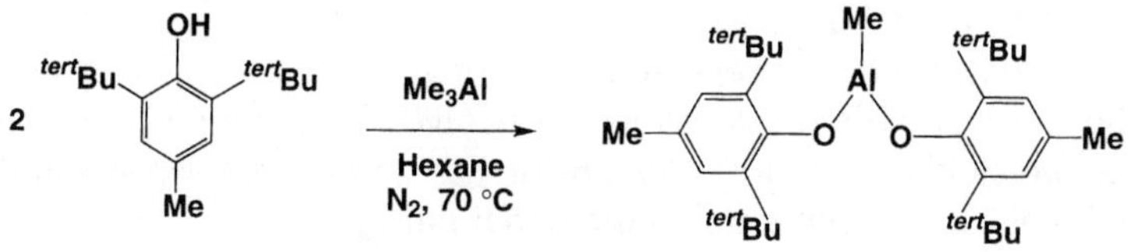

Scheme 14.5

Equipment

- Magnetic stirrer
- One-necked, round-bottomed flask (50 mL)
- Three-way stopcock
- Teflon-coated magnetic stirring bar (1 × 0.4 cm)
- All-glass syringes (1 and 20 mL) with a needle-lock luer
- Needles
- Source of dry nitrogen (preferably from a nitrogen line)

Materials

- 2, 6-Di-*tert*-butyl-4-methylphenol[a] (FW 220.4), 2.75 g, 12.5 mmol   **irritant**
- Trimethylaluminium[b] (FW 72.1), 0.6 mL, 6.25 mmol   **pyrophoric, air- and moisture-sensitive**
- Hexane (FW 86.2)[c] 10 mL   **flammable, irritant**

1. Clean all glassware, syringes, needles, and stirring bar and dry for at least 5 h in a 110°C electric oven before use.

2. Put recrystallized 2, 6-di-*tert*-butyl-4-methylphenol (2.75 g, 12.5 mmol) and a Teflon-coated magnetic stirring bar in a one-necked, round-bottomed flask, and equip the neck of the flask with a three-way stopcock using a Demnum grease.

3. Support the apparatus using a clamp and a stand with a heavy base, and connect the apparatus to a vacuum/nitrogen line via the three-way stopcock.

4. Dry the above apparatus under vacuum (10^{-2} mm Hg) for 1 h, then back-fill the apparatus with nitrogen. Repeat to a total of three times. Do not heat the apparatus, so as to avoid sublimation of 2, 6-di-*tert*-butyl-4-methylphenol.

5. Charge the apparatus containing 2, 6-di-*tert*-butyl-4-methylphenol with distilled hexane (10 mL), using a syringe through the three-way stopcock in a nitrogen stream, to dissolve the phenol.

6. Support a flask, attached to a three-way stopcock, containing distilled trimethylaluminium under nitrogen, using a clamp and a stand with a heavy base, and connect the flask to a nitrogen line via the three-way stopcock.

Protocol 1. *Continued*

7. Fill a syringe with trimethylaluminium (0.6 mL, 6.25 mmol) from the flask through the three-way stopcock in a nitrogen stream, and add the reagent in the syringe, through the three-way stopcock in a nitrogen stream, dropwise at 0 °C to the reaction apparatus containing the hexane solution of 2, 6-di-*tert*-butyl-4-methylphenol.

8. Stir the content of the above apparatus magnetically at room temperature under nitrogen. The reaction mixture evolves methane gas and produces white precipitates in the initial 5–10 min. Stir the suspension for an additional 2 h at room temperature under nitrogen.

9. Warm the suspension at 70 °C under nitrogen until the precipitate completely dissolves. Allow the clear solution to stand overnight at room temperature, to give white crystals.

10. Remove a supernatant, liquid phase with a syringe from the above apparatus in a nitrogen stream, and add distilled hexane (5 mL) by syringe to the residue and stir the mixture for a while. Repeat this procedure to a total of three times in order to wash the crystals.

11. Dry the crystals under reduced pressure (10^{-2} mm Hg) for 1 h at room temperature to produce methylaluminium bis(2, 6-di-*tert*-butyl-4-methylphenolate) (**18**) in 64% yield (1.9 g, 4.0 mmol).

[a] Recrystallize commercial 2, 6-di-*tert*-butyl-4-methylphenol from hexane.
[b] Distil trimethylaluminium fractionally under reduced pressure in a nitrogen atmosphere.
[c] Distil hexane after refluxing over sodium wire under nitrogen.

Step 3. Polymerization of methyl methacrylate (MMA) (Scheme 14.6)

Caution! Carry out all procedures in a well-ventilated hood, and wear disposable vinyl or latex gloves and chemical-resident safety goggles.

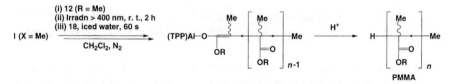

Scheme 14.6

Equipment

- Magnetic stirrer
- One-necked, round-bottomed flask (100 mL)
- Three-way stopcock
- Beaker (500 mL)
- Teflon-coated magnetic stirring bars (1 × 0.4 cm, 2 × 0.6 cm)
- Needles

- All-glass syringes (2, 20, and 30 mL) with a needle-lock luer
- Source of dry nitrogen (preferably from a nitrogen line)
- 300-W Xenon arc lamp equipped with thermal- and UV-cutoff filters
- Iced water bath

14: Lewis acid-assisted anionic polymerizations

Materials

- (5, 10, 15, 20-Tetraphenylporphinato)aluminium methyl
  ((TPP)AlMe; **1**, X = Me) (FW 654.7) 0.655 g, 1 mmol in
  dichloromethane (40 mL) (prepared in Step 1)   **air- and moisture-sensitive, light-sensitive**
- Methylaluminium bis(2, 6-di-*tert*-butyl-4-methylphenoxide)
  (**18**) (FW 480.4), 1.44 g, 3.0 mmol in dichloromethane
  (10 mL) (prepared in step 2)   **air- and moisture-sensitive**
- Methyl methacrylate (MMA; **12**, R = Me)^a (FW 100.1),
  20 mL, 200 mmol   **flammable, corrosive**
- Methanol (FW 32.0), 500 mL   **flammable, toxic**
- Benzene (FW 78.1) 200 mL   **flammable, toxic**

1. Support the apparatus containing a dichloromethane solution (40 ml) of **1** (X = Me) (0.655 g, 1 mmol; Step 1, using a clamp and a stand with a heavy base, and connect the apparatus to a nitrogen line via the three-way stop-cock.

2. Support a flask, attached to a three-way stopcock, containing distilled MMA (**12** R = Me) under nitrogen, using a clamp and a stand with a heavy base, and connect the flask to a nitrogen line via the three-way stopcock.

3. Fill a syringe with **12** (R = Me) (20 mL, 200 mmol) from the flask through the three-way stopcock in a nitrogen stream, and add the reagent in the syringe, through the three-way stopcock in a nitrogen stream, dropwise to the reaction apparatus containing the dichloromethane solution of **1** (X = Me).

4. Stir the contents of the reaction apparatus magnetically at room temperature, and expose for 2 h to a 300-W xenon arc light through thermal- and UV-cutoff filters from a distance of 10 cm. During this period, the solution turns from greenish purple to dark reddish purple, characteristic of an enolatoaluminium porphyrin.

5. Stop the irradiation, and put the reaction apparatus in an iced water bath.

6. Support a flask, attached to a three-way stopcock, containing a dichloromethane solution (0.3 M) of **18** (Step 2) under nitrogen, using a clamp and a stand with a heavy base, and connect the flask to a nitrogen line via the three-way stopcock.

7. Fill a syringe with the dichloromethane solution (10 mL) of **18** (1.4 g, 3 mmol) from the flask through the three-way stopcock in a nitrogen stream, and add the solution in the syringe dropwise to the reaction apparatus containing the above photoirradiated reaction mixture of **1** (X = Me) and **12** (R = Me) while stirring magnetically under nitrogen. The polymerization proceeds with a considerable evolution of heat, and reaches 100% monomer conversion within a few seconds.

8. After 30 s, add methanol (1 mL, 24.7 mmol) to the reaction apparatus containing the polymerization mixture, and pour the contents into a large volume of methanol (500 mL) with vigorous stirring magnetically, to give a precipitate.

Protocol 1. *Continued*

9. Collect the precipitate, dissolve it in benzene (200 mL), and freeze-dry to give a poly(methyl methacrylate) (M_n = 21500, M_w/M_n = 1.10) in 90% yield (18 g) based on the charged monomer.

[a] Distil **12** (R = Me) after stirring with calcium hydride or more vigorously with trimethylaluminium under nitrogen.

By changing the mole ratio of **12** (R = Me) to **1** (X = Me), the molecular weight of the polymer can be controlled over a wide range up to 10^6, retaining the narrow MWD (M_w/M_n = 1.05–1.2).[28] The present procedure is also applicable to the living anionic polymerization of other methacrylates (**12**; R = ethyl, isopropyl, n-butyl, isobutyl, benzyl, and dodecyl,[29] and methacrylonitrile (**13**).[30]

Protocol 2.
Accelerated synthesis of a narrow MWD poly(δ-valerolactone).
Polymerization of δ-valerolactone (5) initiated with (5, 10, 15, 20-tetraphenylporphinato)aluminium methoxide ((TPP)AlOMe; 1, X = OMe) in the presence of methylaluminium bis)2, 6-diphenylphenoxide) (19)[31]

The procedure consists of three steps; (1) preparation of (5, 10, 15, 20-tetraphenylporphinato)aluminium methoxide ((TPP)AlOMe; **1**, X = OMe), (2) preparation of methylaluminium bis(2, 6-diphenylphenoxide) (**19**) and (3) polymerization of δ-valerolactone (**5**).

Step 1. Preparation of (5, 10, 15, 20-tetraphenylporphinato)aluminium methoxide ((TPP)AlOMe; **1**, X = OMe) (Scheme 14.7)

Caution! Carry out all procedures in a well-ventilated hood, and wear disposable vinyl or latex gloves and chemical-resistant safety goggles.

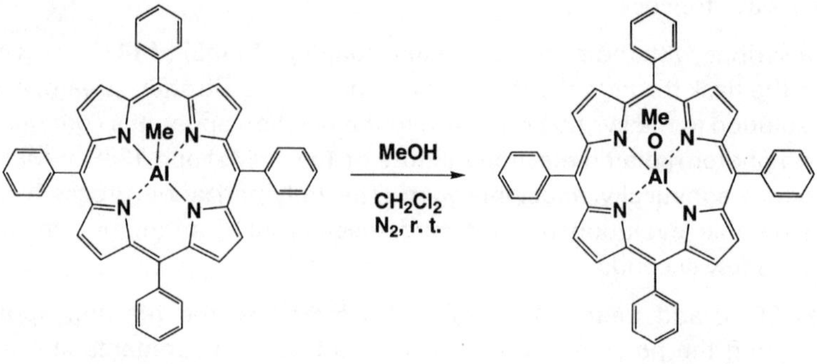

1 (X = Me)   1 (X = OMe)

Scheme 14.7

Equipment

- Magnetic stirrer
- One-necked, round-bottomed flask (50 ml)
- Three-way stopcock
- Teflon-coated magnetic stirring bar (1 × 0.4 cm)

- All-glass syringes (10 and 30 mL) with a needle-lock luer
- Needles
- Source of dry nitrogen (Preferably from a nitrogen line)

Materials

- Methylaluminium 5, 10, 15, 20-tetraphenylporphine ((TPP)AlMe; **1**, X = Me)[a] (FW 654.7), 0.065 g, 0.1 mmol in dichloromethane (4 mL) (for preparation, see Protocol 1, Step 1)
- Methanol (FW 32.0),[a] 1 mL, 24.6 mmol

air- and moisture-sensitive, light-sensitive

flammable, toxic

1. Support the reaction apparatus, containing a dichloromethane solution (4 mL) of **1** (X = Me) (0.065 g, 0.1 mmol), under nitrogen (see Protocol 1, Step 1), using a clamp and a stand with a heavy base, and connect the apparatus to a vacuum/nitrogen line via the three-way stopcock.

2. Support a flask, attached to a three-way stopcock, containing distilled methanol under nitrogen, using a clamp and a stand with a heavy base, and connect the flask to a nitrogen line via the three-way stopcock.

3. Fill a syringe with methanol (1 mL, 24.6 mmol) from the flask through the three-way stopcock in a nitrogen stream, and add the reagent in the syringe, through the three-way stopcock in a nitrogen stream, to the reaction apparatus containing the dichloromethane solution of **1** (X = Me) at room temperature.

4. Stir the contents of the reaction apparatus magnetically for 15 h at room temperature under nitrogen. The solution gradually turns from greenish purple to bright reddish purple, characteristic of a (porphinato)aluminium alkoxide.

5. Connect the reaction apparatus to a vacuum line, and remove volatile fractions from the contents under reduced pressure (10^{-2} mm Hg), and back-fill the apparatus with nitrogen, leaving (TPP)AlOMe (**1**, X = OMe) as a reddish purple powder, which displays the appropriate ^{1}H NMR (in CDCl$_3$).

[a] Distil methanol after refluxing over magnesium ribbon under nitrogen.

Step 2. Preparation of methylaluminium bis(2, 6-diphenylphenoxide) (**19**) (Scheme 14.8)

Caution! Carry out all procedures in a well-ventilated hood, and wear disposable vinyl or latex gloves and chemical-resistant safety goggles.

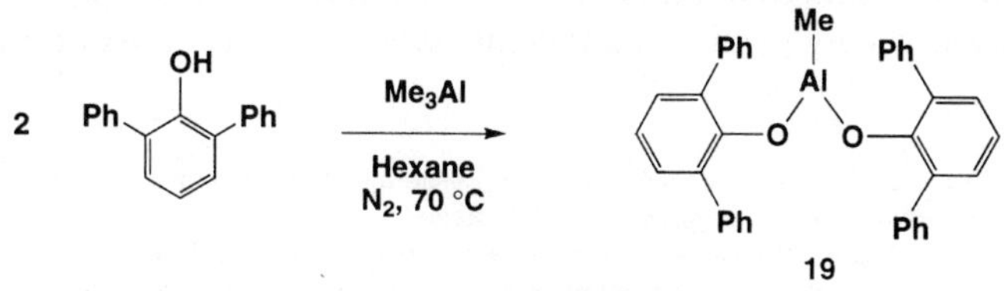

Scheme 14.8

Protocol 2. *Continued*

Equipment

- Magnetic stirrer
- One-necked, round-bottomed flask (5 mL)
- Three-way stopcock
- Teflon-coated magnetic stirring bar (1 × 0.4 cm)

- All-glass syringes (0.5 and 20 mL) with a needle-lock luer
- Needles
- Source of dry nitrogen (preferably from a nitrogen line)

Materials

- Trimethylaluminium[a] (FW 72.09), 0.29 mL, 3 mmol
- 2, 6-Diphenylphenol[b] (FW 246.3), 1.5 g, 6 mmol
- Dichloromethane (FW 84.0,[c] 10 mL

pyrophoric, air- and moisture-sensitive
irritant
irritant, toxic

1. Clean all glassware, syringes, needles, and stirring bar and dry for at least 5 h in a 110 °C electric oven before use.

2. Put recrystallized 2, 6-diphenylphenol (1.5 g, 6 mmol) and a Teflon-coated magnetic stirring bar in a one-necked, round-bottomed flask, and equip the neck of the flask with a three-way stopcock using a Demnum grease (Daikin).

3. Support the apparatus using a clamp and a stand with a heavy base, and connect the apparatus to a vacuum/nitrogen line via the three-way stopcock.

4. Dry the above apparatus under vacuum (10^{-2} mm Hg) for 1 h, then back-fill the apparatus with nitrogen. Repeat to a total of three times. Do not heat the apparatus, so as to avoid sublimation of 2, 6-diphenylphenol.

5. Using a syringe, add dichloromethane (10 mL) to the apparatus containing 2, 6-diphenylphenol through the three-way stopcock in a nitrogen steam, to dissolve the phenol.

6. Support a flask, attached to a three-way stopcock, containing distilled trimethylaluminium under nitrogen, using a clamp and a stand with a heavy base, and connect the flask to a nitrogen line via the three-way stopcock.

7. Fill a syringe with trimethylaluminium (0.29 mL, 3 mmol) from the flask through the three-way stopcock in a nitrogen stream, and add the reagent in the syringe, through the three-way stopcock in a nitrogen stream, dropwise to the reaction apparatus containing the dichloromethane solution of 2, 6-diphenylphenol at 0 °C.

8. Stir the contents of the apparatus at room temperature under nitrogen. The reaction mixture evolves methane gas during the initial 5–10 min. Stir the solution for an additional 2 h at room temperature under nitrogen. Use the resulting pale-yellow solution containing **19** for the subsequent polymerization (Step 3).

---

[a] Distil trimethylaluminium fractionally under reduced pressure in a nitrogen atmosphere.
[b] Recrystallize commercial 2, 6-diphenylphenol from hexane.
[c] Wash commercial dichloromethane successively with concentrated sulfuric acid/water/aqueous NaHCO₃/water in order to remove stabilization agents, and fractionally distil after refluxing over calcium hydride under nitrogen.

14: Lewis acid-assisted anionic polymerizations

Step 3. Polymerization of δ-valerolactone **(5)** (Scheme 14.9)

Caution! Carry out all procedures in a well-ventilated hood, and wear disposable vinyl or latex gloves and chemical-resistant safety goggles.

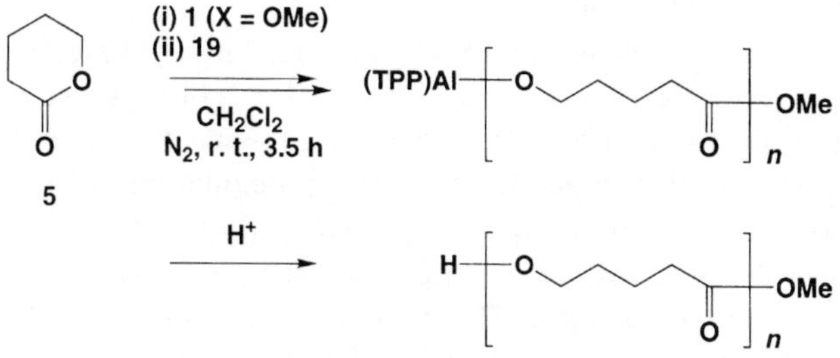

Scheme 14.9

Equipment

- Magnetic stirrer
- One-necked, round-bottomed flask (100 mL)
- Three-way stopcock
- Teflon-coated magnetic stirring bars (1 × 0.4 cm, 2 × 0.6 cm)
- All-glass syringes (2, 5, and 20 mL) with a needle-lock luer
- Needles
- Source of dry nitrogen (preferably from a nitrogen line)

Materials

- (5, 10, 15, 20-Tetraphenylporphinato)aluminium methoxide ((TPP)AlOMe; **1**, X = OMe)[a] (FW 670.7), 0.067 g, 0.1 mmol (prepared in Step 1)   **moisture-sensitive, light-sensitive**
- Methylaluminium bis(2, 6-diphenylphenoxide) **(19)** (FW 532.6) 0.16 g, 0.3 mmol in dichloromethane (10 mL) (prepared in Step 2)   **air- and moisture-sensitive irritant**
- δ-Valerolactone **(5)**[a] (FW 100.1), 1.86 mL, 20 mmol
- Dichloroethane[b] (FW 98.96), 4 mL, 50 mmol   **flammable, toxic**
- methanol (FW 32.0), 500 mL   **flammable, toxic**
- Benzene (FW 78.1), 200 mL   **flammable, toxic**

1. Support the apparatus containing **1** (X = OMe) (0.067 g, 0.1 mmol; Step 1) under nitrogen, using a clamp and a stand with a heavy base, and connect the apparatus to a nitrogen line via the three-way stopcock.

2. Using a syringe add distilled dichloroethane (4 mL) through the three-way stopcock in a nitrogen stream, to give a clear, bright reddish purple solution.

3. Support a flask, attached to a three-way stopcock, containing distilled **5** under nitrogen, using a clamp and a stand with a heavy base, and connect the flask to a nitrogen line via the three-way stopcock.

4. Fill a syringe with **5** (1.86 mL, 20 mmol) from the flask through the three-way stopcock in a nitrogen stream, and add the reagent in the syringe, through the three-way stopcock in a nitrogen stream, dropwise to the reaction apparatus containing the dichloroethane solution of **1** (X = OMe) upon stirring magnetically at room temperature.

Protocol 2. *Continued*

5. Support a flask, attached to a three-way stopcock, containing a dichloromethane solution (0.3 M) of **19** under nitrogen (prepared in Step 2), using a clamp and a stand with a heavy base, and connect the apparatus to a nitrogen line via the three-way stopcock.
6. Fill a syringe with the dichloromethane solution (1 mL) of **19** (0.16 g, 0.3 mmol) from the flask through the three-way stopcock in a nitrogen stream, and add the solution in the syringe, through the three-way stopcock in a nitrogen stream, to the reaction apparatus containing the mixture of **1** (X = OMe) and **5** in dichloroethane upon stirring magnetically at room temperature.
7. Stir the mixture in the reaction apparatus magnetically for 3.5 h at room temperature under nitrogen. The polymerization reaches 74% monomer conversion to give a viscous, partially solidified solution.
8. Pour the polymerization mixture in a large volume of methanol (500 mL) with vigorous stirring, to give a precipitate.
9. Collect the precipitate, dissolve it in benzene (200 mL), and freeze-dry, to give a poly(δ-valerolactone) (M_n = 24000, M_w/M_n = 1.13) in 67% yield (1.3 g) based on the charged monomer.

[a] Distil **5** after stirring with calcium hydride under nitrogen.
[b] Wash commercial dichloroethane successively with concentrated sulfuric acid/water/aqueous NaHCO$_3$/water in order to remove stabilization agents, and fractionally distil after refluxing over calcium hydride under nitrogen.

By changing the mole ratio of **5** to **1** (X = OMe), the molecular weight of the polymer can be controlled over a wide range up to 10^5, retaining the narrow MWD (M_w/M_n <1.2). Use of **18** as accelerator results in broadening of MWD at a high conversion (>70%), due to undesired transesterification of the produced polymer caused by **18**.[31]

Protocol 3.
Accelerated synthesis of a narrow MWD polyoxetane. Polymerization of oxetane (3) initiated with tetraethylammonium chloride (Et$_4$NCl) in the presence of methylaluminium bis(2, 6-di-*tert*-butyl-4-methylphenolate) (18).[24] (Scheme 14.10)

Caution! Carry out all procedures in well-ventilated hood, and wear disposable vinyl or latex gloves and chemical-resistant safety goggles.

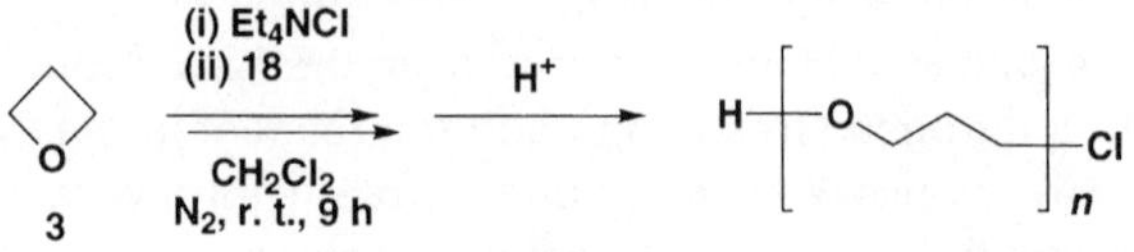

Scheme 14.9

14: Lewis acid-assisted anionic polymerizations

Equipment

- Magnetic stirrer
- One-necked, round-bottomed flask (50 mL)
- Three-way stopcock
- Teflon-coated magnetic stirring bar (1 × 0.4 cm)
- Needles

- All-glass syringes (1, 2, and 3 mL) with a needle-lock luer
- Source of dry nitrogen (preferably from a nitrogen line)
- Glass column for chromatography (50 × 4 cm)

Materials

- Tetraethylammonium chloride (Et$_4$NCl)^a (FW 165.7), 0.0166 g, 0.1 mmol   **hygroscopic, irritant**
- Methylaluminium bis(2, 6-di-*tert*-butyl-4-methylphenolate) (**18**)^b (FW 480.7), (0.14 g, 0.3 mmol) in dichloromethane solution (0.3 M, 1 mL) (for preparation, see Protocol 1, Step 2)   **air- and moisture-sensitive**
- Dichloromethane,^{b} 2 mL   **irritant, toxic**
- Oxetane (**3**) (FW 58.1),^{c} 0.65 mL, 10 mmol   **flammable**
- Silica gel (C300) 150 g

1. Clean all glassware, syringes, needles, and stirring bar and dry for at least 5 h in a 110 °C electric oven before use.

2. Put recrystallized tetraethylammonium chloride (Et$_4$NCl) (0.0166 g) and a Teflon-coated magnetic stirring bar in a one-necked, round-bottomed flask, and equip the neck of the flask with a three-way stopcock using a Demnum grease (Daikin).

3. Support the above apparatus using a clamp and a stand with a heavy base, and dry the content under vacuum (10^{-2} mm Hg) for 1 h, then back-fill the apparatus with nitrogen. Repeat to a total of three times. Do not heat the apparatus, so as to avoid decomposition of Et$_4$NCl.

4. Using a syringe add dichloromethane (2 mL) through the three-way stopcock in a nitrogen stream, to dissolve Et$_4$NCl.

5. Support a flask, attached to a three-way stopcock, containing a dichloromethane solution of **18** (0.3 M) under nitrogen (Protocol 1, Step 2), using a clamp and a stand with a heavy base, and connect the flask to a nitrogen line via the three-way stopcock.

6. Fill a syringe with the dichloromethane solution (1 mL) of **18** (0.14 g, 0.3 mmol) from the flask through the three-way stopcock in a nitrogen stream, and add the solution in the syringe, through the three-way stopcock in a nitrogen stream, dropwise to the reaction apparatus containing the dichloromethane solution of Et$_4$NCl while stirring magnetically at room temperature.

7. Likewise, add **3** (0.65 mL, 10 mmol) to the above reaction apparatus, and stir the contents for 9 h at room temperature. The polymerization reaches 100% monomer conversion, to give a viscous solution.

8. Evaporate the polymerization mixture to dryness with a rotary evaporator. Separate the residue by silica gel column chromatography with ethyl acetate–hexane (4:1) as eluant. Evaporate the fraction containing polymeric

Protocol 3. *Continued*

products with a rotary evaporator, to leave a polyoxetane as a colourless viscous liquid (M_n = 5200, M_w/M_n = 1.12) in 80% yield (0.46 g) based on the monomer charged.

[a] Recrystallize commercial Et$_4$NCl from acetone–hexane.
[b] Wash commercial dichloromethane successively with concentrate sulfuric acid/water/aqueous NaHCO$_3$/water in order to remove stabilization agents, and distil fractionally after refluxing over calcium hydride under nitrogen.
[c] Distil commercial **3** over sodium wire under reduced pressure in a nitrogen stream at room temperature.

By changing the mole ratio of **3** to Et^4 NCl, the molecular weight of the polymer can be controlled up to 20 000, retaining the narrow MWD (M_w/M_n = 1.09–1.15).[24] No polymerization takes place when the mole ratio **18**:Et$_4$NCl is less than unity.

References

1. Szwarc, M. *Nature* **1956**, *178*, 1168–1170.
2. Szwarc, M.; Levy, M.; Milkovich, R. *J. Am. Chem. Soc.* **1956**, *78*, 2656–2657.
3. Aida, T.; Inoue, S. *Acc. Chem. Res.* **1996**, *29*, 39–48.
4. Aida, T. *Prog. Polym. Sci.* **1994**, *19*, 469–528.
5. Aida, T.; Inoue, S. *Macromolecules* **1981**, *14*, 1162–1166.
6. Yasuda, T.; Aida, T.; Inoue, S. *Macromolecules* **1984**, *17*, 2217–2222.
7. Aida, T.; Sanuki, T.; Inoue, S. *Macromolecules* **1985**, *18*, 1049–1055.
8. Aida, T.; Ishikawa, M.; Inoue, S. *Macromolecules* **1986**, *19*, 8–11.
9. Kuroki, M.; Nashimoto, S.; Aida, Y.; Inoue, S. *Macromolecules* **1988**, *21*, 3114–3115.
10. Hosokawa, Y.; Kuroki, M.; Aida, T.; Inoue, S. *Macromolecules* **1991**, *24*, 824–829.
11. Aida, T.; Inoue, S. *J. Am. Chem. Soc.*, **1985**, *107*, 1358–1364.
12. Kuroki, M.; Aida, T.; Inoue, S. *J. Am. Chem. Soc.*, **1987**, *109*, 4737–4738.
13. Sugimoto, H.; Aida, T.; Inoue, S. *Macromolecules* **1990**, *23*, 2869–2875.
14. Watanabe, Y.; Aida, T.; Inoue, S. *Macromolecules* **1991**, *24*, 3970–3972.
15. Kuroki, M.; Watanabe, T.; Aida, T.; Inoue, S. *J. Am. Chem. Soc.* **1990**, *112*, 5639–5640.
16. Yamamoto, H.; Maruoka, K.; Furuta, K.; Naruse, Y. *Pure Appl. Chem.* **1988**, *61*, 419–422.
17. Yamamoto, H.; Maruoka,K. *J. Syn. Org. Soc. Jpn* **1993**, *51*, 1074–1086.
18. Inoue, S.; Aids, T. *CHEMTECH*, **1994**, *24*, 28–35.
19. Aida, T.; Metalloporphyrin catalysis. In *Catalysis in precision polymerization* (ed S. Kobayashi). Wiley, New York, pp 310–322.
20. Sugimoto, H.; Aida, T.; Inioue, S. *Macromolecules* **1994**, *27*, 3672–3674.
21. Sugimoto, H.; Aida, T.; Inoue, S. *Macromolecules* **1993**, *26*, 4751–4755.
22. Takeuchi, D.; Watanabe, Y.; Aida, T.; Inoue, S. *Macromolecules* **1995**, *28*, 651–652.
23. Sugimoto, H.; Kawamura, C.; Kuroki, M.; Aida, T.; Inoue, S. *Macromolecules* **1994**, *27*, 2013–2018.

24. Takeuchi, D.; Aida, T. *Macromolecules* **1996**, *29*, 8096–8100.
25. Akatsuka, M.; Aida, T.; Inoue, S. *Macromolecules* **1995**, *28*, 1320–1322.
26. Kitayama, T.; Zhang, Y.; Hatada, K. *Polym. J.* **1994**, *26*, 868–872.
27. Adler, A. D. *J. Org. Chem.* **1967**, *32*, 476–477.
28. Sugimoto, H.; Kuroki, M.; Watanabe, T.; Kawamura, C.; Aida, T.; Inoue, S. *Macromolecules* **1993**, *26*, 3403–3410.
29. Sugimoto, H.; Saika, M.; Hosokawa, Y.; Aida, T.; Inoue, S. *Macromolecules* **1996**, *29*, 3359–3369.
30. Adachi, T.; Sugimoto, H.; Aida, T.; Inoue, S. *Macromolecules* **1992**, *25*, 2280–2281.
31. Isoda, M.; Sugimoto, H.; Aida, T.; Inoue, S. *Macromolecules* **1997**, *30*, 57–62.

Index

Note: Bold type denotes chapter extent

Index

Index

Index

Index

Friedel–Crafts reaction 2, 3
  hafnium triflate catalysis 180–1
  scandium triflate catalysis 185–9
  ytterbium triflate catalysis 214
Fries rearrangement 180–1, 186
fumarates 15–16
furan 42, 43–4

GABOB 116, 117
gadolinium compounds 214, 218, 219
C-glycoside rearrangement 179–80
glycosyl acetates 160
O-glycosylation 160, 177–8
glycosyl fluoride 177–8
glyoxylate-ene reaction 101–2, 104, 106, 107–8
glyoxylates 99–100, 121, 122, 125, 230
gold(I) reagents 169–73
  bisphosphine complex 169, 170, 171, 172–3
Grignard reagents 6, 96, 97–9

hafnocene complexes 177
  activation of carbon–fluorine bond 177–8
  activation of ethers 179–80
  carbonyl activation 180–1
haloalkanes: activation 225–7
α-haloketones 204
Henry (nitroaldol) reaction 217, 218–19
hexamethyldisiloxane 160
hexanal 163–4
HMG-Co A inhibitors 121
holmium triflate 215
homoallyl alcohols 53, 163, 192
homoallyl ethers 163
Hosomi–Sakurai reaction, *see*
    Sakurai–Hosomi reaction
hydrogen cyanide 130
hydrophosphonylation 217, 218
hydrosilylation 35, 199
threo-β-hydroxy-α-amino acids 169, 171
β-hydroxy-α-amino aldehydes/ketones 169
β-hydroxy aldehydes 26
2-(1-hydroxycyclohexyl)-1-phenyl-1-
    propanone 207–8
β-hydroxy esters 52
α-hydroxy esters 101, 102–3
hydroxy-pyridines 78

imines
  asymmetric reactions with chiral boron
    reagents 57–61
  hydrophosphonylation 217, 218
imino groups 79
iodomethane complexes 225–6
iodotrimethylsilane 159, 160–2, 163–7
(*R*)-(–)-ipsdienol 103, 104

iron complexes
  ferrocenium ion/ferrocene 228, 229
  iron-bisoxazoline 230, 231, 232–3
  iron-bisphosphite 234, 235–6
  iron-bissulfoxide 230, 231
isocarbacycline analogues 105
isocyanides 169–73
α-isocyano Weinreb amide 169

Kaminsky polymerization 182
ketene bis(trialkylsilyl) acetals 153, 154
ketene silyl acetals (KSA)
  aldol and aldol-type reactions 52, 71, 118,
    119, 208–9, 210, 238–42
    with imines 33–4, 58, 59
  asymmetric protonation 153, 154
  Michael reactions 208, 209, 210, 213–14
α-ketoesters 140–2
ketones
  acetalization 160, 189
  cyanosilylation 215
  and diethylzinc 78
  ether formation 164–5
  hydrosilation 35
  and KSA 209
  reduction 203–4
  see also in particular aldol and aldol-type
    reactions; alkylation; allylation; *named
    ketones*
kinetic optical resolution 106–7
KSA, *see* ketene silyl acetals

β-lactone 73–4, 117
lanthanide(III) reagents **203–23**
  alkoxides 220
  hydride complexes 220
  isopropoxides 219–20
  polymerization reactions 220–1
lanthanum(III) reagents 217–19
Lewis acid-assisted chiral Brønsted acid
    (LBA) 153–4
Lewis acid-assisted high-speed living anionic
    polymerization 246–60
Lewis bases 152
ligand modification/redistribution 2, 94–6
lithium alkynides 11
lithium carbenoids 11
lithium diallylcuprate 12
lithium enolates 204
lithium perchlorate 65–7, 71
living polymerization 221, 246
lupin alkaloids 199, 200
lutetium compounds 215, 220

MAD/MAT 6–10, 15–16
magnesium bromide 65, 67–8, 71–3

267

Index